Oxford Lecture Series in
Mathematics and its Applications 24

Series Editors
John Ball Dominic Welsh

OXFORD LECTURE SERIES
IN MATHEMATICS AND ITS APPLICATIONS

1. J. C. Baez (ed.): *Knots and quantum gravity*
2. I. Fonseca and W. Gangbo: *Degree theory in analysis and applications*
3. P. L. Lions: *Mathematical topics in fluid mechanics, Vol. 1: Incompressible models*
4. J. E. Beasley (ed.): *Advances in linear and integer programming*
5. L. W. Beineke and R. J. Wilson (eds): *Graph connections: Relationships between graph theory and other areas of mathematics*
6. I. Anderson: *Combinatorial designs and tournaments*
7. G. David and S. W. Semmes: *Fractured fractals and broken dreams*
8. Oliver Pretzel: *Codes and algebraic curves*
9. M. Karpinski and W. Rytter: *Fast parallel algorithms for graph matching problems*
10. P. L. Lions: *Mathematical topics in fluid mechanics, Vol. 2: Compressible models*
11. W. T. Tutte: *Graph theory as I have known it*
12. Andrea Braides and Anneliese Defranceschi: *Homogenization of multiple integrals*
13. Thierry Cazenave and Alain Haraux: *An introduction to semilinear evolution equations*
14. J. Y. Chemin: *Perfect incompressible fluids*
15. Giuseppe Buttazzo, Mariano Giaquinta and Stefan Hildebrandt: *One-dimensional variational problems: an introduction*
16. Alexander I. Bobenko and Ruedi Seiler: *Discrete integrable geometry and physics*
17. Doina Cioranescu and Patrizia Donato: *An introduction to homogenization*
18. E. J. Janse van Rensburg: *The statistical mechanics of interacting walks, polygons, animals and vesicles*
19. S. Kuksin: *Hamiltonian partial differential equations*
20. Alberto Bressan: *Hyperbolic systems of conservation laws: the one-dimensional Cauchy problem*
21. B. Perthame: *Kinetic formulation of conservation laws*
22. A. Braides: *Gamma-convergence for beginners*
23. Robert Leese and Stephen Hurley: *Methods and algorithms for radio channel assignment*
24. Charles Semple and Mike Steel: *Phylogenetics*

Phylogenetics

Charles Semple and Mike Steel

Department of Mathematics and Statistics, University of Canterbury
Allan Wilson Centre for Molecular Ecology and Evolution, New Zealand

OXFORD
UNIVERSITY PRESS

OXFORD
UNIVERSITY PRESS
Great Clarendon Street, Oxford OX2 6DP

Oxford University Press is a department of the University of Oxford.
It furthers the University's objective of excellence in research, scholarship,
and education by publishing worldwide in

Oxford New York

Auckland Bangkok Buenos Aires Cape Town Chennai
Dar es Salaam Delhi Hong Kong Istanbul Karachi Kolkata
Kuala Lumpur Madrid Melbourne Mexico City Mumbai Nairobi
São Paulo Shanghai Taipei Tokyo Toronto

Oxford is a registered trade mark of Oxford University Press
in the UK and in certain other countries

Published in the United States
by Oxford University Press Inc., New York

First published 2003

A catalogue record for this title is available from the British Library

Library of Congress Cataloging in Publication Data

Semple, Charles.
Phylogenetics / Charles Semple and Mike Steel.
(Oxford lecture series in mathematics and its applications; 24)
Includes bibliographical references (p.).
1. Cladistic analysis–Mathematics 2. Evolutionary genetics–Mathematics I. Steel, M. A. II. Title
III. Series.

QH83 .S45 2003 572.8′38′0151–dc21 2002027406

ISBN 0 19 850942 1 (alk. paper)

Typeset by Newgen Imaging Systems (P) Ltd., Chennai, India
Printed in Great Britain
on acid-free paper by Biddles Ltd, Guildford & King's Lynn

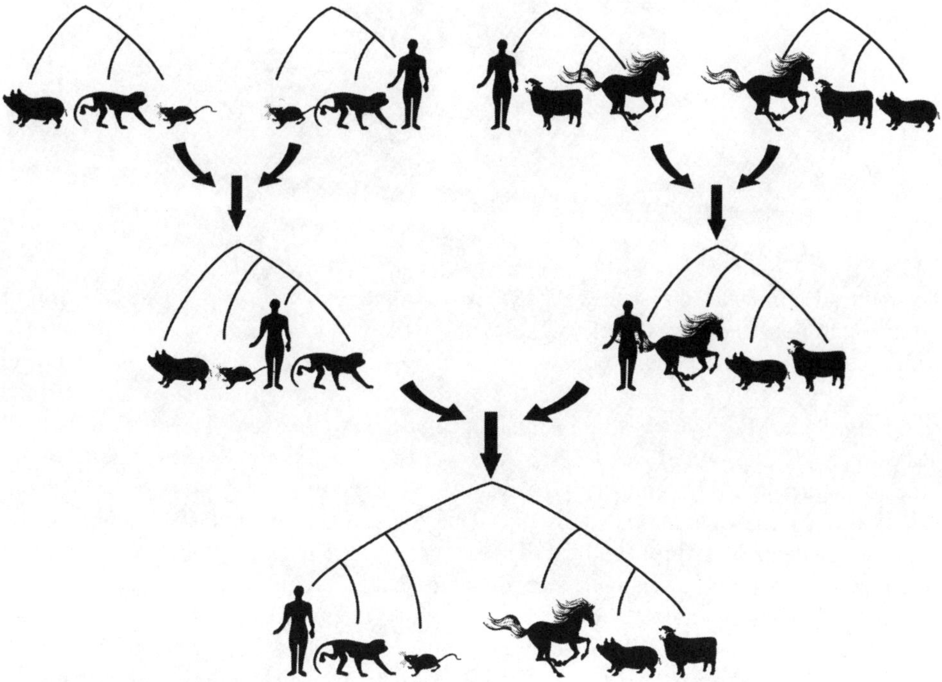

Preface

We possess no pedigrees or armorial bearings; and we have to discover and trace the many diverging lines of descent in our natural genealogies, by characters of any kind which have long been inherited.

Charles Darwin (1872)

This book deals with a fundamental problem that has been of interest since Charles Darwin first proposed the theory of evolution. Namely, how can one use the present-day characteristics of a group of species to infer, in their evolution from a common ancestor, the historical relationships between these species? Typically, these historical relationships are represented by an evolutionary (phylogenetic) tree and such representations were already suggested by Darwin in the nineteenth century (see Fig. 0.1). Determining this tree for different groups of species, or groups of populations, is fundamental to many questions in evolutionary biology as well as for related areas such as conservation genetics and epidemiology. In addition to the discrete structure of the phylogenetic tree, biologists are also interested in many related questions such as how long ago did the various lineages diverge, what are the processes involved in speciation or in the evolution of the characteristics being compared, and what assumptions are required in order to correctly reconstruct the underlying phylogenetic tree?

Over time these questions have been studied from many perspectives, particularly as the types of data available have increased. Initially, comparisons of the physical characteristics of species, such as their morphology or physiology, gave clues to their evolutionary relationships. However, there are processes that can mislead simplistic inferences. For example, the same characteristic can evolve independently in unrelated species (convergent evolution), or a characteristic can evolve and later disappear (reverse transitions). Moreover, using just morphological data, it is difficult to infer evolutionary relationships between very distantly related species or, at the other extreme, organisms that are so similar that they share all the characteristics under study.

From the biologist's perspective, the field was revolutionized by the arrival of molecular data. This began with protein sequences in the late 1960s, genetic (DNA and RNA) sequences in the late 1970s, and most recently whole genome data. The abundance of these data has led to the resolution of many outstanding problems in biology, along with extensive revision in what had previously been believed. Two examples of such revisions—both involving human evolution, but on differing time scales—is the molecular data suggesting that human and chimpanzee are more closely related than either is to gorilla, and that all humans are descended from a migration out of Africa approximately

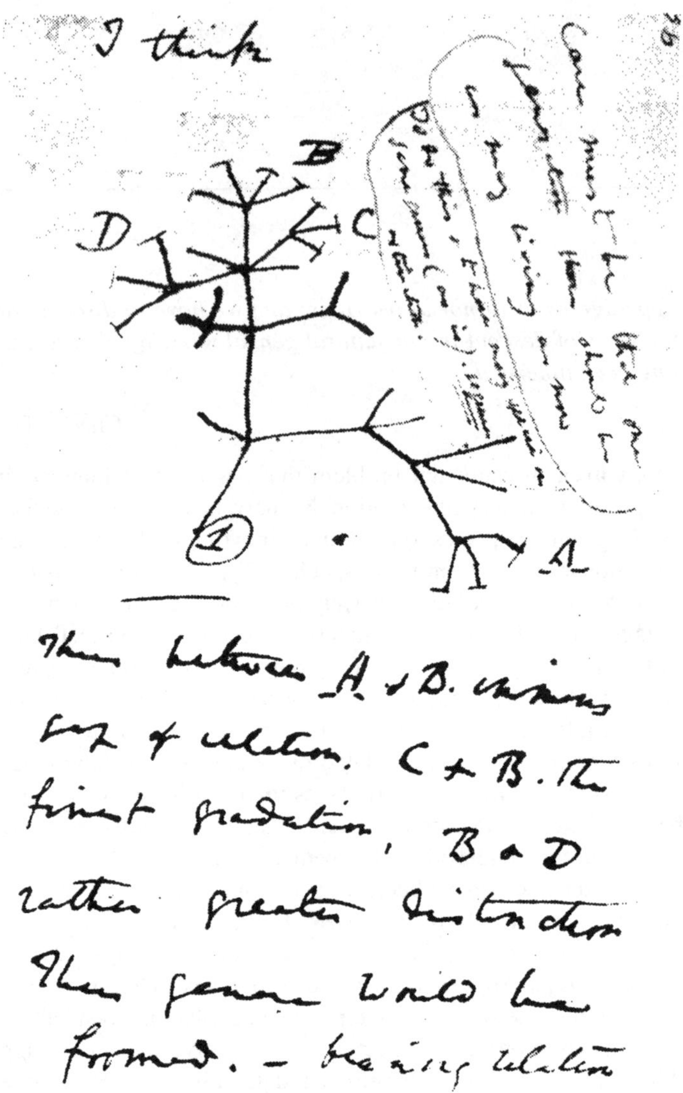

Figure 0.1 The first diagram by Charles Darwin (*First notebook on transmutation of species* 1837) of an evolutionary tree. The ancestral species is at position '1' with the extant species denoted by endpoints (and letters) on certain pendant edges, the remaining pendant edges represent extinctions.

200 000 years ago. Yet many problems still remain, and new questions have arisen as a result of investigations into molecular data.

With this blossoming of activity has come the need for better mathematical foundations, improved computational approaches, and more powerful statistical analyses.

Today, the field of *phylogenetics*—the reconstruction and analysis of phylogenetic trees and networks—is a flourishing area of interaction between mathematics, statistics, computer science, and biology.

For the biologist, there are now many phylogenetic software packages available, and a steady growth in the publication and use of phylogenetic techniques and methods. Applications of these techniques extend well beyond 'reconstructing the past', and the methods are applied in areas such as epidemiology to investigate the origins, relationships, and future development of viruses such as influenza and HIV. Other areas where phylogenetic methods have found applications include ecology (for classifying new species), medicine, and some quite different areas of classification such as linguistics and cognitive psychology.

Our interest in this book is the mathematical foundations of phylogenetics. These foundations date back at least to the pioneering work by Peter Buneman, David Sankoff, and others in the early 1970s. Curiously, Buneman's early paper (1971) dealt not with biology but rather with reconstructing the copying history of manuscripts. The data-driven expansion of phylogenetics during the 1980s and 1990s has led to the need for further mathematical development. For example, many of the traditional techniques developed for analysing the earlier types of data were applied directly to these new data, often with little concern for their appropriateness. Increasingly, researchers asked how such methods would perform under simple models of character evolution, and whether better methods could be devised. With such questions came the need for a better understanding of the underlying theory required to study and analyse phylogenetic trees.

Our aim in this graduate-level text is to outline the key mathematical concepts that underpin phylogenetics. These concepts involve, almost exclusively, discrete mathematics (particularly graph theory) and probability theory. To date much of the theory that underlies this field is scattered widely across the mathematical, statistical, computer science, and biological literature. We describe the central results, and provide proofs of the fundamental theorems where these are illuminating. We also describe how the definitions and approaches relate to the biological applications, and indicate how algorithms can be derived from the results.

A brief outline of the contents is as follows. In Chapter 1, we provide some necessary graph-theoretic background, and give three brief examples of how graphs have been applied in biology. In Chapter 2, we introduce phylogenetic trees, and their mathematically convenient generalization *X-trees*. In Chapter 3, we describe and exploit Buneman's fundamental equivalence between X-trees and a certain type of set system. In Chapters 4–6, we study several concepts of *compatibility*, which, informally speaking, expresses whether the historical relationships described by the data can be represented by a single tree. In Chapter 7, we investigate how distance measures between species can be represented by a tree with weighted edges. In Chapter 8, we consider models of character evolution and describe how these can be used to reconstruct trees from sequence data.

Our presentation includes exercises and, apart from Chapter 1, at least one specialized topic at the end of each chapter. The results in these specialized topics are not required in the non-specialist sections (and so may be omitted for a graduate-level course); however, they are referred to in later specialist topic sections.

This book is intended to present a reasonably self-contained overview of an expanding field, rather than a comprehensive survey. Also, we have concentrated on the underlying mathematics rather than on computational aspects as these have been dealt with in other recent books. We have also attempted to make our approaches sufficiently general so as to be applicable to a wide variety of past, present, and likely future data. Although our selection of topics is doubtlessly influenced by our particular perspectives and interests, we hope we have at least touched on most of the central concepts that underlie the wide variety of methods currently in use.

Finally, we wish to thank Laura Sessions for some proof-reading, the New Zealand Marsden Fund for financial support, and Brigitte, Cheryl, and the three babies that arrived (to join the fourth) during the writing of this book.

Contents

Preliminaries

This section introduces some notation and outlines some background concepts.

Notation

Throughout this book, X denotes a non-empty finite set. The sets of integers, rational numbers, real numbers, non-negative real numbers, and positive real numbers will be denoted by $\mathbb{Z}$, $\mathbb{Q}$, $\mathbb{R}$, $\mathbb{R}^{\geq 0}$, and $\mathbb{R}^{>0}$, respectively. The collection of subsets of X is denoted by 2^X.

A *partition* of X is a decomposition of X into a set of non-empty *blocks* such that every element of X is in exactly one block. For example, if $X = \{a, b, c, d, e\}$, then $\{\{a, c, d\}, \{b\}, \{e\}\}$ is a partition of X with blocks $\{a, c, d\}$, $\{b\}$, and $\{e\}$. Furthermore, if $A_1, A_2, \ldots, A_k$ are the blocks of a partition of X, we often write $A_1|A_2|\cdots|A_k$ in place of $\{A_1, A_2, \ldots, A_k\}$. Note that, in writing a partition of X in this way, the ordering of the blocks is arbitrary.

For sets X and Y, a function $f : X \to Y$, subsets $A \subseteq X$ and $B \subseteq Y$, and an element $y \in Y$, we denote the set $\{f(x): x \in A\}$ by $f(A)$, the set $\{x \in X: f(x) = y\}$ by $f^{-1}(y)$, and the set $\{x \in X: f(x) \in B\}$ by $f^{-1}(B)$, respectively.

Background concepts

Partial orders. A *partial order* on a set S is a relation $\leq$ on S such that, for all $x, y, z \in S$, the following hold:

(i) $x \leq x$;
(ii) if $x \leq y$ and $y \leq x$, then $x = y$;
(iii) if $x \leq y$ and $y \leq z$, then $x \leq z$ (*transitivity*).

If, in addition, a partial order $\leq$ on S satisfies the property that, for every pair of elements $x, y \in S$ either $x \leq y$ or $y \leq x$, then $\leq$ is said to be a *total order*. A total order is also called a *linear order*. For example, taking $S = \mathbb{R}$ the usual less-than-or-equal-to relation is a total order. However, for the collection of all subsets of a set, the relation of set inclusion is a partial order but not a total order.

For a set S on which a partial order $\leq$ has been defined, a subset S' of S has a *least upper bound* if there is an element s of S such that $s' \leq s$ for all $s' \in S'$ and, whenever $t \in S$ is an element satisfying this last condition, $s \leq t$. Observe that a least upper bound for S' need not exist, but if there is one, then it is easily seen that it is unique. The *greatest lower bound* of a subset of a partially ordered set is defined similarly. An element m of S is *maximal* if there is no element s of S such that $m < s$.

Metrics. A *metric* on a set S is a function $d : S \times S \to \mathbb{R}^{\geq 0}$ such that, for all $x, y, z \in S$, the following hold:

(i) $d(x, y) = 0$ if and only if $x = y$ (*definiteness*);
(ii) $d(x, y) = d(y, x)$;
(iii) $d(x, z) \leq d(x, y) + d(y, z)$ (*triangle inequality*).

The pair (S, d) is called a *metric space*. If condition (i) is weakened to

$$d(x, x) = 0 \quad \text{for all } x \in S,$$

then d is said to be a *pseudometric*. Familiar examples of metrics are the real line with $d(x, y) = |x - y|$ and the collection of all subsets of a finite set with $d(A, B) = |A \cup B| - |A \cap B|$. An example of a pseudometric that is not a metric is the Euclidean plane with $d((x_1, y_1), (x_2, y_2)) = |x_1 - x_2|$.

The *discrete metric* on a set S is the metric d defined, for all $x, y \in S$, by

$$d(x, y) = \begin{cases} 1, & \text{if } x \neq y, \\ 0, & \text{otherwise.} \end{cases}$$

Groups. A *group* $\mathcal{G}$ is an ordered pair $(S, *)$ consisting of a set S and a binary operation $*$ on S such that the following hold:

(i) $*$ is associative;
(ii) there is an element $1_{\mathcal{G}}$ in S such that $1_{\mathcal{G}} * x = x * 1_{\mathcal{G}} = x$ for all $x \in S$;
(iii) for each $x \in S$, there is an element x^{-1} in S such that $x^{-1} * x = x * x^{-1} = 1_{\mathcal{G}}$.

It is straightforward to show that a group $\mathcal{G}$ has a unique choice for $1_{\mathcal{G}}$ called the *identity element*. A group is *Abelian* if its binary operation is commutative. For example, the set of integers together with addition is an Abelian group. A group $(S', *)$ is a *subgroup* of a group $(S, *)$ if $S' \subseteq S$. Given two groups $(S, *)$ and $(S_1, \circ)$, a function $f : S \to S_1$ is a *homomorphism* from the group $(S, *)$ into $(S_1, \circ)$ if $f(x * y) = f(x) \circ f(y)$ for all $x, y \in S$. If, in addition, f is a bijection, then f is an *isomorphism*, and the two groups are said to be *isomorphic*. Throughout this book we multiply elements of $\mathcal{G}$ from left to right.

Enumeration. Here, we describe some notation concerning two aspects of combinatorial enumeration, namely generating functions and asymptotics. For an odd positive integer m, we denote by $m!!$ the product $m \times (m - 2) \times \cdots \times 1$, and we let $\binom{n}{k}$ denote $n!/(k!(n - k)!)$, the number of subsets of $\{1, \ldots, n\}$ of size k.

Generating functions are a convenient algebraic tool in enumeration. If $f(x) = \sum_{n\geq 0} a_n x^n$, we say that $f(x)$ is the *ordinary generating function* for the numbers a_n, and let $[x^n]f(x)$ denote the coefficient a_n of x^n in $f(x)$. Similarly, if $f(x, y)$ is a power series in x and y, then $[x^n y^k]f(x, y)$ denotes the coefficient of $x^n y^k$ in $f(x, y)$. If $f(x) = \sum_{n\geq 0} a_n(x^n/n!)$, we say that $f(x)$ is the *exponential generating function* for the numbers a_n. For all positive integers a and non-negative integers n, we will make occasional use of the identity

$$[x^n](1 - bx)^{-a} = b^n \binom{a + n - 1}{n}, \tag{0.1}$$

where b is an arbitrary real number.

Now suppose $f(n)$ and $g(n)$ are real-valued functions defined on the positive integers. If $\lim_{n\to\infty}[f(n)/g(n)] = 1$, then $f(n)$ and $g(n)$ are said to be *asymptotically equivalent*, in which case we write $f(n) \sim g(n)$. Furthermore, if there exists a constant $c > 0$ and an integer m such that $|f(n)| \leq c|g(n)|$ for all $n \geq m$, then we write $f(n) = O(g(n))$. If, for every $\epsilon > 0$, there exists a value of m such that $|f(n)| < \epsilon|g(n)|$ for all $n > m$, we write $f(n) = o(g(n))$.

For approximating factorials, we will use *Stirling's formula*

$$n! \sim \sqrt{2\pi}\, e^{-n} n^{n+(1/2)}.$$

Complexity theory. In applying algorithms to large-scale problems, it is particularly useful if the running time of the algorithm does not grow too quickly with the size of the input. A *polynomial–time algorithm* is one in which the number of steps in the algorithm grows no faster than some polynomial in the parameters associated with the input. If the running time is $f(n)$ where $f(n) = O(n^k)$, then we say that the algorithm runs in $O(n^k)$ time. Such algorithms tend to be highly prized particularly when k is small.

There is a large class of decision problems that have the property that if any one of the problems in the class has a polynomial–time algorithm to solve it, then so does every other problem in the class. This is the class of 'NP-complete' problems. A related class of 'NP-hard' problems consists of more general 'search' problems. In the latter class, a polynomial–time solution to any one of the problems would imply that all NP-complete problems have a polynomial–time algorithm. Since it is widely believed that NP-complete problems do not have a polynomial–time algorithm for their solution, showing that a decision problem is NP-complete, or a search problem is NP-hard, is regarded as strong evidence that there is no polynomial–time algorithm for this problem. For further details, we refer the interested reader to Garey and Johnson (1979).

Elementary (discrete) probability theory. Given a countable set S and a function $P : S \to \mathbb{R}^{\geq 0}$ satisfying $\sum_{s\in S} P(s) = 1$, we obtain a corresponding real-valued function $\mathbb{P}$ defined on the subsets of S by setting, for all subsets A of S, $\mathbb{P}(A) = \sum_{x\in A} P(x)$. Note that $0 \leq \mathbb{P}(A) \leq 1$ and $\mathbb{P}(S - A) = 1 - \mathbb{P}(A)$. In the language of probability theory, S is a *sample space*, A is an *event* in S, P is a *probability distribution* on S, and $\mathbb{P}(A)$ is the *probability of* A (induced by P). If S is finite and $P(s) = 1/|S|$ for all $s \in S$, we say that P is the *uniform* probability distribution on S.

Let S be a sample space, and let A and B be events in S. Provided $\mathbb{P}(B) > 0$, the ratio

$$\mathbb{P}(A|B) = \frac{\mathbb{P}(A \cap B)}{\mathbb{P}(B)} \tag{0.2}$$

is called the *conditional probability* of A given B. Observe that $0 \leq \mathbb{P}(A|B) \leq 1$. In practice, conditional probabilities are often prescribed for a particular problem and so eqn (0.2) is useful for computing $\mathbb{P}(A \cap B)$.

For a sample space S, we have the following useful elementary results:

(i) (*Law of total probability*) If $B_1, \ldots, B_k$ partition S and A is an event in S, then

$$\mathbb{P}(A) = \sum_{i=1}^{k} \mathbb{P}(A|B_i)\mathbb{P}(B_i).$$

(ii) (*Bayes' rule*) If A and B are events in S, and $\mathbb{P}(A), \mathbb{P}(B) > 0$, then

$$\mathbb{P}(B|A) = \frac{\mathbb{P}(A|B)\mathbb{P}(B)}{\mathbb{P}(A)}.$$

The value of $\mathbb{P}(A)$ on the right-hand side of this last expression is often evaluated by the law of total probability.

(iii) (*The product rule*) If $A_1, \ldots, A_k$ are events in S, then

$$\mathbb{P}(\cap_{i=1}^{k} A_i) = \mathbb{P}(A_1)\mathbb{P}(A_2|A_1)\mathbb{P}(A_3|A_1 \cap A_2) \cdots \mathbb{P}(A_k| \cap_{j<k} A_j).$$

A family $A_1, A_2, \ldots, A_k$ of events in a sample space is *mutually independent* if, for every non-empty subset H of $\{1, 2, \ldots, k\}$,

$$\mathbb{P}(\cap_{i \in H} A_i) = \prod_{i \in H} \mathbb{P}(A_i).$$

Similarly, if B is an event in the sample space and $\mathbb{P}(B) > 0$, a family $A_1, A_2, \ldots, A_k$ of events are *conditionally independent given* B if, for all subsets H of $\{1, 2, \ldots, k\}$,

$$\mathbb{P}(\cap_{i \in H} A_i|B) = \prod_{i \in H} \mathbb{P}(A_i|B).$$

A (*discrete*) *random variable* ξ is a function from a sample space S into a set W, and it is usual to write $\mathbb{P}(\xi = t)$ for $\mathbb{P}(\xi^{-1}(t))$. If $W = \mathbb{R}$, then the *mean* and *variance* of ξ are defined by

$$\mu = \mathbb{E}(\xi) = \sum_t t\mathbb{P}(\xi = t)$$

and

$$\sigma^2 = \mathbb{V}(\xi) = \sum_t (t - \mu)^2 \mathbb{P}(\xi = t),$$

respectively, where the summation is over the range of possible values that ξ takes with positive probability.

1

Graphs and their role in biology

1.1 Graphs

Graphs play a fundamental role in phylogenetics. This section introduces some notation and concepts of graph theory.

A *graph* G is an ordered pair (V, E) consisting of a non-empty set V of *vertices* and a multiset E of *edges* each of which is an element of $\{\{x, y\}: x, y \in V\}$. Unless otherwise stated, all graphs will have a finite set of vertices. An edge that joins a vertex to itself is a *loop* and edges that join the same distinct pair of vertices are called *parallel edges*. A *simple graph* is a graph that has no loops and no parallel edges. Unless otherwise stated, we will assume that all graphs in this book are simple.

If $e = \{u, v\}$ is an edge of a graph G, then u and v are *adjacent* or *neighbours*, and e is said to be *incident* with u and v. The vertices u and v are the *ends* or *end vertices* of e. For example, in Fig. 1.1, v_2 and v_6 are adjacent, but v_2 and v_5 are not adjacent; and the end vertices of e_7 are v_5 and v_6. We denote the set of vertices and edges of G by $V(G)$ and $E(G)$, respectively.

Let v be a vertex of a graph G. The *degree* of v, denoted $d(v)$, is the number of edges in G that are incident with v. A vertex of degree zero is said to be *isolated*. A vertex of degree one is called a *pendant* vertex, and an edge incident with a pendant vertex is called a *pendant* edge. In Fig. 1.1, v_3 has degree 4, v_7 is an isolated vertex, and v_1 is a pendant vertex. Since each edge of a graph is incident with exactly two vertices,

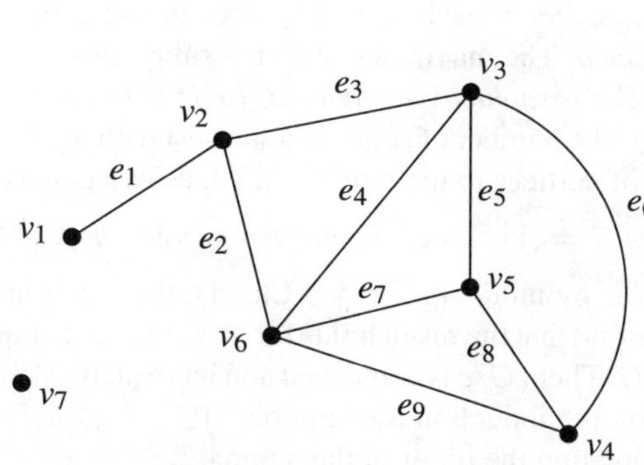

Figure 1.1 A graph.

it follows that, in summing the degrees of the vertices of a graph, we count each edge exactly twice. This observation gives the well-known equality in Lemma 1.1.1.

Lemma 1.1.1. *Let $G = (V, E)$ be a graph. Then,*

$$\sum_{v \in V} d(v) = 2|E|.$$

A graph H is a *subgraph* of a graph G if $V(H)$ and $E(H)$ are subsets of $V(G)$ and $E(G)$, respectively. If V' is a non-empty subset of $V(G)$, then the subgraph of G that has vertex set V' and edge set consisting of those edges of G that have both ends in V' is the subgraph of G *induced* by V', and is denoted by $G[V']$.

Two basic operations in graph theory are deletion and contraction. Let e be an edge of a graph G. Then $G \backslash e$ denotes the graph obtained from G by removing or *deleting* e. The graph obtained from G by identifying the ends of e and then deleting e, denoted G/e, is said to be obtained from G by *contracting* e. For a subset E' of the edge set of G, the graph obtained from G by deleting (resp. contracting) each of the edges in E' is denoted by $G \backslash E'$ (resp. G/E'). Note that both resulting graphs are well defined. If v is a vertex of G, then $G \backslash v$ denotes the graph obtained from G by removing v and all edges incident with v. We say $G \backslash v$ is obtained from G by *deleting* v.

Two graphs $G_1 = (V_1, E_1)$ and $G_2 = (V_2, E_2)$ are *isomorphic* if there is a bijection $\psi : V_1 \rightarrow V_2$ that preserves adjacency; that is, $\{u, v\} \in E_1$ precisely if $\{\psi(u), \psi(v)\} \in E_2$. The map ψ is called a *graph isomorphism.*

A *clique* is a graph in which each pair of distinct vertices is joined by one edge. A clique is also known as a *complete graph.* A clique with k vertices is called a *k-clique.* For example, the vertices in $\{v_3, v_4, v_5, v_6\}$ induce a 4-clique of the graph shown in Fig. 1.1.

A *path* in a graph G is a sequence of distinct vertices $v_1, v_2, \ldots, v_k$ such that, for all $i \in \{1, 2, \ldots, k-1\}$, v_i and v_{i+1} are adjacent. If, in addition, v_1 and v_k are adjacent, then the subgraph of G whose vertex set is $\{v_1, v_2, \ldots, v_k\}$ and whose edge set is $\{\{v_k, v_1\}\} \cup \{\{v_i, v_{i+1}\}: i \in \{1, \ldots, k-1\}\}$ is a *cycle.* A cycle with k vertices is called a *k-cycle.*

A graph G is *connected* if each pair of vertices in G can be joined by a path; otherwise, G is *disconnected.* The maximal connected subgraphs of a graph G are called the *components* of G. The (*graphical*) *distance* $d_G(u, v)$ between two vertices u and v in a connected graph G is the number of edges in a shortest path joining u and v. Lemma 1.1.2 relates the number of vertices to the number of edges in a connected graph.

Lemma 1.1.2. *Let $G = (V, E)$ be a connected graph. Then, $|V| \leq |E| + 1$.*

Proof. The proof is by induction on $|V|$. Clearly, the result holds for $|V| \leq 2$. Now, assume that $|V| \geq 3$ and that the result holds for all connected graphs with fewer vertices. Let e be an edge of G. Then, G/e is connected and has exactly $|V|-1$ vertices and $|E|-1$ edges. Therefore, by the induction assumption, $|V| - 1 \leq |E| - 1 + 1 = |E|$. Hence $|V| \leq |E| + 1$, completing the proof of the lemma. □

A further graph operation that will be used in this book is that of subdivision. Replacing an edge e of a graph by a path consisting of three vertices so that the end vertices of e are the end vertices of the path is referred to as *subdividing* e. In subdividing an edge, the vertex of the path that is not an end vertex of e is the vertex that *subdivides* e. A graph H is a *subdivision* of a graph G if G can be obtained from H by repeatedly subdividing edges.

We end this section with a brief discussion on directed graphs. A *directed graph* or *digraph* D is an ordered pair (V, A) consisting of a non-empty set V of vertices and a multiset A of *arcs* each of which is an element of $\{(x, y) \colon x, y \in V\}$. If $a = (u, v)$ is an element of A, then u is the *tail* of a and v is the *head* of a. The arc a is said to be *directed from* u *to* v. In general, the terminology for directed graphs is similar to that for graphs. For example, a *directed path* is a sequence of distinct vertices $v_1, v_2, \ldots, v_k$ such that, for all $i \in \{1, 2, \ldots, k-1\}$, there is an arc (v_i, v_{i+1}). A digraph is *acyclic* if there are no directed cycles.

Let $D = (V, A)$ be a digraph and let v be an element of V. The *in-degree* (resp. *out-degree*) of v is the number of arcs of D whose head (resp. tail) is v.

A digraph D is an *orientation* of a graph G if D can be obtained from G by simply specifying a direction on each of the edges of G. Furthermore, a graph G is the *underlying* graph of a digraph D if G can be obtained from D by replacing each arc with an edge having the same end vertices.

Lastly, a partial order $\leq$ on a set S induces an associated directed graph, called the *cover digraph* associated with $(S, \leq)$. This graph has vertex set S and, for all $x, y \in S$, an arc (x, y) precisely if $y \leq x$ and there is no element $z \in S$ such that $y < z < x$.

1.2 Trees and forests

In this section, we introduce trees, which are the basic graphical structures that we will be dealing with in this book.

A *tree* $T = (V, E)$ is a connected graph with no cycles. A *forest* is a disjoint union of trees or, equivalently, a graph with no cycles. A vertex of T of degree at most one is called a *leaf*. A vertex of T that is not a leaf is called an *interior* vertex. An edge of T is *interior* if both of its ends are interior vertices. We denote the set of interior vertices and interior edges of T by $\mathring{V}$ and $\mathring{E}$, respectively. A connected subgraph of T is a *subtree* of T.

A tree is *binary* if every interior vertex has degree three. Binary trees are sometimes called *cubic*, *ternary*, or *trivalent* trees, but we will use the word 'binary' to keep the terminology closer to the biology. A tree is a *path graph* if all vertices have degree at most two. Clearly, a path graph has exactly two vertices of degree one.

Trees can be characterized amongst the class of connected graphs via an equality based on the number of vertices and the number of edges.

Theorem 1.2.1. *Let $G = (V, E)$ be a graph. Then, the following are equivalent:*

(i) G is a tree;

(ii) for any two vertices v_1 and v_2 in V there exists a unique path in G from v_1 to v_2;

(iii) G is connected and $|V| = |E| + 1$.

Proof. The equivalence of (i) and (ii) is left as an exercise. Here, we establish the equivalence of (i) and (iii). The proof that (i) implies (iii) is by induction on $|V|$. Evidently, this implication holds for $|V| \leq 2$. Let $|V| \geq 3$ and suppose that the implication holds for all graphs with fewer vertices. Then, as $|V| \geq 3$, G has a leaf v with a single incident edge. Now, $G \backslash v = (V', E')$ is a tree with one less vertex than G. Therefore, by the induction assumption, $|V'| = |E'| + 1$ and so $|V| = |E| + 1$ as required.

Now assume that (iii) holds, but G is not a tree. Then, as G is connected, there is an edge e of G contained in a cycle of G. Clearly, $G \backslash e = (V, E')$ is connected. Therefore, by Lemma 1.1.2, $|V| \leq |E'| + 1$, and so, as $|E'| = |E| - 1$, we get $|V| < |E| + 1$; a contradiction. Hence G is a tree. □

Combining Lemma 1.1.1 and the equivalence of (i) and (iii) in Theorem 1.2.1, we get Corollary 1.2.2.

Corollary 1.2.2. *Let T be a tree with vertex set V. Then,*

$$\sum_{v \in V} d(v) = 2|V| - 2.$$

The proofs of the next three propositions are left as exercises.

Proposition 1.2.3. *Let $T = (V, E)$ be a tree with $|V| \geq 3$, and let n_1 and n_2 denote the number of vertices of T of degree one and degree two, respectively. Then*

(i) $|\mathring{E}| \leq n_1 + n_2 - 3$.
(ii) Suppose that $n_2 = 0$. Then, $|\mathring{E}| = n_1 - 3$ if and only if T is binary.

Proposition 1.2.4. *Suppose that T_1 and T_2 are trees, and that ψ and ψ' are graph isomorphisms from the vertex set of T_1 into the vertex set of T_2. If ψ and ψ' agree on the leaves of T_1, then $\psi = \psi'$.*

Two distinct leaves of a tree are said to form a *cherry* if they are adjacent to a common vertex.

Proposition 1.2.5. *Let T be a tree having no vertices of degree two.*

(i) If T has at least three leaves, then T has at least two distinct pairs of cherries.
(ii) If T has at least four leaves, then T has at least two disjoint pairs of cherries.

A *rooted* tree is a tree that has exactly one distinguished vertex called the *root* which we denote by the letter ρ. Rooted trees play an important role throughout this book.

For a rooted tree T, a natural partial order $\leq_T$ on the vertex set V of T is obtained by setting $v_1 \leq_T v_2$ precisely if the path from the root of T to v_2 includes v_1. If this is the case, we say that v_2 is a *descendant of* v_1 and that v_1 is an *ancestor of* v_2. A vertex of a rooted tree is both a descendant and ancestor of itself. Consistent with this terminology, we will always draw a rooted tree with the root at the top of the figure and oriented so as to respect the ancestor–descendant relationship.

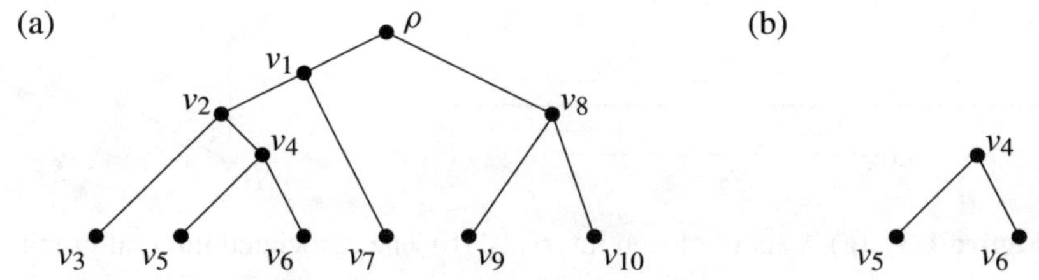

Figure 1.2 (a) A rooted tree. (b) A rooted subtree.

Let v be a vertex of a tree T. A rooted tree $T' = (V', E')$ with root vertex ρ' is a *rooted subtree of T lying below v* if T' is a subgraph of T such that V' is the set of descendants of ρ' in T, and ρ' is a proper descendant of v. These concepts are illustrated in Example 1.2.6.

Example 1.2.6. Let T be the rooted tree shown in Fig. 1.2(a). Then, $v_1 \leq_T v_4$ and so v_4 is a descendant of v_1. Furthermore, the rooted tree shown in Fig. 1.2(b) is a rooted subtree of T lying below v_1.

We end this section by describing two operations on a tree $T = (V, E)$. Suppose that v is a degree-two vertex of T. Let e be an edge of T incident with v. The tree T/e is said to be obtained from T by *suppressing* v. Furthermore, for a subset V' of V, we let $T(V')$ denote the minimal connected subgraph of T that contains the vertices in V' and we say that $T(V')$ is the *minimal subtree of T induced by* V'. Analogously, one can define the operation of suppression for rooted trees. Moreover, if T is now a rooted tree with root vertex ρ and V' is a subset of $V(T)$, then the *minimal rooted subtree of T induced by* V' is the minimal connected subgraph T' of T that contains the vertices in V' and for which the vertex closest to ρ in T' is distinguished.

1.3 Intersection graphs

Historically, intersection graphs have played a significant role in combinatorial biology. Their role is highlighted in Section 1.4 and particularly in Chapter 4.

Let $\mathcal{S} = \{S_1, S_2, \ldots, S_k\}$ be a family of sets. The *intersection graph* of $\mathcal{S}$, denoted $\text{int}(\mathcal{S})$, is the graph that has vertex set $\mathcal{S}$ and an edge between S_i and S_j precisely if $S_i \cap S_j \neq \emptyset$, where i and j are distinct elements of $\{1, 2, \ldots, k\}$. The following result established by Marczewski (1945) is easily proved and left as an exercise.

Proposition 1.3.1. *For any graph G, there is a graph H and a collection $\mathcal{S}$ of vertex sets of subgraphs of H so that G is isomorphic to* $\text{int}(\mathcal{S})$.

Two classes of intersection graphs that will be important to us arise from finite closed intervals of the real line and from subtrees of a tree. A graph that is isomorphic to the intersection graph of a collection of finite closed intervals of the real line is said to be an *interval graph*. For example, the graph in Fig. 1.3(b) is an interval graph as it

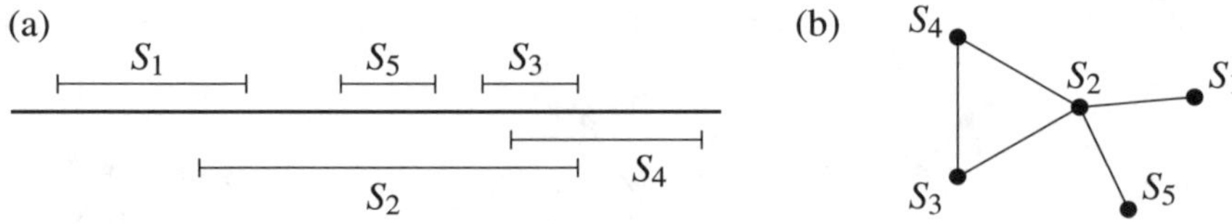

Figure 1.3 (a) A set of closed intervals. (b) The associated interval graph.

is the intersection graph of the collection of closed intervals displayed in Fig. 1.3(a). Equivalently, a graph is an interval graph if it is isomorphic to the intersection graph of a collection of vertex sets of connected subgraphs of a path graph. A graph that is isomorphic to the intersection graph of a collection of vertex sets of subtrees of a given tree is said to be a *subtree intersection graph.*

Characterizations of subtree intersection graphs have been given in a number of different ways. The main reason for this is that the class of subtree intersection graphs coincides with the class of chordal graphs. Three characterizations of subtree intersection graphs are given in Theorem 1.3.2.

A graph G is *chordal*, also called *triangulated*, if every induced subgraph of G that is a cycle has at most three edges. Equivalently, a graph is chordal if every cycle with at least four vertices has an edge (called a 'chord') connecting two non-consecutive vertices in the cycle. Observe that every vertex-induced subgraph of a chordal graph is chordal. A *chordalization* (also called a *triangulation*) of a graph $G = (V, E)$ is a graph $G' = (V, E')$ with the properties that G' is chordal and $E \subseteq E'$.

A vertex of a graph G is *simplicial* if its neighbours together with itself induce a clique of G. Thus, a simplicial vertex of a graph is in a unique maximal clique. We say that G has a *perfect elimination ordering* if the vertices of G can be ordered, $v_1, v_2, \ldots, v_k$ say, so that, for each $i \in \{1, 2, \ldots, k\}$, v_i is a simplicial vertex of the subgraph of G induced by $\{v_i, \ldots, v_k\}$.

The equivalences in Theorem 1.3.2 are due, in various parts, to Buneman (1974*a*), Fulkerson and Gross (1965), Gavril (1974), Rose (1970), and Walter (1978).

Theorem 1.3.2. *Let G be a graph. Then the following statements are equivalent:*

(i) G is chordal;
(ii) G is a subtree intersection graph;
(iii) G has a perfect elimination ordering;
(iv) there exists a tree T whose vertex set $\mathcal{K}$ *is the set of maximal cliques of G and, for each vertex v in G, the subgraph of T induced by the elements of* $\mathcal{K}$ *containing v is a subtree of T.*

We refer to the tree described in (iv) of Theorem 1.3.2 as a *maximal clique tree representation of G*. The equivalence of (i) and (ii) in the statement of Theorem 1.3.2 is illustrated in Example 1.3.3.

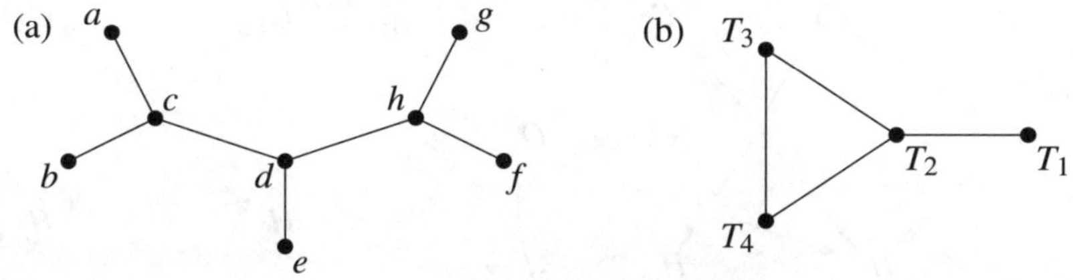

Figure 1.4 (a) A tree. (b) A subtree intersection graph.

Example 1.3.3. Let T denote the tree shown in Fig. 1.4(a). Let T_1, T_2, T_3, T_4, denote the subtrees of T induced by the vertex sets $V_1 = \{a, b, c\}$, $V_2 = \{b, c, d\}$, $V_3 = \{d, e\}$, and $V_4 = \{d, f, g, h\}$, respectively. The subtree intersection graph of $\{T_1, T_2, T_3, T_4\}$ is shown in Fig. 1.4(b).

As an interval graph is isomorphic to the intersection graph of a collection of vertex sets of connected subgraphs of a path graph, we have the following consequence of the equivalence between (i) and (ii) in Theorem 1.3.2.

Corollary 1.3.4. *All interval graphs are chordal.*

There are polynomial–time algorithms for recognizing chordal and interval graphs, and also for realizing them as the appropriate type of intersection graphs. For a survey of these algorithms see McKee and McMorris (1999).

An important concept in computational graph theory is the notion of tree-width, which can be defined in several equivalent ways. Given a graph H, let $\omega(H)$ denote the size of the largest clique in H. The *tree-width* of a graph $G = (V, E)$, denoted $tw(G)$, is

$$tw(G) = \min\{\omega(H) - 1\colon H = (V, E') \text{ is chordal and } E \subseteq E'\}.$$

Many intractable computational problems on graphs have polynomial–time solutions if a bound is placed on the tree-width of the graph. For details of such solutions, we refer the reader to Downey and Fellows (1999).

1.4 Applications of graphs in biology

In this section, we present a brief sample of three applications of graph theory in biology. In the next chapter, we introduce a fourth application: the theory of phylogenetic trees, which is central to the rest of the book.

Food webs and competition graphs. A *food web* is an acyclic digraph $D = (V, A)$, where the sets V and A have the biological interpretation that V is a set of species and $(v, w) \in A$ if species v feeds or preys on species w. The *competition graph* associated with a food web $D = (V, A)$, denoted $C(D)$, is the graph (V, E), where $\{v, v'\} \in E$ if there exists some $w \in V$ such that both (v, w) and (v', w) are elements of A. In particular,

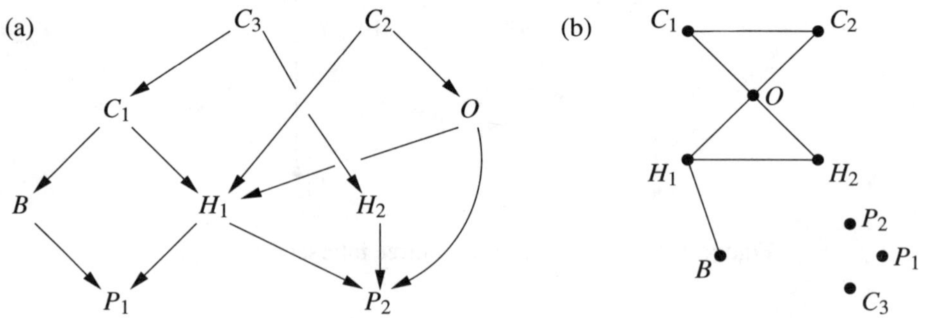

Figure 1.5 (a) A food web and (b) its associated competition graph.

the edges of a competition graph display which pairs of species are in competition for some shared food resource.

Example 1.4.1. A food web for a set of species is shown in Fig. 1.5(a), where C_1, C_2, and C_3 are carnivores, H_1 and H_2 are herbivores, B is a bird, O is an omnivore, and P_1 and P_2 are plants. The associated competition graph is shown in Fig. 1.5(b).

Proposition 1.4.2 is easily established and is left as an exercise.

Proposition 1.4.2.

(i) Every competition graph has at least one isolated vertex.

(ii) Let $G = (V, E)$ be a graph. Then, for some graph $G' = (V', E)$ with $V \subseteq V'$ and $|V'| - |V| \leq |E|$, there exists an acyclic digraph D such that $C(D)$ is isomorphic to G'.

Referring to Proposition 1.4.2(ii), the smallest number of additional isolated vertices required to make G a competition graph is called the *competition number* for G. Computing this number is an NP-hard problem.

Curiously, competition graphs that arise in biology are typically interval graphs, even though a randomly generated graph is unlikely to be an interval graph. This is of interest to biologists as it suggests that competition can often be modelled by a one-dimensional 'niche space'. It also raises an interesting problem, namely to characterize exactly those acyclic digraphs D for which $C(D)$ is an interval graph. Although there has been much work on this problem and many partial results, the general problem remains unsolved. For further details, see McKee and McMorris (1999).

Interval graphs and genome mapping. A central task in computational biology is to reconstruct genes based on information about short overlapping fragments of DNA. Here, we think of DNA sequences as comprising a string of letters chosen from the four-letter alphabet $\{A, C, G, T\}$. Indeed, Benzer (1959), in one of the first papers to deal with interval graphs, addressed a much more fundamental question: based on information about which fragments of DNA overlap, is the data consistent with the hypothesis that the gene is a linear arrangement of letters? Mathematically, this question is tantamount to asking whether the intersection graph of the fragments $S_1, S_2, \ldots, S_k$ forms an interval

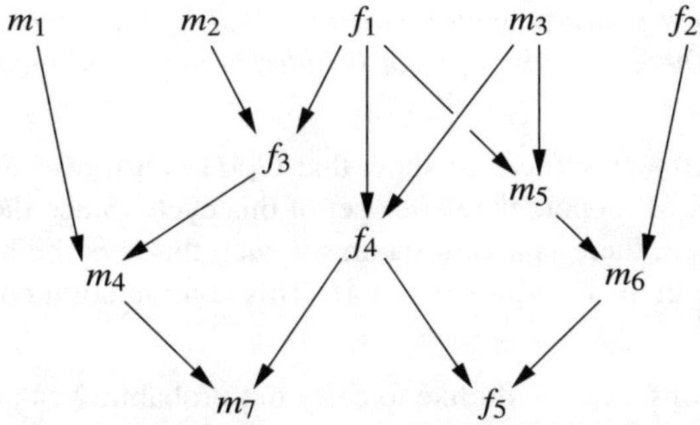

Figure 1.6 A pedigree.

graph. Subsequently, interval graphs have played an important role in the assembly of sequences from subsequences.

A generalization of interval graphs for the problem of 'physical mapping' of DNA are 'probe' interval graphs. Formally, a graph $G = (V, E)$ is a *probe interval graph* if V can be partitioned into two subsets P and N, and each vertex $v \in V$ can be assigned an interval $I(v)$ in such a way that $\{u, v\}$ is an edge of G precisely if both $I(u) \cap I(v) \neq \emptyset$ and either u or v (or both) lie in P. By setting $N = \emptyset$, we see that every interval graph is a probe interval graph. In the physical mapping problem, the DNA sequence is broken into fragments, a subset of which forms the set P of probes while the complement of this subset forms the set N. An edge between $p \in P$ and $n \in N$ denotes that probe p overlaps clone n.

Graphs and digraphs are also widely used to study other problems related to DNA including 'sequence alignment' and 'sequencing by hybridization'. In the latter problem, the reconstruction of a sequence, given the set of l-tuples of letters present in the sequence as consecutive subsequences, can be reduced to determining an 'Eulerian' path in an associated digraph. Waterman (1995) discusses this issue in further detail.

Pedigree (di)graphs. A further application of graph theory in biology is the representation and analysis of sexually reproducing populations through time, according to their parent–offspring history. The following brief overview follows Thomas (1993).

A *pedigree* is an acyclic digraph in which the vertex set is partitioned into two subsets M and F so that each vertex either has in-degree zero, or has exactly one incoming arc from a vertex in M and exactly one incoming arc from a vertex in F.

The usual interpretation here is that the members of M and F are the male and female individuals, respectively, and that each arc is directed from a parent to a child. We will denote the set of vertices of in-degree zero by I and will assume that the digraph is finite, in which case I is necessarily non-empty. The set I can be regarded as the initial 'founders' in the pedigree.

Example 1.4.3. A pedigree is shown in Fig. 1.6 with $F = \{f_1, f_2, \ldots, f_5\}$ and $M = \{m_1, m_2, \ldots, m_7\}$. The founders of this pedigree are m_1, m_2, f_1, m_3, and f_2.

Lemma 1.4.4. *For a pedigree with bipartiton $\{M, F\}$ of the vertex set, let G be the underlying graph. Then, the subgraphs of G induced by M and induced by F are both forests.*

Proof. By symmetry, it suffices to show that $G[M]$ is a forest. Suppose that $G[M]$ contains a cycle. Let M' denote the vertex set of this cycle. Since there are no directed cycles in the pedigree, there is a vertex v in M' such that v is the head of two arcs in the pedigree with both arcs having tails in M. This contradiction completes the proof. □

For various reasons, one would like to carry out probability calculations on a pedigree. In general, for such calculations, it is computationally useful to triangulate it, so that one can construct a perfect elimination ordering and thereby carry out calculations recursively. The computational complexity of the calculations depend critically on the size of the largest clique in the triangulated graph, so triangulating a pedigree while keeping this number small is particularly important.

Thomas (1993) determined a special case where such a triangulation can be achieved. For a pedigree with bipartition $\{M, F\}$ of the vertex set, an *augmented pedigree graph G* is a graph that can be obtained from the underlying graph of a pedigree by adding new edges that join a vertex in M with a vertex in F. It follows from Lemma 1.4.4 that the subgraphs of G induced by M and induced by F are both forests. A *proper k-colouring* of a graph G is an assignment of colours to the vertices of G so that adjacent vertices are assigned distinct colours and at most k colours are used.

Proposition 1.4.5.

(i) *Every augmented pedigree graph can be* 4*-coloured.*

(ii) *If a pedigree can be triangulated by only adding edges which have an end vertex in M and an end vertex in F, then the largest clique in the resulting augmented pedigree graph has size at most* 4.

Proof. To prove (i), let G be an augmented pedigree graph. Since any tree can be 2-coloured by distinguishing an arbitrary vertex v_0 and assigning to each vertex v of the tree colour $i \in \{0, 1\}$ depending on whether the number of edges separating v_0 and v has parity i (modulo 2), we can properly 2-colour the graphs $G[M]$ and $G[F]$. By choosing distinct colours, this induces a 4-colouring of G, thus establishing (i). The second part now follows by observing that, for all $k > 4$, a k-clique cannot be 4-coloured. □

It follows from Proposition 1.4.5 that any pedigree satisfying the condition described in (ii) has tree-width at most 3.

1.5 Exercises

1. Prove that a graph $G = (V, E)$ is a tree if and only if G is connected but $G \backslash e$ is disconnected for all $e \in E$.
2. Prove the equivalence of (i) and (ii) in Theorem 1.2.1.

3. Let T be a tree. Show that, for any three vertices v_1, v_2, and v_3 of T, there is a unique vertex in the intersection of the (unique) paths joining v_1 to v_2, v_1 to v_3, and v_2 to v_3.
4. Let $\{V_1, V_2, \ldots, V_k\}$ be a collection of vertex sets of subtrees of a tree. Prove that the elements of this collection satisfy the following *Helly property*: if $V_i \cap V_j \neq \emptyset$ for all $i, j \in \{1, 2, \ldots, k\}$, then $\bigcap_{i=1}^{k} V_i \neq \emptyset$.
5. A vertex v of a tree $T = (V, E)$ is a *centroid* if each component of $T \backslash v$ has at most $(1/2)|V|$ vertices. Show that a tree has either a single centroid or two adjacent centroids, in which case $|V|$ is even.
6. Prove Proposition 1.2.3.
7. Prove Proposition 1.2.4 by first establishing the result for the special case T_1 and T_2 are both path graphs.
8. Show that every graph can be realized as the intersection graph of some family of sets (see Proposition 1.3.1).
9. Let T be a rooted tree with n leaves, and suppose that each non-leaf vertex of T (including the root) has exactly two adjacent descendant vertices. Let $n_1(v), n_2(v)$ denote the numbers of leaves in the two maximal rooted subtrees of T lying below v. Prove that

$$\sum_{v \in \mathring{V}(T)} n_1(v) n_2(v) = \binom{n}{2}.$$

10. Find a chordal graph that is not an interval graph.
11. Let $\{V_1, V_2, \ldots, V_k\}$ be a partition of the vertex set of a graph G such that, for all i, $G[V_i]$ can be r_i-coloured. Suppose there exists a chordalization of G that introduces only edges that join vertices in different blocks of this partition. Prove that

$$tw(G) \leq -1 + \sum_{i=1}^{k} r_i.$$

12. Prove both parts of Proposition 1.4.2.
13. Let G be a chordal graph with at least one isolated vertex. Prove that G is the competition graph of an acyclic digraph.
14. Prove that all (finite) pedigrees can be 3-coloured. Furthermore, construct examples to show that
    (i) not every pedigree can be 2-coloured, and
    (ii) not every augmented pedigree graph can be 3-coloured.
15. Can the pedigree in Fig. 1.6 be triangulated by introducing only edges that join individuals of the opposite gender?

2

X-trees and phylogenetic trees

'Phylogenetic trees' provide a standard graphical representation of evolutionary relationships in biology. However, from a mathematical perspective, it is natural to consider a slightly more general class of objects called 'X-trees'. Briefly, an X-tree is a finite tree in which some vertices (including all vertices of degree one or two) are labelled by disjoint subsets of a set X. Although this definition may seem somewhat arbitrary, we will see in the next chapter that X-trees correspond naturally to a suitably structured set system. A phylogenetic tree is an X-tree for which only the leaves of the tree are labelled, and by singleton subsets of X.

2.1 X-trees

In this section, we introduce the concept of an X-tree.

Definition 2.1.1. An *X-tree* $\mathcal{T}$ is an ordered pair $(T; \phi)$, where T is a tree with vertex set V and $\phi : X \to V$ is a map with the property that, for each $v \in V$ of degree at most two, $v \in \phi(X)$. An X-tree is also called a *semi-labelled tree* (*on* X). Figure 2.1(a) shows an X-tree, where $X = \{1, 2, \ldots, 8\}$.

Let $\mathcal{T}$ be a semi-labelled tree $(T; \phi)$. The tree T is called the *underlying tree* of $\mathcal{T}$ and the map ϕ is called the *labelling map* of $\mathcal{T}$. We shall sometimes write $T(\mathcal{T})$ for

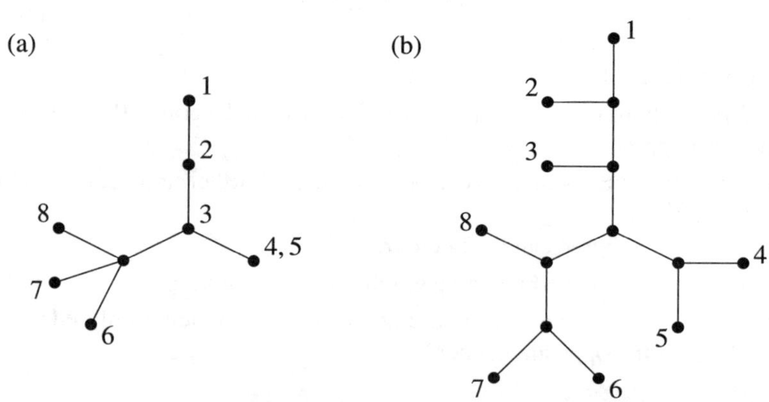

Figure 2.1 (a) An X-tree. (b) A binary phylogenetic X-tree.

$\mathcal{T}$ and $\phi(\mathcal{T})$ for ϕ, particularly when several semi-labelled trees are being considered. The domain of ϕ is called the *label set of* $\mathcal{T}$ and is denoted $\mathcal{L}(\mathcal{T})$. For convenience, and provided there is no ambiguity, we often refer to the vertices and edges of T as the vertices and edges of $\mathcal{T}$, respectively. In this case, we write $V(\mathcal{T})$ for $V(T)$ and $E(\mathcal{T})$ for $E(T)$.

For a collection $\mathcal{P}$ of semi-labelled trees, we let $\mathcal{L}(\mathcal{P}) = \bigcup_{\mathcal{T} \in \mathcal{P}} \mathcal{L}(\mathcal{T})$ and call $\mathcal{L}(\mathcal{P})$ the *label set of* $\mathcal{P}$.

Two X-trees $\mathcal{T}_1 = (T_1; \phi_1)$ and $\mathcal{T}_2 = (T_2; \phi_2)$, where $T_1 = (V_1, E_1)$ and $T_2 = (V_2, E_2)$, are *isomorphic* if there exists a bijection $\psi : V_1 \to V_2$ which induces a bijection between E_1 and E_2, and satisfies $\phi_2 = \psi \circ \phi_1$, in which case ψ is unique. We write $\mathcal{T}_1 \cong \mathcal{T}_2$ if $\mathcal{T}_1$ is isomorphic to $\mathcal{T}_2$. Throughout this book, we regard isomorphic X-trees as being equivalent. For a fixed set X, we denote the collection of all X-trees, up to equivalence, by $\mathcal{P}(X)$.

Definition 2.1.2. A *phylogenetic* (X-) *tree* $\mathcal{T}$ is an X-tree $(T; \phi)$ with the property that ϕ is a bijection from X into the set of leaves of T. A phylogenetic X-tree is also called a *phylogenetic tree* (*on* X). If, in addition, every interior vertex of T has degree three, $\mathcal{T}$ is a *binary phylogenetic* (X-) *tree*.

If $\mathcal{T} = (T; \phi)$ is a phylogenetic X-tree, we can view X as the set of leaves of the underlying tree T, and so we frequently denote the leaves of T by the elements of X since ϕ is implicitly determined. Figure 2.1(b) shows a binary phylogenetic tree on $\{1, 2, \ldots, 8\}$.

Some authors (e.g. Barthélemy and Guénoche 1991) have used different terminology, referring to phylogenetic X-trees as 'free, separated X-trees', and to binary phylogenetic X-trees as 'free, separated, ternary X-trees'. In this book, we will keep the terminology closer to the biology.

Two frequently used properties of binary phylogenetic trees are contained in Propositions 2.1.3 and 2.1.4. The first proposition follows immediately from Proposition 1.2.3.

Proposition 2.1.3. *Let* $\mathcal{T}$ *be a binary phylogenetic* X*-tree and let* $n = |X|$*. Then, for all* $n \geq 2$*,* $\mathcal{T}$ *has* $2n - 3$ *edges and* $n - 3$ *interior edges.*

Let $B(n)$ denote the collection of all binary phylogenetic trees with label set $\{1, 2, \ldots, n\}$ and let $b(n) = |B(n)|$. If $n \in \{1, 2\}$, then $b(n) = 1$. For all $n \geq 3$, the following proposition provides a well-known formula for $b(n)$, a result that dates back to Schröder (1870).

Proposition 2.1.4. *For all* $n \geq 3$,

$$b(n) = 1 \times 3 \times 5 \times \cdots \times (2n - 5) = \frac{(2n-4)!}{(n-2)!2^{n-2}}.$$

Proof. The proof is by induction on n. Since $b(3) = 1$, the result holds for $n = 3$. Suppose that the result holds for $k = n - 1$, where $k \geq 3$. Let $\psi : B(k+1) \to B(k)$ be the map that is defined by setting $\psi(\mathcal{T})$ to be the binary phylogenetic tree in $B(k)$ that is obtained from $\mathcal{T}$ by deleting the leaf labelled $k + 1$ and its incident edge, and then suppressing the resulting degree-two vertex. Clearly, ψ is onto. Furthermore, it follows by Proposition 2.1.3 that each binary phylogenetic tree in $B(k)$ is the image of $2k-3$ trees in $B(k+1)$. Therefore, by the induction assumption, $b(k+1) = (2k-3)b(k)$, and so

$$b(n) = 1 \times 3 \times 5 \times \cdots \times (2n-5),$$

as required. A routine check shows that this last expression is also equal to $(2n-4)!/[(n-2)!2^{n-2}]$. □

Recall that if m is an odd integer, $m!!$ denotes the product $m \times (m-2) \times (m-4) \times \cdots \times 1$. Using this notation, $b(n) = (2n-5)!!$. The function $b(n)$ grows quite rapidly with n as illustrated by the fact that $b(10) = 2\,027\,025$ and $b(20) \approx 2 \times 10^{20}$. Furthermore, applying Stirling's formula to the second formula for $b(n)$ in Proposition 2.1.4 gives the asymptotic equivalences

$$b(n) \sim \frac{1}{\sqrt{\pi}} 2^{n-2} n! n^{-5/2} \sim \frac{1}{2\sqrt{2}} \left(\frac{2}{e}\right)^n n^{n-2}.$$

The rate of growth in the number of phylogenetic trees is an important issue for the reconstruction of such trees from various types of data. Often we wish to select a 'best' tree under some criterion, and clearly this will be computationally impossible by searching all possible trees when n is large. Indeed, this is a major reason for developing new mathematical techniques aimed at finding optimal (or near optimal) trees without having to carry out an entire search of all trees.

2.2 Rooted *X*-trees

In this section, we extend the notion of X-trees and phylogenetic trees to rooted X-trees and rooted phylogenetic trees, respectively. Furthermore, we highlight some of the fundamental interplays between X-trees and rooted X-trees.

Definition 2.2.1. A *rooted X-tree $\mathcal{T}$* is an ordered pair $(T;\phi)$, where $T = (V, E)$ is a rooted tree with root vertex ρ, and $\phi : X \to V$ is a map with the property that, for each $v \in V - \{\rho\}$ of degree at most two, $v \in \phi(X)$. A rooted X-tree is also called a *rooted semi-labelled tree* (*on X*). An example of a rooted semi-labelled tree is shown in Fig. 2.2.

Let $\mathcal{T}$ be a rooted semi-labelled tree $(T;\phi)$. Analogous to the definitions given in Section 2.1, the tree T and the map ϕ are called the underlying tree of $\mathcal{T}$ and the labelling map of $\mathcal{T}$, respectively. Again, we shall sometimes use the notation $T(\mathcal{T})$ for T and $\phi(\mathcal{T})$ for ϕ. The label set of $\mathcal{T}$, denoted $\mathcal{L}(\mathcal{T})$, is the domain of ϕ.

The concepts of a phylogenetic tree and a binary phylogenetic tree also have analogues in the rooted tree setting.

Definition 2.2.2. A *rooted phylogenetic* (X-) *tree $\mathcal{T}$* is a rooted X-tree $(T;\phi)$ such that

(i) ϕ is a bijective map from X into the set of leaves of T and
(ii) the root has degree at least two.

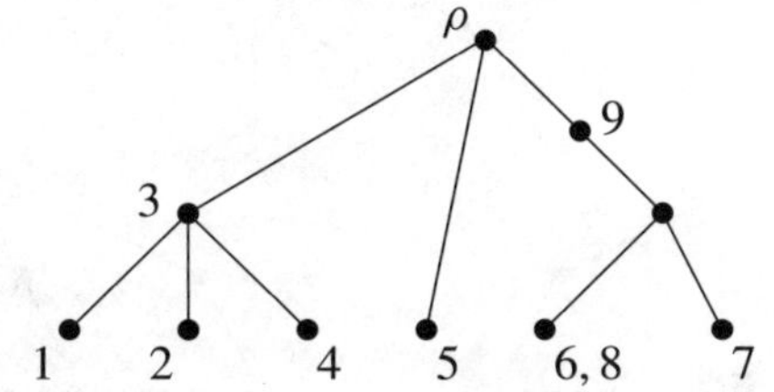

Figure 2.2 A rooted semi-labelled tree on $\{1, 2 \ldots, 9\}$.

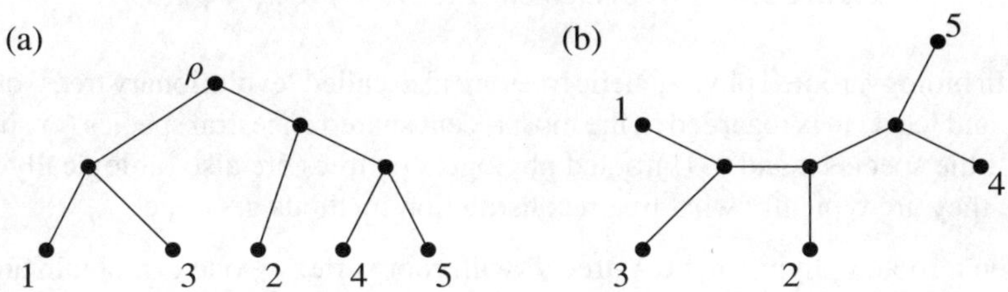

Figure 2.3 (a) A rooted binary phylogenetic tree $\mathcal{T}$ with root ρ. (b) The unrooted phylogenetic tree $\mathcal{T}^{-\rho}$.

If, in addition, every interior vertex of T has degree three except for the root which has degree two, then $\mathcal{T}$ is a *rooted binary phylogenetic* (X-) *tree*.

For technical reasons, we also consider the tree consisting of a single-root vertex labelled by the element of a singleton set to be a rooted binary phylogenetic tree. A rooted phylogenetic tree on $\{1, 2, 3, 4, 5\}$ is shown in Fig. 2.3(a).

We can regard a rooted tree as a digraph by viewing each edge as an arc directed away from the root. Such a tree is a rooted binary tree precisely if, under the digraph interpretation, each vertex has out-degree zero or two, thereby justifying the adjective 'binary'.

Let $\mathcal{T} = (T; \phi)$ be a rooted X-tree and let A be a subset of X. We will refer to the unique vertex of T that is the greatest lower bound of the set $\{\phi(x): x \in A\}$ under the partial order $\leq_T$ as the *most recent common ancestor* of A in $\mathcal{T}$. In case $A = \{x, y\}$, we will denote the most recent common ancestor of A as $\text{lca}_{\mathcal{T}}(x, y)$ or, more briefly, $\text{lca}(x, y)$. For example, in Fig. 2.3(a), we have $\text{lca}(2, 3) = \rho$.

Phylogenetic trees and evolutionary biology. Phylogenetic trees and their rooted counterparts provide a convenient representation for evolutionary relationships in biology. In particular, for a rooted phylogenetic tree $\mathcal{T}$ on X, we view the edges of $\mathcal{T}$ as being directed away from the root ρ, and then regard $\mathcal{T}$ as describing the evolution of the set X of extant species that label the leaves of $\mathcal{T}$ from a common hypothetical ancestral species at ρ; the other interior vertices of $\mathcal{T}$ correspond to further hypothetical ancestral species. Alternatively, we can view the interior vertices of $\mathcal{T}$ as corresponding to past speciation

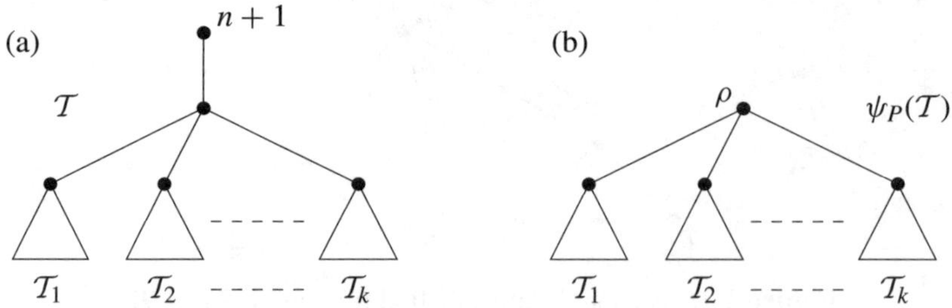

Figure 2.4 (a) An element $\mathcal{T}$ of $P(n+1)$. (b) $\psi_P(\mathcal{T})$.

events. In biology, rooted phylogenetic trees are also called 'evolutionary trees' or 'cladograms', and $\mathrm{lca}(x, y)$ is regarded as the most recent shared ancestral species (or speciation event) of the species x and y. Unrooted phylogenetic trees are also biologically relevant because they are typically what tree reconstruction methods generate.

Given a rooted phylogenetic X-tree $\mathcal{T}$ with root vertex ρ, one can obtain an associated unrooted phylogenetic X-tree, denoted $\mathcal{T}^{-\rho}$, as follows. If the degree of ρ in the underlying tree T is at least three, we simply take $\mathcal{T}^{-\rho}$ as $\mathcal{T}$ without distinguishing vertex ρ. On the other hand, if ρ has degree two, we suppress ρ to obtain $\mathcal{T}^{-\rho}$. Figure 2.3 illustrates a rooted phylogenetic tree and its associated unrooted phylogenetic tree.

For all $n \geq 2$, let $P(n)$ denote the collection of phylogenetic trees with label set $\{1, 2, \ldots, n\}$. Furthermore, for all $n \geq 1$, let $RP(n)$ and $RB(n)$ denote the collections of rooted phylogenetic tree and rooted binary phylogenetic trees, respectively, with label set $\{1, 2, \ldots, n\}$. For all $n \geq 1$, we now describe a natural bijection between $P(n+1)$ and $RP(n)$, and between $B(n+1)$ and $RB(n)$.

Let $\psi_P : P(n+1) \rightarrow RP(n)$ be the map that deletes from a tree in $P(n+1)$ the leaf labelled $n+1$ and its incident edge, and then roots the resulting tree at the remaining end-vertex of this edge. This map is illustrated in Fig. 2.4. Let ψ_B denote the restriction of ψ_P to $B(n+1)$. The straightforward proof of Proposition 2.2.3 is left to the reader.

Proposition 2.2.3. *For all $n \geq 1$,*

(i) $\psi_P : P(n+1) \rightarrow RP(n)$ is a bijection and
(ii) $\psi_B : B(n+1) \rightarrow RB(n)$ is a bijection.

Combining Proposition 2.1.4 with Proposition 2.2.3(ii), we get the following result.

Corollary 2.2.4. *For all $n \geq 2$,*

$$|RB(n)| = (2n-3)!! = \frac{(2n-2)!}{(n-1)!2^{n-1}}.$$

For all $n \geq 2$, the value $(2n-3)!!$ in Corollary 2.2.4 is also the number of ways to pair $2n-2$ points into $n-1$ groups each of size two. Equivalently, it is the number of 'perfect matchings' of a complete graph on $2n-2$ vertices. In fact, there is a natural bijection between $RB(n)$ and the set of matchings of $2n-2$ points into $n-1$ pairs. Such a bijection

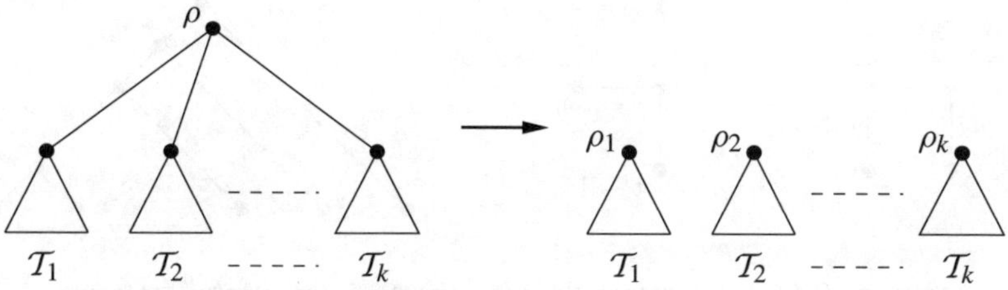

Figure 2.5 The standard decomposition of a rooted phylogenetic tree.

has been described by Diaconis and Holmes (1998) and, in a more general setting, by Erdös and Székely (1989).

Let $\mathcal{T} = (T; \phi)$ and $\mathcal{T}' = (T'; \phi')$ be rooted phylogenetic trees on X and X', respectively, where $X' \subseteq X$. We say that $\mathcal{T}'$ is a *rooted phylogenetic subtree* of $\mathcal{T}$ if T' is a rooted subtree of T lying below some vertex and $\phi' = \phi|X'$. A natural way to decompose a rooted phylogenetic tree $\mathcal{T}$ is to associate with $\mathcal{T}$ its set of maximal rooted phylogenetic subtrees. We call this the *standard decomposition* and it underlies a variety of inductive arguments, algorithms based on dynamic programming, and techniques for tree enumeration. Figure 2.5 illustrates this decomposition.

As an application of this decomposition, we now describe a recursive enumeration of $RB(n)$. Although this may seem excessive, since we already have a simple explicit formula for $|RB(n)|$, the approach is useful for other reasons. In particular, it provides the generating function for $|RB(n)|$ that will be used later, and it suggests a corresponding result for $|RP(n)|$ (see Exercise 2).

For all $n \geq 2$, each tree in $RB(n)$ decomposes as in Fig. 2.5 into two rooted binary phylogenetic trees with label sets X_1 and X_2 such that $1 \leq |X_1| \leq n - 1$, and X_1 and X_2 partitions $\{1, 2, \ldots, n\}$. Furthermore, the number of rooted binary phylogenetic trees in $RB(n)$ that partitions $\{1, 2, \ldots, n\}$ in this way is

$$\frac{1}{2}\binom{n}{i}|RB(i)||RB(n-i)|,$$

where $i = |X_1|$. The factor of $\frac{1}{2}$ appears as the order of X_1 and X_2 is irrelevant. Summing over i, we obtain the recursion

$$|RB(n)| = \frac{1}{2}\sum_{i=1}^{n-1}\binom{n}{i}|RB(i)||RB(n-i)|. \tag{2.1}$$

The recursion (2.1) leads to a useful relationship for the exponential generating function for $RB(n)$. Let $R(x) = \sum_{n \geq 1} |RB(n)|(x^n/n!)$. It is easily checked that (2.1)

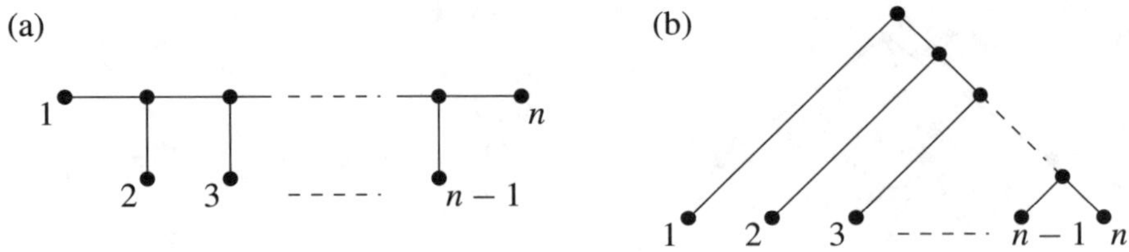

Figure 2.6 (a) A caterpillar tree. (b) A rooted caterpillar tree.

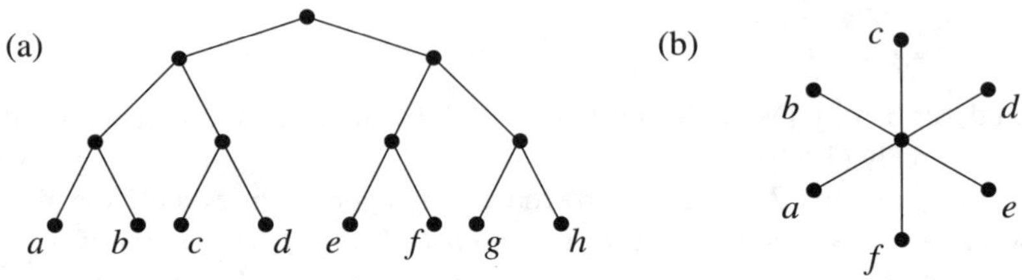

Figure 2.7 (a) A balanced tree of height three. (b) A star tree.

implies the quadratic equation

$$R(x) = x + \tfrac{1}{2}R(x)^2. \tag{2.2}$$

Solving this quadratic equation for $R(x)$ and choosing the root whose power series has non-negative coefficients, we get $R(x) = 1 - \sqrt{1-2x}$.

We end this section by describing some families of phylogenetic trees that are used in examples and in the proofs of certain results.

Caterpillar trees, balanced trees, and stars. For all $n \geq 1$, a *caterpillar* on n leaves is any binary phylogenetic tree for which the induced subtree on the interior vertices forms a path graph. An example of a caterpillar on n leaves is shown in Fig. 2.6(a). The rooted analogue, the *rooted caterpillar*, is any rooted phylogenetic tree for which the induced subtree on the interior vertices is a path graph with the root at one end of the path. An example of a rooted caterpillar on n leaves is shown in Fig. 2.6(b).

A *balanced tree* of height $h \geq 0$ is a rooted binary phylogenetic tree, with $n = 2^h$ leaves, each of which is separated from the root by exactly h edges. Figure 2.7(a) shows a balanced tree of height three. A phylogenetic tree with no interior edges, and therefore with a single interior vertex that is adjacent to all the leaves, is called a *star tree*. A star tree on six leaves is shown in Fig. 2.7(b).

2.3 Ranked trees and the coalescent process

Let $\mathcal{T}$ be a rooted phylogenetic tree and let T denote the underlying tree of $\mathcal{T}$. Let h be an injective map from the set $\mathring{V}$ of interior vertices of T into the set $\{1, 2, \ldots, |\mathring{V}|\}$ that

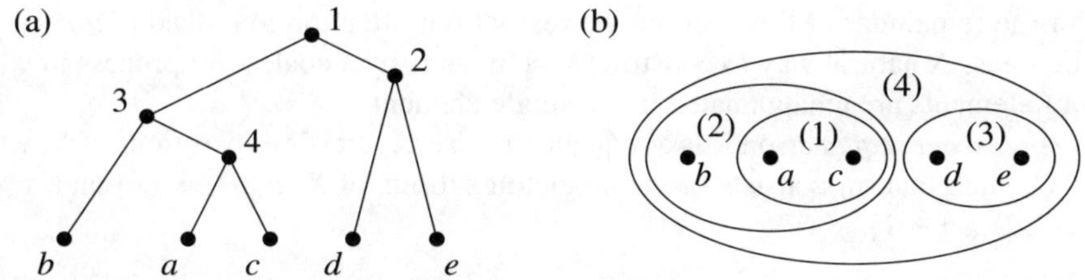

Figure 2.8 (a) A ranked phylogenetic tree. (b) A coalescent sequence on $\{a, b, c, d, e\}$.

satisfies the following property:

$$\text{if } v_1 \leq_{\mathcal{T}} v_2, \quad \text{then } h(v_1) \leq h(v_2). \tag{2.3}$$

The pair $(\mathcal{T}, h)$ is called a *ranked phylogenetic tree* and the map h is a *rank function* for $\mathcal{T}$.

In biology, as the interior vertices of a rooted phylogenetic tree $\mathcal{T}$ correspond to speciation events, a natural choice for a rank function h for $\mathcal{T}$ would be to assign each element in $\mathring{V}(\mathcal{T})$ their position in the order of the occurrence of these events. For this reason, the pair $(\mathcal{T}, h)$ is sometimes called a 'labelled history' or a 'ranked dendogram'.

Example 2.3.1. Consider the phylogenetic tree $\mathcal{T}$ shown in Fig. 2.8(a) with label set $\{a, b, c, d, e\}$. Each interior vertex of $\mathcal{T}$ is assigned a distinct element of $\{1, 2, 3, 4\}$ and this assignment satisfies (2.3). Hence, $\mathcal{T}$ together with this assignment is a ranked phylogenetic tree.

Observe that every rooted phylogenetic tree admits a rank function, but this function may not be unique. The next proposition, due to Knuth (1997), allows us to count the number of rank functions associated with a rooted phylogenetic tree.

Proposition 2.3.2. *Let $\mathcal{T}$ be a rooted phylogenetic tree and let $\mathring{V}$ denote the set of interior vertices of $\mathcal{T}$. For each $v \in \mathring{V}$, let λ_v denote the number of elements of $\mathring{V}$ that are descendants of v. Then, the number of rank functions for $\mathcal{T}$ is*

$$\frac{|\mathring{V}|!}{\prod_{v \in \mathring{V}} \lambda_v}.$$

Proof. Consider the partially ordered set $\mathring{V}$ under $\leq_{\mathcal{T}}$. In the language of poset theory, the number of rank functions for $\mathcal{T}$ corresponds to the number of linear extensions of this poset. In general, if one has an n-element poset P and, for all $x \in P$, we let λ_x denote the cardinality of $\{y \in P \colon x \leq y\}$, then the number of linear extensions of P is exactly $n!/\prod_{x \in P} \lambda_x$, provided the associated cover digraph is a rooted tree with the root being the unique minimal element of the poset (see Stanley 1997, p. 312). □

For the remainder of this section, we restrict our attention to ranked *binary* phylogenetic trees. A natural way to construct such trees is by a coalescent process in which pairs of elements are amalgamated into a single element.

A *coalescent sequence* on X is a sequence $\pi_0, \pi_1, \ldots, \pi_{n-1}$ of partitions of X, where $n = |X|$, such that π_0 consists of the singleton subsets of X, $\pi_{n-1} = \{X\}$ and, for all $i \in \{1, 2, \ldots, n-1\}$,

$$\pi_i = (\pi_{i-1} - \{A_{i-1}, B_{i-1}\}) \cup \{A_{i-1} \cup B_{i-1}\}$$

for some distinct blocks A_{i-1} and B_{i-1} of π_{i-1}.

Example 2.3.3. For $X = \{a, b, c, d, e\}$, a coalescent sequence on X is $\pi_0 = \{\{a\}, \{b\}, \{c\}, \{d\}, \{e\}\}, \pi_1 = \{\{a, c\}, \{b\}, \{d\}, \{e\}\}, \pi_2 = \{\{a, b, c\}, \{d\}, \{e\}\}, \pi_3 = \{\{a, b, c\}, \{d, e\}\}$, $\pi_4 = \{X\}$. Pictorially, this sequence can be displayed as shown in Fig. 2.8(b).

Based on a coalescent sequence on X, the *coalescent process* constructs a particular ranked binary phylogenetic X-tree as follows. Let $\pi_0, \pi_1, \ldots, \pi_{n-1}$ be a coalescent sequence on X. Define $\mathcal{P}_0$ to be the collection of rooted binary phylogenetic trees that consists of a single vertex labelled by a single element of X. Thus, $|\mathcal{P}_0| = |X|$. In general, for all $i \in \{1, 2, \ldots, n-1\}$, define $\mathcal{P}_i$ to be the collection of rooted binary phylogenetic trees that are obtained from $\mathcal{P}_{i-1}$ by

(i) replacing the rooted binary phylogenetic trees on A_{i-1} and B_{i-1} with the rooted binary phylogenetic tree whose maximal rooted phylogenetic subtrees are the rooted binary phylogenetic trees with label sets A_i and B_i, and
(ii) denoting the root of the newly created tree by v_i.

Let $\mathcal{T}$ denote the unique element of $\mathcal{P}_{n-1}$ and let h be the map from $\{v_1, v_2, \ldots, v_{n-1}\}$ into $\{1, 2, \ldots, n-1\}$ defined by $h(v_i) = n - i$ for all i. The pair $(\mathcal{T}, h)$ is the particular ranked binary phylogenetic tree associated with the coalescent sequence $\pi_0, \pi_1, \ldots, \pi_{n-1}$ on X. As an example, the ranked binary phylogenetic tree shown in Fig. 2.8(a) is the tree associated with the coalescent sequence given in Example 2.3.3.

It can be shown that every ranked binary phylogenetic X-tree is associated with exactly one coalescent sequence on X. Furthermore, if π_{i-1} is a partition in a coalescent sequence on X, then $|\pi_{i-1}| = n+1-i$ and so there are $\binom{n+1-i}{2}$ possible choices for which pair of elements in π_{i-1} are amalgamated to obtain π_i. Consequently, the total number of ranked binary phylogenetic X-trees is $\prod_{i=1}^{n-1} \binom{n+1-i}{2}$. This establishes the following result from Murtagh (1984).

Proposition 2.3.4. *The number of ranked binary phylogenetic trees on a fixed label set of size n is*

$$\frac{n!(n-1)!}{2^{n-1}}.$$

2.4 Tree shapes

If we take a phylogenetic tree and ignore the labels on the leaves, we obtain a *tree shape*. Formally, a tree shape can be viewed as representing an equivalence class of phylogenetic trees. In particular, two phylogenetic trees $\mathcal{T}_1$ and $\mathcal{T}_2$ with equal label sets are *shape equivalent* if $T(\mathcal{T}_1)$ is isomorphic to $T(\mathcal{T}_2)$. This concept extends naturally to the rooted tree setting: two rooted phylogenetic trees $\mathcal{T}_1$ and $\mathcal{T}_2$ with equal label sets are shape equivalent if there is an isomorphism between $T(\mathcal{T}_1)$ and $T(\mathcal{T}_2)$, which maps the root of the first tree to the root of the second. The resulting equivalence class of rooted phylogenetic trees is called a *rooted tree shape*. We may also add the word 'binary' to tree shape if the underlying phylogenetic trees are (rooted or unrooted) binary phylogenetic trees.

For the class of unrooted binary phylogenetic trees on n leaves, there is exactly one tree shape for each $n \in \{2,3,4,5\}$, but for $n = 6$ there are two tree shapes (Fig. 2.9). For the class of rooted binary phylogenetic trees on n leaves, there is one rooted tree shape for $n = 3$, but two rooted tree shapes for $n = 4$ (Fig. 2.10). This prompts two natural enumeration problems both for the rooted and unrooted settings:

(i) given a fixed n, determine the number of tree shapes for phylogenetic trees on n leaves;

(ii) for a tree shape τ, count the number of phylogenetic trees on a given label set with shape τ.

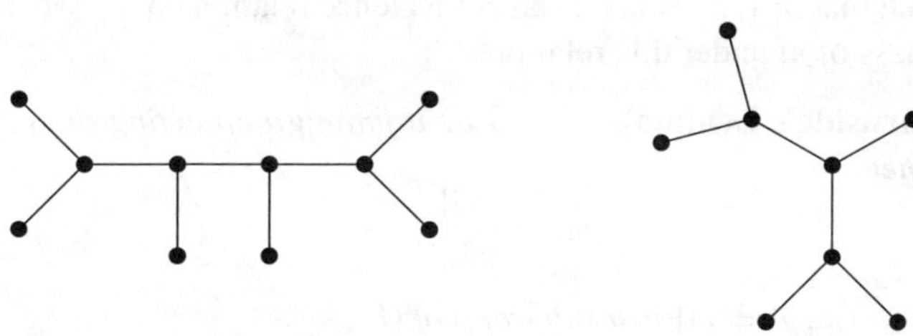

Figure 2.9 The two tree shapes for $B(6)$.

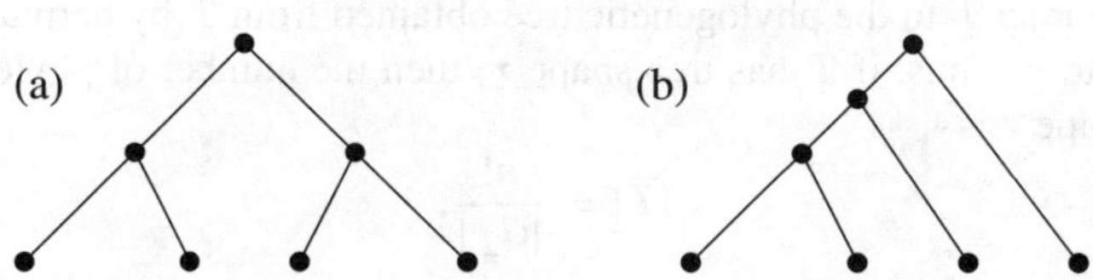

Figure 2.10 The two rooted tree shapes for $RB(4)$.

Regarding (i), there are the following results for rooted binary tree shapes. Let r_n denote the number of rooted binary tree shapes on n leaves, and let

$$r(x) = \sum_{n\geq 1} r_n x^n = x + x^2 + x^3 + 2x^4 + 3x^5 + \cdots$$

be the ordinary generating function for $r(x)$. Using the standard decomposition of rooted phylogenetic trees illustrated in Fig. 2.5, Harding (1971) showed that

$$r(x) = x + \tfrac{1}{2}(r^2(x) + r(x^2)).$$

Using analytical techniques, this last equation can be used to establish the asymptotic equivalence

$$r_n \sim ab^n n^{-3/2},$$

where $a = 0.31877\ldots$ and $b = 2.4832\ldots$ (see Harding 1971 for further details).

In the case of unrooted binary phylogenetic trees, the corresponding generating function for the number of tree shapes can be expressed in terms of $r(x)$ (for details see Dobson 1974).

Problem (ii) is easier to answer, but it requires a short excursion into the theory of group actions on sets. An *action of a group* $\mathcal{G}$ on a set M is a map $M \times \mathcal{G} \to M$ such that, for all $m \in M$ and for all $g_1, g_2 \in \mathcal{G}$,

(i) $(m, 1_{\mathcal{G}}) = m$ and
(ii) $(m, g_1 g_2) = ((m, g_1), g_2)$.

It is straightforward to show that the relation $\sim$ defined by $m_1 \sim m_2$, if there is an element g of $\mathcal{G}$ such that $(m_1, g) = m_2$, is an equivalence relation on M. We denote by $\overline{m}$ the equivalence class of m under this relation.

Lemma 2.4.1 (Burnside's Lemma). *Let $\mathcal{G}$ be a finite group acting on a finite set M and let $m \in M$. Then,*

$$|\overline{m}| = \frac{|\mathcal{G}|}{|\mathcal{G}_m|},$$

where $\mathcal{G}_m = \{g \in \mathcal{G}\colon (m, g) = m\}$ is a subgroup of $\mathcal{G}$.

In our context, M is the collection of either all phylogenetic trees or all rooted phylogenetic trees with label set $\{1, \ldots, n\}$, while $\mathcal{G}$ is the (symmetric) group of all $n!$ permutations of $\{1, \ldots, n\}$. For an element g of $\mathcal{G}$ and an element $\mathcal{T}$ of M, the action of g on $\mathcal{T}$ is to map $\mathcal{T}$ to the phylogenetic tree obtained from $\mathcal{T}$ by permuting the label set according to g. Thus, if $\mathcal{T}$ has tree shape τ, then the number of phylogenetic trees having tree shape τ is

$$|\overline{\mathcal{T}}| = \frac{n!}{|\mathcal{G}_{\mathcal{T}}|}, \tag{2.4}$$

where $\mathcal{G}_{\mathcal{T}}$ is the collection of permutations of $\{1, \ldots, n\}$ that leaves $\mathcal{T}$ unchanged. Hendy *et al.* (1984) provide the following formulae for determining $|\mathcal{G}_{\mathcal{T}}|$.

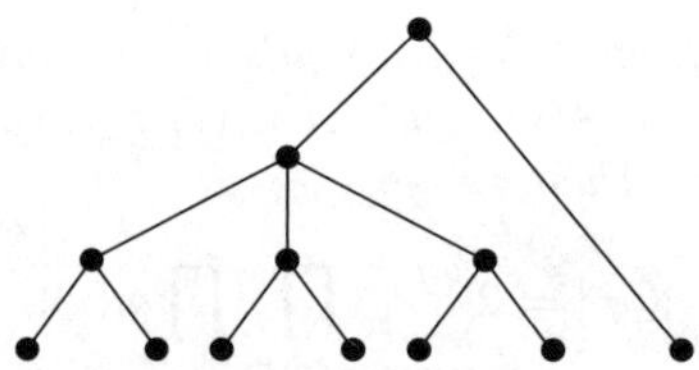

Figure 2.11 A rooted tree shape.

Proposition 2.4.2. *Let $\mathcal{T}$ be a rooted phylogenetic tree. For each interior vertex v of $\mathcal{T}$, let $D(v)$ denote the collection of maximal rooted phylogenetic subtrees that lie below v. Now, let us impose the rooted shape equivalence relation on $D(v)$ and let $n_1(v), n_2(v), \ldots$ denote the sizes of the resulting equivalence classes. Then,*

$$|\mathcal{G}_{\mathcal{T}}| = \prod_{v \in \mathring{V}(\mathcal{T})} \prod_i n_i(v)!$$

To illustrate Proposition 2.4.2, consider a rooted phylogenetic tree $\mathcal{T}$ having the shape shown in Fig. 2.11. Applying Proposition 2.4.2 and considering the interior vertices as one moves down the tree,

$$|\mathcal{G}_{\mathcal{T}}| = (1!)^2 \times (3!) \times (2!) \times (2!) \times (2!) = 48.$$

For a rooted binary phylogenetic tree $\mathcal{T}$, the formula for $|\mathcal{G}_{\mathcal{T}}|$ is simpler. An interior vertex v of a rooted binary tree shape τ is a *symmetry vertex* if the two maximal rooted subtrees that lie below v have the same shape. Let $s(\tau)$ denote the number of symmetry vertices of a rooted binary tree shape τ. Combining Proposition 2.4.2 and eqn (2.4), we obtain the following result.

Corollary 2.4.3. *For a rooted binary phylogenetic tree $\mathcal{T}$ of shape τ,*

$$|\mathcal{G}_{\mathcal{T}}| = 2^{s(\tau)}.$$

Thus, the number of trees in $RB(n)$ having shape τ is $n!2^{-s(\tau)}$.

For unrooted phylogenetic trees $\mathcal{T}$, there is a corresponding formula for $|\mathcal{G}_{\mathcal{T}}|$. However, it is complicated slightly by the fact that an unrooted tree can also have a symmetry that interchanges two adjacent vertices. More precisely, a *central edge* in an unrooted phylogenetic tree is an edge for which the two rooted subtrees obtained by deleting this edge have the same shape. Note that a phylogenetic tree can have at most one central edge. Let $c(\mathcal{T})$ take the value 1 if $\mathcal{T}$ has a central edge, otherwise set $c(\mathcal{T}) = 0$. The analogue of Proposition 2.4.2 for unrooted phylogenetic trees is the following.

Proposition 2.4.4. *Let $\mathcal{T}$ be a phylogenetic tree. For each interior vertex v of $\mathcal{T}$, let $A(v)$ denote the collection of all maximal rooted phylogenetic subtrees of the rooted* *phylogenetic tree obtained from $\mathcal{T}$ by distinguishing v. Now, let us impose the rooted shape equivalence relation on $A(v)$, and let $n_1(v), n_2(v), \ldots$ denote the sizes of the resulting equivalence classes. Then*

$$|\mathcal{G}_{\mathcal{T}}| = 2^{c(\mathcal{T})} \prod_{v \in \mathring{V}(\mathcal{T})} \prod_i n_i(v)!$$

2.5 Randomly generated trees

In this section, we consider two simple models that randomly generate rooted binary phylogenetic trees. These models have been investigated for a variety of reasons. For example, simple models of speciation (and extinction) are closely related to the models we describe below, and the predictions that such models make about tree shape can be used to test hypotheses concerning speciation. Also, these models are useful for exploring biases in tree reconstruction methods, and for providing prior distributions on trees in tree comparison and Bayesian approaches to tree reconstruction. This section deals with models for generating rooted binary phylogenetic trees; however, if we suppress the root of the generated tree we can also obtain an induced probability distribution on (unrooted) binary phylogenetic trees.

Yule–Harding model. The first model we consider is the *Yule–Harding model*. For all $n \geq 2$, it generates a random element of $RB(n)$ as follows. Beginning with the rooted phylogenetic tree on two leaves that is randomly labelled with a pair of distinct elements from $\{1, \ldots, n\}$, repeat the following procedure until the resulting rooted binary phylogenetic tree has n leaves:

> To the rooted binary phylogenetic tree so far constructed, select a pendant edge randomly and with uniform probability. Subdivide this edge to create a new vertex and add a new leaf adjacent to this vertex. Label this new leaf by selecting with uniform probability one of the labels from $\{1, \ldots, n\}$ that does not yet label a leaf of the tree.

It is easy to check that, for the Yule–Harding model on $RB(4)$, the probability of generating a rooted binary phylogenetic tree that has the shapes shown in Fig. 2.10(a) and (b) is $\frac{1}{3}$ and $\frac{2}{3}$, respectively.

The Yule–Harding model has a variety of attractive mathematical properties and has been widely studied (Harding 1971; Slowinski 1990; Brown 1994; Aldous 1996). In evolutionary biology, the model describes the probability distribution on rooted binary phylogenetic trees under a general process that allows for speciation and extinction, provided a certain exchangeability condition is satisfied. Informally, this condition states that at any time when a speciation or extinction event occurs, it is equally likely to apply to any of the species extant at that moment (see Aldous 1996 for details).

The Yule–Harding model can be characterized in several different ways including the following. First, the Yule–Harding model gives the same probability distribution on $RB(n)$ as that obtained by taking the uniform distribution on the class of ranked binary phylogenetic trees with label set $\{1, \ldots, n\}$, and ignoring the ranking of the interior vertices. Secondly, the Yule–Harding model also gives the same distribution on $RB(n)$ as the 'coalescent model' from population genetics.

Uniform model. A second model for randomly generating rooted binary phylogenetic trees is the *uniform model* (sometimes called the *PDA model* in biology). In this model, each tree in $RB(n)$ is selected with equal probability.

The Yule–Harding and uniform models lead to quite different probability distributions on rooted binary phylogenetic trees. For example, under the uniform model, the probability of generating a rooted binary phylogenetic tree in $RB(4)$ that has the shape in Fig. 2.10(a) is $3/|RB(4)| = \frac{1}{5}$; however, for the Yule–Harding model, the corresponding probability is $\frac{1}{3}$.

The Yule–Harding model tends to favour rooted binary phylogenetic trees that have greater 'balance' between the numbers of descendants of each interior vertex. This is not surprising given the close connection described above between the Yule–Harding model and the uniform distribution on the set of ranked binary phylogenetic trees with label set $\{1, \ldots, n\}$. Clearly, rooted binary phylogenetic trees that have more balanced assignments of species allow for more rankings of the interior vertices. To compare the distribution on the numbers of leaves on either side of the root under the two models, generate a tree in $RB(n)$ under either model and then select, with uniform probability, one of the two maximal rooted phylogenetic subtrees lying below the root. For all $i \in \{1, 2, \ldots, n\}$, let $p_{\text{unif}}(i, n)$ and $p_{\text{Yule}}(i, n)$ denote the probability that this randomly generated tree has exactly i leaves. The following result is from Slowinski (1990).

Theorem 2.5.1. *For all* $i \in \{1, 2, \ldots, n-1\}$,

$$p_{\text{Yule}}(i, n) = \frac{1}{n-1}$$

and

$$p_{\text{unif}}(i, n) = \frac{1}{2}\binom{n}{i}\frac{|RB(i)||RB(n-i)|}{|RB(n)|}.$$

Now, for all $i \geq 1$, let $c_i = [1/(i+1)]\binom{2i}{i}$ denote the i-th *Catalan number.* Theorem 2.5.1 implies that, for all i,

$$\lim_{n \to \infty} p_{\text{unif}}(i, n) = c_{i-1}4^{-i}.$$

Observe that in the case of the uniform model, as i ranges from 1 to 4, we have $\lim_{n \to \infty} p_{\text{unif}}(i, n) = \frac{1}{4}, \frac{1}{16}, \frac{1}{32}, \frac{5}{128}$, respectively, which contrasts sharply with the corresponding flat distribution described by the Yule–Harding model.

We end this section by describing explicit formulae for the probability of generating a rooted binary phylogenetic tree of a particular shape under the two models. Given a rooted binary tree shape τ, let $p_{\text{unif}}(\tau)$ and $p_{\text{Yule}}(\tau)$ denote the probability that the tree generated by the uniform and Yule–Harding models, respectively, has shape τ. Recall that $s(\tau)$ is the number of symmetry vertices of any rooted binary phylogenetic tree having shape τ, c_n is the n-th Catalan number, and, for an interior vertex v in τ, λ_v is the number of interior vertices of τ that are descendants of v. Since τ is binary, λ_v is one less than the number of leaves that lie below v.

Theorem 2.5.2. *Let τ be a rooted binary tree shape with n leaves. Then*

(i) $p_{\text{unif}}(\tau) = 2^{n-1-s(\tau)}/c_{n-1}$ *and*

(ii) $p_{\text{Yule}}(\tau) = 2^{n-1-s(\tau)}/\prod_{v\in\mathring{V}} \lambda_v$, *where $\mathring{V}$ is the set of interior vertices of τ.*

Proof. The formula for $p_{\text{unif}}(\tau)$ is obtained by taking the number of trees in $RB(n)$ having shape τ and dividing it by the total number of trees in $RB(n)$. By Corollary 2.4.3, the first number is $n!2^{-s(\tau)}$ and, by Corollary 2.2.4, the second number is $(2n-2)!/[(n-1)!2^{n-1}]$. The result now follows using straightforward algebra.

To prove (ii), we exploit the observation that the distribution on $RB(n)$ under the Yule–Harding model is precisely the distribution that arises by selecting, with uniform probability, a ranked binary phylogenetic tree and then ignoring the ranking of the interior vertices. Thus, the probability of generating a rooted binary phylogenetic tree of shape τ under this model is the number of ranked binary phylogenetic trees with label set $\{1,\ldots,n\}$ that have shape τ divided by the total number of ranked binary phylogenetic trees with label set $\{1,\ldots,n\}$. By Corollary 2.4.3, the number of trees in $RB(n)$ of shape τ is $n!2^{-s(\tau)}$ and, by Proposition 2.3.2, each of these trees has exactly

$$\frac{(n-1)!}{\prod_{v\in\mathring{V}} \lambda_v}$$

possible rankings. By Proposition 2.3.4, the total number of ranked binary phylogenetic trees with label set $\{1,\ldots,n\}$ is

$$\frac{n!(n-1)!}{2^{n-1}}.$$

Combining these three quantities we obtain the desired result. □

2.6 Tree rearrangement operations

In this section, we introduce three types of tree rearrangement operations on binary phylogenetic trees. These operations are useful in numerous ways for the reconstruction and comparison of phylogenetic trees.

In the following descriptions of these tree rearrangement operations, $\mathcal{T} = (T;\phi)$ is a binary phylogenetic X-tree and $e = \{u, v\}$ is an edge of $\mathcal{T}$. We begin with the least restrictive operation.

(TBR) Let T' be the binary tree obtained from T by deleting e, adding an edge between a vertex that subdivides an edge of one component of $T\backslash e$ and a vertex that subdivides an edge of the other component of $T\backslash e$, and then suppressing any resulting degree-two vertices. If a component of $T\backslash e$ consists of a single vertex, then the added edge is attached to this vertex. The binary phylogenetic X-tree $(T'; \phi')$, where $\phi'(x) = \phi(x)$ for all $x \in X$, is said to be obtained from $\mathcal{T}$ by a single *tree bisection and reconnection* (TBR) operation. The generic form of a TBR operation is illustrated in Fig. 2.12, where e is the 'deleted' edge and f is the 'added' edge.

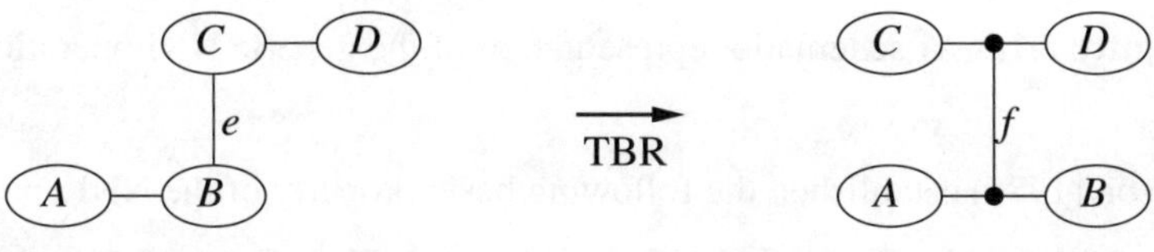

Figure 2.12 A schematic representation of the generic TBR operation.

(SPR) Now, let T' be the binary tree obtained from T by deleting e (*pruning* a subtree), adding an edge between an end vertex u of e and a vertex that subdivides an edge in the component of $T\backslash e$ that does not contain u (*regrafting* the subtree), and then suppressing any resulting degree-two vertices. The binary phylogenetic X-tree $(T'; \phi')$, where $\phi'(x) = \phi(x)$ for all $x \in X$, is said to be obtained from $\mathcal{T}$ by a single *subtree prune and regraft* (SPR) operation. The generic form of an SPR operation is illustrated in Fig. 2.13, where e is the 'deleted' edge and f is the 'added' edge.

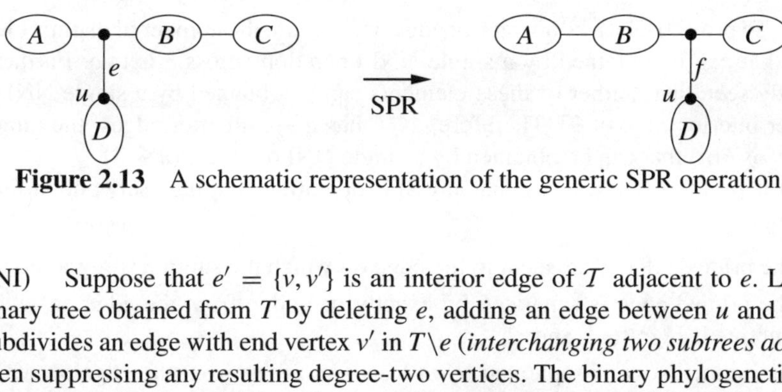

Figure 2.13 A schematic representation of the generic SPR operation.

(NNI) Suppose that $e' = \{v, v'\}$ is an interior edge of $\mathcal{T}$ adjacent to e. Let T' be the binary tree obtained from T by deleting e, adding an edge between u and a vertex that subdivides an edge with end vertex v' in $T\backslash e$ (*interchanging two subtrees across* e'), and then suppressing any resulting degree-two vertices. The binary phylogenetic X-tree $(T'; \phi')$, where $\phi'(x) = \phi(x)$ for all $x \in X$, is said to be obtained from $\mathcal{T}$ by a single *nearest neighbour interchange* (NNI) operation. Figure 2.14 illustrates the generic form of an NNI operation, where e is the 'deleted' edge and f is the 'added' edge.

Clearly, any NNI operation is a particular type of SPR operation and any SPR operation is a particular type of TBR operation. Also, every TBR operation is either a single SPR operation or the composition of two SPR operations. Furthermore, each of the operations is reversible; that is, if a binary phylogenetic tree $\mathcal{T}'$ is obtained from $\mathcal{T}$ by a particular tree rearrangement operation, then $\mathcal{T}$ can be obtained from $\mathcal{T}'$ by using the same type of operation.

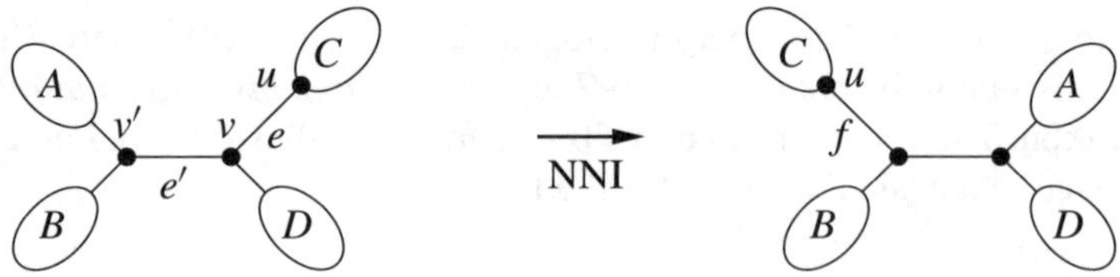

Figure 2.14 A schematic representation of the generic NNI operation.

Robinson (1971) established the following basic property of the NNI operation.

Proposition 2.6.1. *Let $\mathcal{T}$ and $\mathcal{T}'$ be elements of $B(n)$. Then, $\mathcal{T}'$ can be obtained from $\mathcal{T}$ by a sequence of* NNI *operations and, therefore, by a sequence of* SPR *(or* TBR*) operations.*

For an element $\mathcal{T}$ of $B(n)$, the next theorem determines the number of elements of $B(n)$ that can be obtained by a single NNI operation on $\mathcal{T}$ (Robinson 1971) and a single SPR operation on $\mathcal{T}$ (Allen and Steel 2001).

Theorem 2.6.2. *Let $\mathcal{T}$ be an element of $B(n)$. Then, the number of elements of $B(n)$, apart from $\mathcal{T}$, that can be obtained by*

(i) a single NNI *operation on $\mathcal{T}$ is $2(n-3)$ and*
(ii) a single SPR *operation on $\mathcal{T}$ is $2(n-3)(2n-7)$.*

Proof. To prove (i), let e be an interior edge of $\mathcal{T}$. Then, the number of distinct elements of $B(n)$ that can be obtained by a single NNI operation across e is two. Furthermore, it is easily seen that neither of these elements can be obtained by a single NNI across any other interior edge of $\mathcal{T}$. Therefore, as $\mathcal{T}$ has $n-3$ interior edges, the number of elements of $B(n)$ that can be obtained by a single NNI on $\mathcal{T}$ is $2(n-3)$.

For the proof of (ii), we note that any SPR operation is of precisely one of the three following types:

(a) the pruned edge is adjacent to the edge on which the subtree is regrafted;
(b) the pruned edge is separated by exactly one edge from the edge on which the subtree is regrafted; and
(c) the pruned edge is separated by at least two edges from the edge on which the subtree is regrafted.

For a type (a) operation, the resulting element of $B(n)$ is $\mathcal{T}$. For type (b), the single SPR operation on $\mathcal{T}$ corresponds to a single NNI operation on $\mathcal{T}$. Thus, by part (i) of this theorem, there are $2(n-3)$ elements of $B(n)$ resulting from a single type (b) SPR operation on $\mathcal{T}$.

Now consider type (c). Every SPR operation on $\mathcal{T}$ corresponds to an ordered pair of distinct edges of T, where the first component is the pruned edge and the second component is the edge on which the subtree is regrafted. One can easily check that two distinct type (c) SPR operations on $\mathcal{T}$ result in two distinct elements of $B(n)$. Therefore, to count the number of distinct elements of $B(n)$ resulting from a type (c) SPR operation on $\mathcal{T}$, we need to count the number of such ordered pairs arising from type (a) and type (b) SPR operations on $\mathcal{T}$, and then subtract these from $(2n-3)(2n-4)$, which is the total number of such ordered pairs.

The number of ordered pairs corresponding to a type (a) SPR operation on $\mathcal{T}$ is $6(n-2)$, since there are six such pairs for each of the $n-2$ interior vertices. Now a type (b) SPR operation corresponds to an NNI operation across an interior edge. Therefore, the number of ordered pairs corresponding to a type (b) SPR operation on $\mathcal{T}$ is $8(n-3)$, since there are eight such pairs for each of the $n-3$ interior edges. Therefore, the number of distinct elements of $B(n)$ resulting from a type (c) SPR operation on $\mathcal{T}$ is

$$(2n-3)(2n-4) - [6(n-2) + 8(n-3)] = 4(n-3)(n-4).$$

Combining (a)–(c), we deduce that the total number of distinct elements of $B(n)$, apart from $\mathcal{T}$, resulting from a single SPR operation on $\mathcal{T}$ is

$$0 + 2(n-3) + 4(n-3)(n-4) = 2(n-3)(2n-7).$$

□

Observe in the statement of Theorem 2.6.2 that the number of elements of $B(n)$ that can be obtained by a single NNI (or SPR) operation on $\mathcal{T}$ is independent of the shape of $\mathcal{T}$. However, in the case of a single TBR operation on $\mathcal{T}$ the number of such elements is dependent on the shape of $\mathcal{T}$. Nevertheless, one can easily show that this number is $O(n^3)$ regardless of the shape of $\mathcal{T}$.

Each of the three types of tree rearrangement operation induces a metric on $B(n)$. Under this metric, the distance between any two binary phylogenetic trees $\mathcal{T}, \mathcal{T}' \in B(n)$ is the minimum number of operations (of the given type) required to convert $\mathcal{T}$ into $\mathcal{T}'$. An interesting problem is to determine the 'diameter' of this metric space under these three metrics, where the diameter of a metric space (S, d) is $\max\{d(x, y): x, y \in S\}$. It can be shown that for the NNI operation, the diameter of $B(n)$ is bounded above and below by functions that grow at the rate of $n \log n$ (for details see Li *et al.* 1996). For the SPR and TBR operations, the diameter of $B(n)$ is bounded above and below by functions that grow linearly with n (for details see Allen and Steel 2001).

2.7 Reticulate evolution and co-phylogeny

A single tree is sometimes overly simplistic for representing evolutionary relationships in biology and other areas of classification. We now briefly discuss some of the mechanisms that cause this to occur.

First, the evolution of a species may not be purely tree-like. Sometimes two ancestral species hybridize to produce offspring that have genetic material from both species. This occurs, for example, in certain plants and fish. In such a case it may be more reasonable to represent the evolutionary picture either by a set of trees as in Fig. 2.15(a) or by a digraph as in Fig. 2.15(b). A second mechanism is the possible conflict between gene trees and species trees as we describe next.

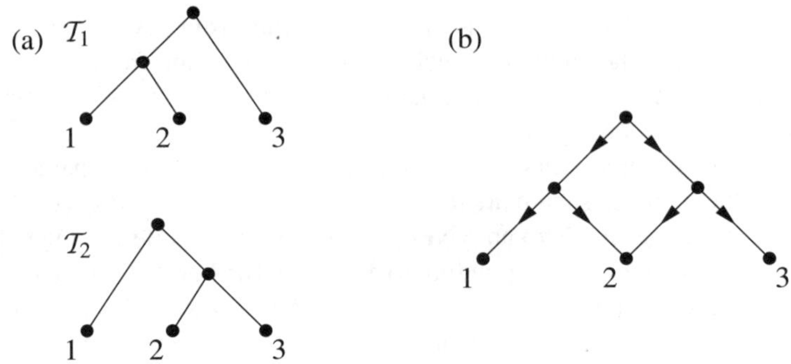

Figure 2.15 (a) $\mathcal{T}_1$ and $\mathcal{T}_2$. (b) A digraph representing $\mathcal{T}_1$ and $\mathcal{T}_2$.

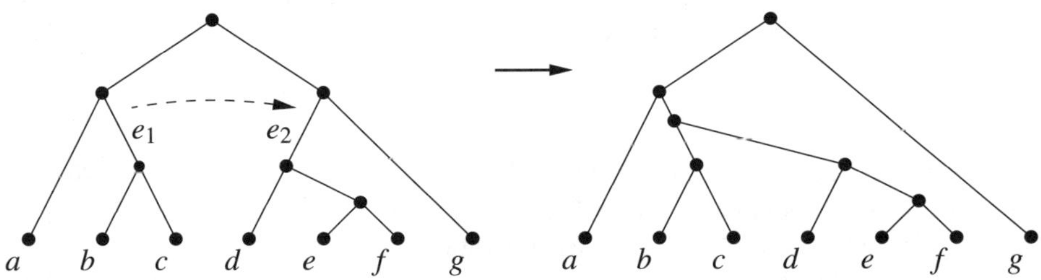

Figure 2.16 Representation of horizontal gene transfer by an SPR operation.

Gene trees versus species trees. When a tree does adequately describe the history of a collection of species, the evolution of individual genes carried by each of these species is sometimes described more effectively by a different tree. One mechanism for this to occur is the process called 'horizontal gene transfer', where genetic material from one species is inserted into the genetic makeup of another species. If this insertion is associated with the movement of a gene from an (ancestral) species corresponding to edge e_1 of the tree $\mathcal{T}$ to a contemporaneous species corresponding to edge e_2 of $\mathcal{T}$, then the resulting evolution of the gene is that obtained from $\mathcal{T}$ by an SPR (subtree prune and regraft) operation. Specifically, the subtree descendant from e_2 is regrafted onto the edge e_1 as illustrated in Fig. 2.16. In this way, the 'gene trees' for different genes of the same set of species can suggest different evolutionary relationships between the species.

Gene trees may also differ from species trees because of a process called 'lineage sorting', in which a population of individuals carry one of several copies of a gene. If two speciation events occur within a relatively short time period, then the relationships between the different copies of the gene that become fixed in the subsequent species may not reflect the order of speciation. This is illustrated in Fig. 2.17(a) where the solid lines describe the evolution of a particular copy of a gene resulting in the phylogenetic tree shown in Fig. 2.17(b). In contrast, the species relationships support a different phylogenetic tree as indicated by the dashed lines in Fig. 2.17(a).

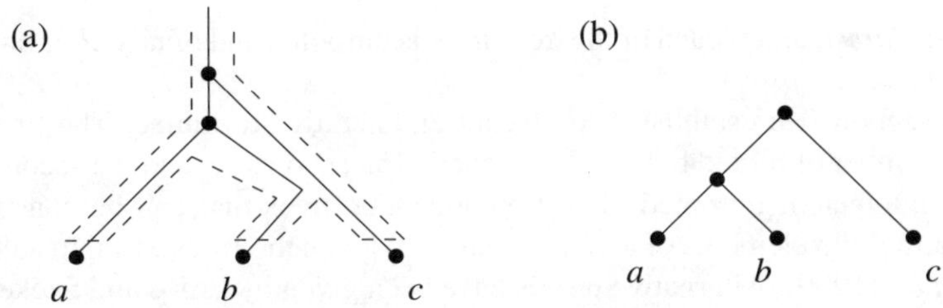

Figure 2.17 The effect of lineage sorting.

A gene tree may also differ from a species tree for other reasons. For example, when the elements of X correspond to genetic sequences of populations of the same sexually reproducing species, then the trees of descent of different portions of the sequence may differ by SPR operations due to a process called 'recombination'. A further mechanism that can cause a gene tree to differ from a species tree is 'gene duplication and loss'.

Note that our discussion here has been in terms of the actual species trees and gene trees, rather than those that are estimated from data (by techniques we will explore in subsequent chapters). Consequently, a further reason for differences between an estimated species tree and an estimated gene tree is that either estimated tree may incorrectly describe the true history.

In view of all of these mechanisms, it should be clear that the trees for different genes of the same set of species can suggest different evolutionary relationships between the species. An interesting question is how to reconstruct a 'species tree' from several conflicting gene trees.

It is also useful to describe the evolution of a gene (given the information provided by the gene tree) in terms of the species tree. This approach is sometimes called *co-phylogeny* and is familiar to those who study the joint evolution of hosts and parasites. In this case, the analogue of horizontal gene transfer is simply called 'host switching' as it corresponds to an ancestral parasite moving onto a different host. In this more general setting, the host tree consists of species that 'carry' the (ancestral and extant) objects of an associate tree (such as the parasite or gene tree). However, these associate objects may sometimes move across the tree or undergo lineage sorting. A graph-theoretic approach to this co-phylogeny problem has been described by Charleston (1998). Such approaches can also be extended to deal with multiple associate trees and to estimate a host tree from a collection of associate trees.

2.8 Specialist topic: phylogenetic forests

A *phylogenetic forest* $\mathcal{F}$ on X is a collection $\{\mathcal{T}_1, \mathcal{T}_2, \ldots, \mathcal{T}_k\}$ phylogenetic trees—each of which may be either rooted or unrooted—such that $\{\mathcal{L}(\mathcal{T}_1), \mathcal{L}(\mathcal{T}_2), \ldots, \mathcal{L}(\mathcal{T}_k)\}$ is a partition of X. If each of the trees in $\mathcal{F}$ are binary, we call $\mathcal{F}$ a *binary phylogenetic forest*. Furthermore, if each of the trees in $\mathcal{F}$ is rooted and binary, $\mathcal{F}$ is a *rooted binary phylogenetic forest*, and if each of the trees in $\mathcal{F}$ is unrooted and binary, $\mathcal{F}$ is an *unrooted binary phylogenetic forest*.

In this section, we establish two elegant enumerative formulae. The first formula counts the number of rooted binary phylogenetic forests on a set and the second formula counts the number of unrooted binary phylogenetic trees that can be constructed by 'amalgamating' the elements of a given binary phylogenetic forest. Both results are due to Carter *et al.* (1990). Alternative proofs have been given by Erdös and Székely (1993) and Steel (1993). Both formulae will be used in Section 5.6.

To establish the first formula, we use the following tool from enumerative combinatorics.

The Lagrange inversion formula. Frequently in enumeration problems, one can show that an associated generating function $w(t)$ satisfies an equation of the form $w(t) = t\phi(w(t))$ for some power series ϕ. In such cases, the Lagrange inversion formula (see Goulden and Jackson 1983) provides an explicit description of the coefficients of $w(t)$. Recall that $[x^n]g(x)$ denotes the coefficient of x^n in the power series $g(x)$.

Suppose we have two (formal) power series $\phi(\lambda) = \sum_{i\geq 0} c_i\lambda^i$ where $c_0 \neq 0$ and $f(\lambda) = \sum_{i\geq 0} d_i\lambda^i$. Then, there is a unique power series $w(t)$ such that $w(t) = t\phi(w(t))$ and, for each $n > 0$,

$$[t^n]f(w(t)) = \frac{1}{n}[\lambda^{n-1}]f'(\lambda)\phi^n(\lambda), \tag{2.5}$$

where $f'(\lambda) = \sum_{i\geq 1} id_i\lambda^{i-1}$ denotes the (formal) derivative of f. Equation (2.5) is known as the *Lagrange inversion formula*.

Counting phylogenetic forests. Let $N(n, k)$ denote the number of rooted binary phylogenetic forests on a set X of size n and consisting of k trees. Clearly, for all $n \geq 1$, $N(n, k) \neq 0$ if and only if $1 \leq k \leq n$, $N(n, 1) = |RB(n)|$, and $N(n, n) = 1$. Furthermore, by considering the standard decomposition of a rooted binary phylogenetic tree, it is easily seen that $N(n, 2) = |RB(n)|$ for all $n \geq 2$. Surprisingly, there is an explicit formula for $N(n, k)$.

Proposition 2.8.1. *For every pair of positive integers n and k with $1 \leq k \leq n$,*

$$N(n, k) = \frac{(2n-k-1)!}{(n-k)!(k-1)!2^{n-k}}.$$

Proof. Our proof follows Carter *et al.* (1990), and relies on the Lagrange inversion formula. As in Section 2.2, let $R(x)$ denote the exponential generating function for $RB(n)$, that is $R(x) = \sum_{n\geq 1} |RB(n)|(x^n/n!)$, and recall [see eqn (2.2)] that $R(x)$ satisfies the quadratic equation $R(x) = x + \frac{1}{2}R(x)^2$. This quadratic equation can be rewritten as

$$R(x) = x\phi(R(x)), \tag{2.6}$$

where $\phi(\lambda) = \left(1 - \frac{1}{2}\lambda\right)^{-1}$.

For a fixed k, consider $n!R(x)^k$ as a power series in x. The coefficient of x^n in this power series is the sum of the terms $(n!/(a_1!\cdots a_k!))\prod_{i=1}^{k} N(a_i, 1)$ over all k-tuples $(a_1, \ldots, a_k)$ of positive integers that sum to n. However, this summation counts the number of k-tuples $(\mathcal{T}_1, \ldots, \mathcal{T}_k)$, where $\{\mathcal{T}_1, \ldots, \mathcal{T}_k\}$ is a rooted binary phylogenetic forest on a set of size n, and this is $k!N(n, k)$. Thus,

$$N(n, k) = n! \times [x^n]f(R(x)),$$

where $f(t) = t^k/k!$. Applying the Lagrange inversion formula to (2.6) with this choice of f,

$$[x^n]f(R(x)) = \frac{1}{n}[\lambda^{n-1}]\frac{\lambda^{k-1}}{(k-1)!}\left(1 - \frac{1}{2}\lambda\right)^{-n}.$$

The right-hand side of the last equation is

$$\frac{1}{n(k-1)!}[\lambda^{n-k}]\left(1 - \frac{1}{2}\lambda\right)^{-n}.$$

It now follows by eqn (0.1) that

$$N(n, k) = n! \times [x^n]f(R(x)) = n! \times \frac{1}{n(k-1)!2^{n-k}}\binom{2n-k-1}{n-k}.$$

Simplifying the right-hand side of this equation gives the equation in the statement of the proposition. □

Building trees from forests. Given a binary phylogenetic forest $\mathcal{F}$ on X with a distinguished unrooted binary phylogenetic tree $\mathcal{T}_1$, we now wish to count the number of unrooted binary phylogenetic X-trees that can be 'built' from the elements of $\mathcal{F}$ by starting with $\mathcal{T}_1$ and successively 'attaching' the remaining trees (in any order) to the unrooted binary phylogenetic tree so far constructed.

Let $\mathcal{T}_1 = (T_1; \phi_1)$ be an unrooted phylogenetic tree on X_1 where $|X_1| \geq 2$ and let $\mathcal{T}_2 = (T_2; \phi_2)$ be either a rooted or an unrooted phylogenetic tree on X_2 such that $X_1 \cap X_2$ is empty and, if $\mathcal{T}_2$ is unrooted, $|X_2| \geq 2$. Let T be the tree obtained from T_1 and T_2 by adding an edge between a vertex that subdivides an edge of T_1 and either the root of T_2 if T_2 is rooted, or a vertex that subdivides an edge of T_2 if T_2 is unrooted. In either case, we view T as an unrooted tree. The phylogenetic tree $(T; \phi)$, where $\phi : X_1 \cup X_2 \rightarrow V(T)$ is defined, for all $x \in X_1 \cup X_2$, by

$$\phi(x) = \begin{cases} \phi_1(x), & \text{if } x \in X_1, \\ \phi_2(x), & \text{if } x \in X_2, \end{cases}$$

is the tree that results from an *edge attachment* by $\mathcal{T}_2$ to $\mathcal{T}_1$. Observe that if $\mathcal{T}_1$ and $\mathcal{T}_2$ are both binary, then $\mathcal{T}$ is also binary.

Example 2.8.2. Let $\mathcal{T}_1$ be the unrooted binary phylogenetic tree and let $\mathcal{T}_2$ be the rooted binary phylogenetic tree shown in Fig. 2.18(a). Noting that $\mathcal{T}_2$ is rooted, an

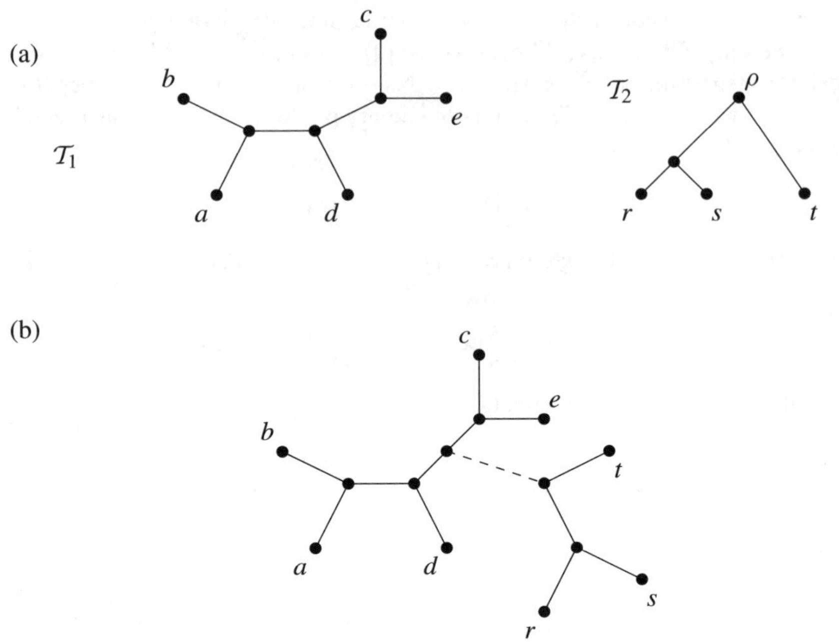

Figure 2.18 (a) $\mathcal{T}_1$ and $\mathcal{T}_2$. (b) An edge attachment by $\mathcal{T}_2$ to $\mathcal{T}_1$.

edge attachment by $\mathcal{T}_2$ to $\mathcal{T}_1$ is shown in Fig. 2.18(b), where the dashed edge is the 'attached' edge.

We now formally describe the elements we wish to count. Let $\mathcal{F}$ be a binary phylogenetic forest on X of size k such that all unrooted binary phylogenetic trees have label set of size at least two, and let $\mathcal{T}_1$ be a distinguished unrooted binary phylogenetic tree in $\mathcal{F}$. An unrooted binary phylogenetic X-tree $\mathcal{T}$ is an *extension of $\mathcal{F}$ with kernel $\mathcal{T}_1$* if there exists a sequence $(\mathcal{T}^1, \mathcal{F}^1), (\mathcal{T}^2, \mathcal{F}^2), \ldots, (\mathcal{T}^k, \mathcal{F}^k)$ of ordered pairs such that $(\mathcal{T}^1, \mathcal{F}^1) = (\mathcal{T}_1, \mathcal{F} - \{\mathcal{T}_1\})$, $(\mathcal{T}^k, \mathcal{F}^k) = (\mathcal{T}, \emptyset)$, and, for all $i \in \{2, 3, \ldots, k\}$, $\mathcal{T}^i$ is obtained from $\mathcal{T}^{i-1}$ by selecting any tree $\mathcal{T}' \in \mathcal{F}^{i-1}$ and applying an edge attachment by $\mathcal{T}'$ to $\mathcal{T}^{i-1}$, in which case $\mathcal{F}^i = \mathcal{F}^{i-1} - \{\mathcal{T}'\}$. Of course, such a sequence may not be unique. We denote the set of all extensions of $\mathcal{F}$ with kernel $\mathcal{T}_1$ by $\mathrm{Ext}(\mathcal{T}_1, \mathcal{F})$.

If all of the trees in $\mathcal{F}$ are unrooted, then $\mathrm{Ext}(\mathcal{T}', \mathcal{F})$ describes the same set of binary phylogenetic X-trees for each $\mathcal{T}' \in \mathcal{F}$. In particular, it is the collection of unrooted binary phylogenetic X-trees $\mathcal{T}$ such that, for each $\mathcal{T}' \in \mathcal{F}$, a subdivision of $T(\mathcal{T}')$ is a subtree of $T(\mathcal{T})$ and collectively these subtrees are disjoint, and $\phi|\mathcal{L}(\mathcal{T}')$ is the labelling map of $\mathcal{T}'$. Thus, in this case, we simply write $\mathrm{Ext}(\mathcal{F})$ for $\mathrm{Ext}(\mathcal{T}_1, \mathcal{F})$.

Theorem 2.8.3. *Let $\mathcal{F} = \{\mathcal{T}_1, \mathcal{T}_2, \ldots, \mathcal{T}_k\}$ be a binary phylogenetic forest on X such that all unrooted binary phylogenetic trees in $\mathcal{F}$, $\mathcal{T}_1, \mathcal{T}_2, \ldots, \mathcal{T}_p$ say, have label set of size* *at least two, where* $|X| = n$ *and* $1 \le p \le k$*. Then,*

$$|\text{Ext}(\mathcal{T}_1, \mathcal{F})| = \frac{b(n)}{b(n-k+2)} \prod_{i=1}^{p} |E_i|,$$

where E_i *is the set of edges of* $\mathcal{T}_i$ *for all* $i \in \{1, 2, \ldots, p\}$*.*

Proof. Using induction on k, we first establish the theorem for $p = 1$. If $k = 1$, then $|\text{Ext}(\mathcal{T}_1)| = 1 = [b(n)/b(n+1)]|E_1|$, and so the result holds for $p = 1$ and $k = 1$. Now suppose that $k \ge 2$, and that the theorem holds for $p = 1$ and all smaller values of k. To establish the induction step, it suffices to show that

$$|\text{Ext}(\mathcal{T}_1, \mathcal{F})| = (2n - 2k + 1)\frac{b(n)}{b(n-(k-1)+2)} \tag{2.7}$$

as

$$\frac{b(n)}{b(n-k+2)} = (2n-2k+1)\frac{b(n)}{b(n-(k-1)+2)}.$$

For each phylogenetic tree $\mathcal{T}$ in $\text{Ext}(\mathcal{T}_1, \mathcal{F})$, consider the element $\mathcal{T}_j$ of $\mathcal{F}$ that is determined as follows. Fix a vertex, v_1 say, of $\mathcal{T}$ that corresponds to a vertex of $\mathcal{T}_1$, and, for each $i \in \{2, 3, \ldots, k\}$, let v_i denote the vertex of $\mathcal{T}$ corresponding to the root of $\mathcal{T}_i$. Then, for some $j \in \{1, 2, \ldots, k-1\}$, v_j is the last vertex in $\{v_1, v_2, \ldots, v_{k-1}\}$ that is on the path in $\mathcal{T}$ from v_1 to v_k. Having determined $\mathcal{T}_j$, now consider the 'placement' of $\mathcal{T}_j$ and $\mathcal{T}_k$ in $\mathcal{T}$. It is easily seen that there is a sequence $(\mathcal{T}^1, \mathcal{F}^1), (\mathcal{T}^2, \mathcal{F}^2), \ldots, (\mathcal{T}^k, \mathcal{F}^k)$ of ordered pairs that constructs the extension $\mathcal{T}$ such that $(\mathcal{T}^1, \mathcal{F}^1) = (\mathcal{T}_1, \mathcal{F} - \{\mathcal{T}_1\})$, $(\mathcal{T}^k, \mathcal{F}^k) = (\mathcal{T}, \emptyset)$, and, for some $i \in \{2, 3, \ldots, k\}$, $\mathcal{T}^i$ is obtained from $\mathcal{T}^{i-1}$ by an edge attachment by $\mathcal{T}_k$ to $\mathcal{T}^{i-1}$, where the edge of $\mathcal{T}^{i-1}$ used for this attachment corresponds to an edge of $\mathcal{T}_j$. Determining j for each element of $\text{Ext}(\mathcal{T}_1, \mathcal{F})$, we distinguish two disjoint cases:

(i) $j \in \{2, 3, \ldots, k-1\}$ and
(ii) $j = 1$.

Consider case (i). For a fixed j, let $\mathcal{F}_j$ denote the binary phylogenetic forest obtained from $\mathcal{F}$ by deleting $\mathcal{T}_k$ and replacing $\mathcal{T}_j$ with a tree $\mathcal{T}_j'$ that is an edge attachment by $\mathcal{T}_k$ to $\mathcal{T}_j$. By the induction assumption, there are

$$\frac{b(n)}{b(n-(k-1)+2)}|E_1|$$

extensions of $\mathcal{F}_j$ with kernel $\mathcal{T}_1$. Since there are $|E_j|$ ways to make an edge attachment by $\mathcal{T}_k$ to $\mathcal{T}_j$, it follows that there are

$$|E_j|\frac{b(n)}{b(n-(k-1)+2)}|E_1|$$

elements $\mathcal{T}$ of $\mathrm{Ext}(\mathcal{T}_1, \mathcal{F})$ such that v_j is the last vertex in $\{v_1, v_2, \ldots, v_{k-1}\}$ that is on the path in $\mathcal{T}$ from v_1 to v_k. Consequently, the total number of extensions of $\mathcal{F}$ with kernel $\mathcal{T}_1$ covered by case (i) is

$$(|E_2| + |E_3| + \cdots + |E_{k-1}|)\frac{b(n)}{b(n-(k-1)+2)}|E_1|. \tag{2.8}$$

Now consider case (ii). Let $\mathcal{F}_1$ denote the binary phylogenetic forest obtained from $\mathcal{F}$ by deleting $\mathcal{T}_k$ and replacing $\mathcal{T}_1$ with a tree $\mathcal{T}_1'$ that is an edge attachment by $\mathcal{T}_k$ to $\mathcal{T}_1$. By the induction assumption, there are

$$\frac{b(n)}{b(n-(k-1)+2)}(|E_1| + |E_k| + 2)$$

extensions of $\mathcal{F}_1$ with kernel $\mathcal{T}_1$. Since there are $|E_1|$ ways to make an edge attachment by $\mathcal{T}_k$ to $\mathcal{T}_1$, it follows that the total number of extensions of $\mathcal{F}$ with kernel $\mathcal{T}_1$ covered by case (ii) is

$$|E_1|\frac{b(n)}{b(n-(k-1)+2)}(|E_1| + |E_k| + 2). \tag{2.9}$$

Adding the totals (2.8) and (2.9), and noting that $|E_1|+|E_2|+\cdots+|E_k|+2 = 2n-2k+1$, we establish (2.7). Hence, Theorem 2.8.3 holds for $p = 1$.

We now use the fact that the theorem holds for $p = 1$, to prove the theorem for all $p > 1$. For all $i \in \{2, 3, \ldots, p\}$, let $\mathcal{T}_i'$ be a rooted binary phylogenetic tree obtained from $\mathcal{T}_i$ by subdividing an edge and distinguishing the subdividing vertex. Then, as Theorem 2.8.3 holds for $p = 1$,

$$|\mathrm{Ext}(\mathcal{T}_1, \{\mathcal{T}_1, \mathcal{T}_2', \ldots, \mathcal{T}_p', \mathcal{T}_{p+1}, \ldots, \mathcal{T}_k\})| = \frac{b(n)}{b(n-k+2)}|E_1|.$$

The union over all such extensions of binary phylogenetic forests constructed this way is equal to $\mathrm{Ext}(\mathcal{T}_1, \mathcal{F})$. Furthermore, these extensions partition $\mathrm{Ext}(\mathcal{T}_1, \mathcal{F})$. It now follows that

$$|\mathrm{Ext}(\mathcal{T}_1, \mathcal{F})| = \frac{b(n)}{b(n-k+2)}\prod_{i=1}^{p}|E_i|$$

as required. □

2.9 Exercises

1. Using generating functions, or otherwise, justify for all $n \geq 2$ the recursion

$$|RB(n)| = \sum_{i=1}^{n-1}\binom{n-1}{i-1}|RB(i)||RB(n-i)|.$$

2. Let $P(x) = \sum_{n\geq 1} |RP(n)|(x^n/n!)$ be the exponential generating function for $RP(n)$. Show that

$$P(x) = x + \sum_{k\geq 2} \frac{P(x)^k}{k!}.$$

3. For all $n \geq 2$, construct an explicit bijection from the set $RB(n)$ into the set of matchings of $\{1, 2, \ldots, 2n-2\}$ into $n-1$ pairs.
4. Prove that, for all $n \geq 4$, the number of caterpillar trees with label set $\{1, 2, \ldots, n\}$ is exactly $\frac{1}{8}n!$.
5. (Felsenstein 1978*a*) Let $p(n,f)$ denote the number of phylogenetic trees with label set $\{1, 2, \ldots, n\}$ and f interior edges. For all $n \geq 3$, establish the recursion

$$p(n,f) = (f+1)p(n-1,f) + (n+f-2)p(n-1,f-1).$$

Using this recursion, deduce that $b(n) = (2n-5)!!$ for all $n \geq 3$.

6. Show that for all n there is a binary tree shape on n leaves for which the number of phylogenetic trees with label set $\{1, 2, \ldots, n\}$ of that shape is an odd number.
7. Under the Yule–Harding model, calculate the probability for all $n \geq 2$ of generating the rooted caterpillar tree shown in Fig. 2.6(b).
8. Prove that for all $n \geq 2$, the probability distribution on $RB(n)$ induced by the Yule–Harding model is the same as that obtained by selecting, with uniform probability, a ranked binary phylogenetic tree with label set $\{1, 2, \ldots, n\}$ and ignoring the ranking of vertices on this tree.
9. Using Proposition 2.8.1, prove that, for all $n \geq 2$ and all $k \geq 2$, the number of binary phylogenetic trees $(\mathcal{T}; \phi)$ with label set $\{1, 2, \ldots, n\}$, and for which there are exactly k edges in the path joining $\phi(1)$ and $\phi(2)$ is

$$\frac{(2n-k-4)!}{(n-k-1)!2^{n-k-1}}.$$

10. An *edge-distinguished tree* is a tree in which exactly one edge is distinguished. By counting the number of edge-distinguished rooted binary phylogenetic trees on n leaves in two different ways, establish the identity

$$\sum_{k=2}^{n} N(n,k)k! = (2n-2)N(n,1).$$

11. For all $n \geq 3$, construct an explicit bijection from the set $B(n)$ into the set of ordered tuples $(\mathcal{T}_1, \mathcal{T}_2, \ldots, \mathcal{T}_k)$, where $\{\mathcal{T}_1, \mathcal{T}_2, \ldots, \mathcal{T}_k\}$ is a rooted binary phylogenetic forest on $\{1, 2, \ldots, n\}$ and $1 \leq k \leq n-2$. Deduce that

$$b(n) = \sum_{k=1}^{n-2} N(n-2,k)k!$$

12. Let $F(n)$ denote the number of rooted phylogenetic forests on $\{1, \ldots, n\}$. Show that $F(n) = 2|RP(n)|$.
13. (Steel and Fu 1995) A *subforest* of a phylogenetic tree $\mathcal{T} = (T; \phi)$ is a collection of edge-disjoint subtrees of T such that all subtrees have at least one edge and each degree-one vertex of every subtree is a leaf of T. Let $\mathcal{T}'$ be a binary phylogenetic tree on n leaves. Prove that the number of subforests of $\mathcal{T}'$ is $F_{2n-2} - 1$, where F_k is the *k-th Fibonacci number* defined, for all $k \geq 0$, by $F_{k+2} = F_k + F_{k+1}$ and $F_0 = F_1 = 1$.

3
Trees and splits

In the first section of this chapter, we state and prove a classical result that we will refer to as the Splits-Equivalence Theorem. This result, established by Peter Buneman in 1971, provides a fundamental equivalence between X-trees and a certain type of collection of bipartitions of X. It has played a particularly important role in the mathematical development of phylogenetics as we shall illustrate throughout the chapter. Much of the material presented here builds on the approach of Bandelt and Dress (1986).

3.1 Splits-Equivalence Theorem

Definition 3.1.1. An X-*split* is a partition of X into two non-empty sets. We denote the X-split whose blocks are A and B by $A|B$.

For an X-split $A|B$, the position of A and B is arbitrary, so we make no distinction between the X-splits $A|B$ and $B|A$.

Associated with every X-tree there is a particular collection of X-splits that we next define. Let $\mathcal{T} = (T; \phi)$ be an X-tree, and let e be an edge of T. Then, $T\backslash e$ consists of two components. If V_1 and V_2 denote the vertex sets of these two components, then $\phi^{-1}(V_1)|\phi^{-1}(V_2)$ is an X-split. This X-split *corresponds to e in* $\mathcal{T}$. Observe that the X-split $\phi^{-1}(V_1)|\phi^{-1}(V_2)$ corresponds to no other edge of T. We denote by $\Sigma(\mathcal{T})$ the collection of X-splits that correspond to the edges of $\mathcal{T}$, and refer to the X-splits in $\Sigma(\mathcal{T})$ as the X-*splits of* $\mathcal{T}$ or *induced by* $\mathcal{T}$.

Example 3.1.2. Consider the X-tree shown in Fig. 3.1, where $X = \{1, 2, \ldots, 8\}$. The X-splits corresponding to the edges e_1 and e_2 are $\{1, 2, 3\}|\{4, 5, 6, 7, 8\}$ and $\{1, 2, 3, 4, 5, 8\}|\{6, 7\}$, respectively.

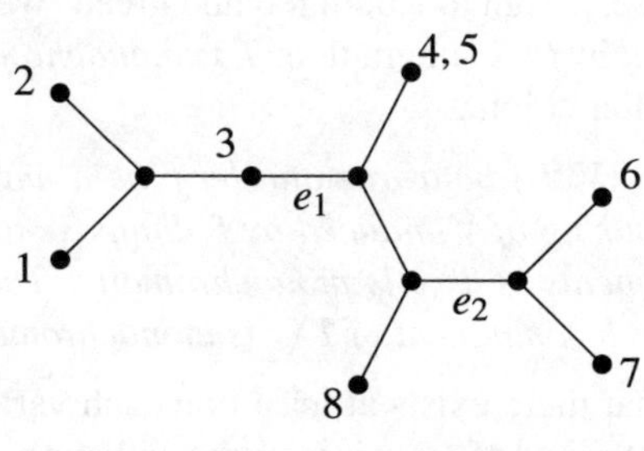

Figure 3.1 $\{1, 2, 3\}|\{4, 5, 6, 7, 8\}$ and $\{1, 2, 3, 4, 5, 8\}|\{6, 7\}$ are induced X-splits.

Definition 3.1.3. A pair of X-splits $A_1|B_1$ and $A_2|B_2$ are *compatible* if at least one of the sets $A_1 \cap A_2$, $A_1 \cap B_2$, $B_1 \cap A_2$, and $B_1 \cap B_2$ is the empty set.

Note that, for two distinct X-splits, this definition is equivalent to requiring that exactly one of the four described intersections is empty.

The justification for Definition 3.1.3 is the following central theorem due to Buneman (1971).

Theorem 3.1.4 (Splits-Equivalence Theorem). *Let Σ be a collection of X-splits. Then, there is an X-tree $\mathcal{T}$ such that $\Sigma = \Sigma(\mathcal{T})$ if and only if the splits in Σ are pairwise compatible. Moreover, if such an X-tree exists, then, up to isomorphism, $\mathcal{T}$ is unique.*

Most of the remainder of this section is devoted to proving the Splits-Equivalence Theorem. The proof given here is somewhat different to Buneman's. However, we will return to his original argument in Section 3.8. We begin the proof with three lemmas, the first of which was established by Robinson and Foulds (1981).

Lemma 3.1.5. *Let $\mathcal{T} = (T; \phi)$ be an X-tree, and let σ_1 and σ_2 be distinct elements of $\Sigma(\mathcal{T})$. Then, X can be partitioned into three sets X_1, X_2, and X_3 such that $\sigma_1 = X_1|(X_2 \cup X_3)$ and $\sigma_2 = (X_1 \cup X_2)|X_3$. Furthermore, the intersection of the vertex sets of the minimal subtrees of T induced by $T(\phi(X_1))$ and $T(\phi(X_3))$ is empty.*

Proof. Let $e_1 = \{u_1, v_1\}$ and $e_2 = \{u_2, v_2\}$ be the unique edges of T corresponding to σ_1 and σ_2, respectively. Evidently, there is a path P in T such that e_1 and e_2 are the first and last edges, respectively, that are traversed by P. Without loss of generality, we may assume that u_1 and u_2 are the initial and terminal vertices of P, respectively. Observe that u_1 and u_2 are distinct, but v_1 and v_2 may not be distinct. Let V_1, V_2, and V_3 denote the vertex sets of the components of $T\backslash\{e_1, e_2\}$ containing u_1, v_1, and u_2, respectively. By choosing, for each $i \in \{1, 2, 3\}$, $X_i = \phi^{-1}(V_i)$, we get the desired result. □

To illustrate Lemma 3.1.5, consider the X-tree shown in Fig. 3.1. Let σ_1 and σ_2 be the X-splits corresponding to the edges e_1 and e_2, respectively. Choosing $X_1 = \{1, 2, 3\}$, $X_2 = \{4, 5, 8\}$, and $X_3 = \{6, 7\}$ provides a partition of X into three sets. Furthermore, $\sigma_1 = X_1|(X_2 \cup X_3)$ and $\sigma_2 = (X_1 \cup X_2)|X_3$.

The next lemma is a general property of trees. Let T be a tree and let f be a function from a finite set Y into the vertex set V of T. Colour the elements of Y either red or green. We assign a colouring to the elements of V in $f(Y)$ in the following way. Let v be an element of $f(Y)$. If all of the elements of $f^{-1}(v)$ are of the same colour, assign that colour to v itself; otherwise, assign to v both red and green. We refer to this colouring as the *colouring of V induced by f*. A subgraph of T is *monochromatic* if all of its coloured vertices are of one particular colour.

Lemma 3.1.6. *Let $T = (V, E)$ be a tree and let f be a mapping from a finite set Y into V. Consider the colouring of V induced by f. Suppose that, for each edge $e \in E$, exactly one of the components of $T\backslash e$ is monochromatic. Then, there exists a unique vertex $v \in V$ for which each component of $T\backslash v$ is monochromatic.*

Proof. We first show that there exists at least one such vertex. For each edge $e \in E$, assign an orientation from the end of e that is incident with the monochromatic component of $T\backslash e$ to the other end of e. Then, there exists in V a vertex v of out-degree zero; otherwise, we would have a directed path of infinite length. Deleting this vertex v produces monochromatic components.

We now show that there can be at most one such vertex. Suppose that distinct vertices v and v' both have the claimed property. Select an edge e in the path connecting v and v'. Then, exactly one of the two components in $T\backslash e$ is not monochromatic. Without loss of generality, this component contains v. However, this contradicts the assumption that each component of $T\backslash v'$ is monochromatic as the component containing v is not monochromatic. This completes the proof of the lemma. □

Lemma 3.1.7. *Let $A|B$ be an X-split. Suppose that $\mathcal{T} = (T; \phi)$ is an X-tree such that $A|B$ is not a split of $\mathcal{T}$, but $A|B$ is compatible with each X-split of $\mathcal{T}$. Then, there exists a unique vertex v of T such that, for each component (V', E') of $T\backslash v$, either $\phi^{-1}(V') \subseteq A$ or $\phi^{-1}(V') \subseteq B$.*

Proof. Colour the elements of A red and the elements of B green, and consider the corresponding colouring of the vertices of T induced by ϕ. Then, for each edge e of T, exactly one of the components of $T\backslash e$ is monochromatic under the colouring of the vertices of T by ϕ. Applying Lemma 3.1.6 with $f = \phi$ and $Y = X$, there exists a unique vertex v of T for which each component of $T\backslash v$ is monochromatic. This implies that $A|B$ satisfies the condition described in the lemma. □

Proof of Splits-Equivalence Theorem. First, suppose that Σ is equal to the set of X-splits induced by an X-tree, and let σ_1 and σ_2 be distinct elements of Σ. By Lemma 3.1.5, there is a partition of X into three sets X_1, X_2, and X_3 such that $\sigma_1 = X_1|(X_2 \cup X_3)$ and $\sigma_2 = (X_1 \cup X_2)|X_3$. Since $X_1 \cap X_3 = \emptyset$, the X-splits σ_1 and σ_2 are compatible, and so the X-splits in Σ are pairwise compatible.

Conversely, suppose that Σ is a pairwise compatible collection of X-splits. We use induction on the cardinality of Σ to simultaneously prove that $\Sigma = \Sigma(\mathcal{T})$ for some X-tree $\mathcal{T}$ and, up to isomorphism, the choice of $\mathcal{T}$ is unique. If $|\Sigma| = 0$, then it is clear that, up to isomorphism, there is a unique X-tree $\mathcal{T}$ such that $\Sigma = \Sigma(\mathcal{T})$, namely the semi-labelled tree consisting of a single vertex labelled X.

Now suppose that $|\Sigma| = k + 1$, where $k \geq 0$, and that the existence and uniqueness properties hold for $|\Sigma| = k$. Let $A|B \in \Sigma$. Since $\Sigma - \{A|B\}$ is pairwise compatible, it follows by our induction assumption that there is, up to isomorphism, a unique X-tree $\mathcal{T}' = (T'; \phi')$ with $\Sigma - \{A|B\} = \Sigma(\mathcal{T}')$. By Lemma 3.1.7, there exists a unique vertex v' of T' such that, for each component (V', E') of $T'\backslash v'$, either $\phi^{-1}(V') \subseteq A$ or $\phi^{-1}(V') \subseteq B$.

Let T be the tree obtained from T' by replacing v' with two new adjacent vertices v_A and v_B, and attaching the subtrees that were incident with v' to the new vertices in such a way that the subtrees consisting of vertices in $\phi'(A)$ and $\phi'(B)$ are attached to v_A and v_B, respectively. Let $\phi : X \to V(T)$ be the map defined as follows:

$$\phi(x) = \begin{cases} \phi'(x), & \text{if } \phi'(x) \neq v', \\ v_A, & \text{if } \phi'(x) = v' \text{ and } x \in A, \\ v_B, & \text{if } \phi'(x) = v' \text{ and } x \in B. \end{cases}$$

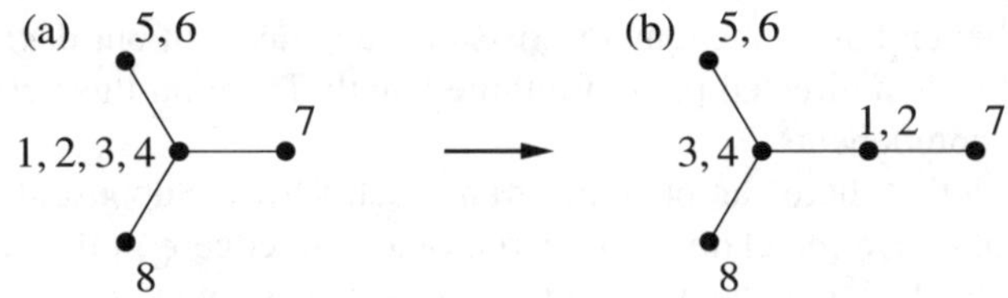

Figure 3.2 One iteration of tree popping.

It is easily checked that $(T;\phi)$ is an X-tree, and that if we denote $(T;\phi)$ by $\mathcal{T}$, then we have $\Sigma = \Sigma(\mathcal{T})$. Moreover, as $\mathcal{T}'$ is the unique X-tree for which $\Sigma - \{A|B\} = \Sigma(\mathcal{T}')$, it is easily seen that, up to isomorphism, $\mathcal{T}$ is the only such X-tree satisfying $\Sigma = \Sigma(\mathcal{T})$. This completes the proof of the Splits-Equivalence Theorem. □

Lemma 3.1.7 provides an efficient algorithm for reconstructing an X-tree $\mathcal{T}$ from $\Sigma(\mathcal{T})$. This method, called *tree popping*, was described by Meacham (1983). In this method, we order the elements of $\Sigma(\mathcal{T})$ arbitrarily, $\sigma_1, \sigma_2, \ldots, \sigma_k$ say, where $k = |\Sigma(\mathcal{T})|$, and then, using this ordering, construct a sequence $\mathcal{T}_0, \mathcal{T}_1, \ldots, \mathcal{T}_k$ of X-trees such that, for all $i \in \{1, 2, \ldots, k\}$, $\Sigma(\mathcal{T}_i) = \{\sigma_1, \sigma_2, \ldots, \sigma_i\}$. Thus, $\Sigma(\mathcal{T}_k) = \Sigma(\mathcal{T})$. This is done by defining $\mathcal{T}_0$ to be the X-tree consisting of a single vertex labelled X and, for all i, defining $\mathcal{T}_i$ to be the X-tree obtained from $\mathcal{T}_{i-1}$ by introducing σ_i as described in the induction step of the proof of thc Splits-Equivalence Theorem.

Example 3.1.8. To illustrate one iteration of tree popping, let $X = \{1, 2, \ldots, 8\}$, and let $\sigma_1 = \{5, 6\}|(X - \{5, 6\})$, $\sigma_2 = \{7\}|(X - \{7\})$, $\sigma_3 = \{8\}|(X - \{8\})$, and $\sigma_4 = \{1, 2, 7\}|\{3, 4, 5, 6, 8\}$ be an ordering of X-splits. Applying the method of tree popping to this ordering, we get after three iterations the X-tree $(T;\phi)$ shown in Fig. 3.1(a). Consider σ_4, and colour the elements of $\{1, 2, 7\}$ red and the elements of $\{3, 4, 5, 6, 8\}$ green. This colouring induces a colouring of the vertices of T via ϕ. The interior vertex, v say, of T is the unique vertex for which each component of $T\backslash v$ is monochromatic. By applying the construction in the proof of the Splits-Equivalence Theorem to this vertex, we get the X-tree shown in Fig. 3.1(b).

An alternative approach to reconstructing semi-labelled trees from their set of splits has been described by Gusfield (1991).

Although a phylogenetic tree may seem a particularly special type of X-tree, the two classes of trees are closely associated. In particular, there is a phylogenetic X-tree associated to each X-tree. We describe this connection now.

Definition 3.1.9. For a finite set X, let $\Sigma_{\text{triv}}(X)$ denote the set of X-splits $A|B$ such that $\min\{|A|, |B|\} = 1$. The elements of $\Sigma_{\text{triv}}(X)$ are called the *trivial* splits of X.

A trivial split of X is compatible with every X-split. Furthermore, in a phylogenetic X-tree $\mathcal{T}$, the trivial splits of X are precisely the X-splits that correspond to the pendant edges of $\mathcal{T}$. Note that, to any X-tree $\mathcal{T}$, we have an associated phylogenetic X-tree $\mathcal{T}'$ defined by $\Sigma(\mathcal{T}') = \Sigma(\mathcal{T}) \cup \Sigma_{\text{triv}}(X)$.

3.2 A partial order on *X*-trees

An immediate application of the Splits-Equivalence Theorem is that it provides a partial order on the set of X-trees. For a fixed set X, consider the collection $\mathcal{P}(X)$ of all X-trees together with the relation $\leq$ where, for elements $\mathcal{T}$ and $\mathcal{T}'$ of $\mathcal{P}(X)$, $\mathcal{T} \leq \mathcal{T}'$ precisely if $\Sigma(\mathcal{T}) \subseteq \Sigma(\mathcal{T}')$. The proof of the following proposition is straightforward and omitted.

Proposition 3.2.1. *Let $\mathcal{P}(X)$ denote the collection of all X-trees. Then*

(i) *$\leq$ is a partial order on $\mathcal{P}(X)$;*
(ii) *the maximal elements of $(\mathcal{P}(X), \leq)$ are the binary phylogenetic X-trees; and*
(iii) *for all subsets $\mathcal{P}$ of $\mathcal{P}(X)$, $\mathcal{P}$ has a unique greatest lower bound in $\mathcal{P}(X)$ under $\leq$; namely the element, $\mathcal{T}$ say, of $\mathcal{P}(X)$ with the property that*

$$\Sigma(\mathcal{T}) = \bigcap_{\mathcal{T}' \in \mathcal{P}} \Sigma(\mathcal{T}').$$

Observe that a subset of $\mathcal{P}(X)$ may not necessarily have an upper bound in $\mathcal{P}(X)$ under $\leq$.

Intuitively, we can view the partial order $\leq$ on $\mathcal{P}(X)$ in the following way. If $\mathcal{T}$ and $\mathcal{T}'$ are elements of $\mathcal{P}(X)$, and $\mathcal{T} \leq \mathcal{T}'$, then $\mathcal{T}$ can be obtained from $\mathcal{T}'$ by contracting edges and amalgamating the label sets of the identified vertices; we say that $\mathcal{T}'$ *refines* $\mathcal{T}$ if $\mathcal{T} \leq \mathcal{T}'$. In Fig. 3.3, each of $\mathcal{T}_1$ and $\mathcal{T}_2$ refine $\mathcal{T}$.

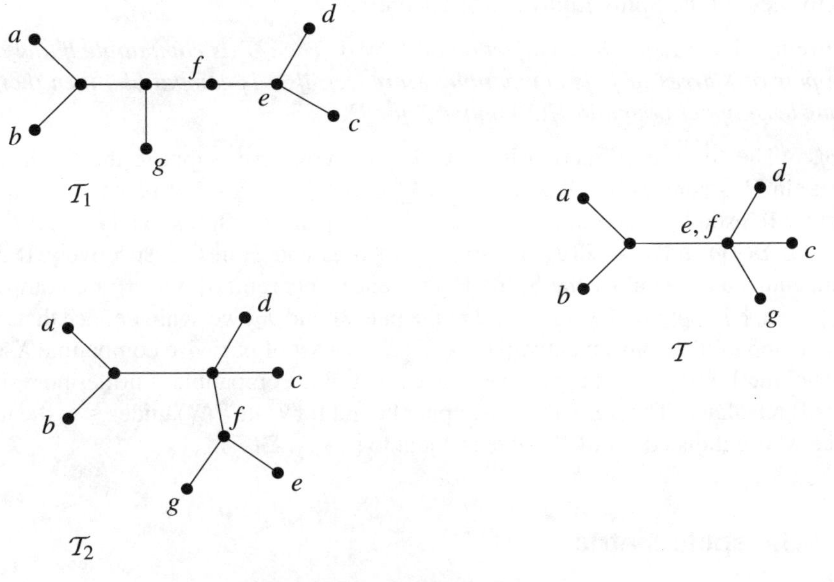

Figure 3.3 $\mathcal{T}_1$ and $\mathcal{T}_2$ refine $\mathcal{T}$.

Remark. Under the partial order $\leq$ on $\mathcal{P}(X)$, we may view vertices of an X-tree of degree at least four as 'unresolved'. In biology, such vertices generally represent uncertainty as to the exact order of speciation (a 'soft polytomy') rather than certainty of simultaneous speciation (a 'hard polytomy').

3.3 Compatibility of X-trees

Definition 3.3.1. For a fixed set X, a *compatible set $\mathcal{P}$ of X-trees* is a subset of $\mathcal{P}(X)$ for which there exists an X-tree in $\mathcal{P}(X)$ that is an upper bound to $\mathcal{P}$ under the partial ordering $\leq$.

Graphically, $\mathcal{P}$ is compatible precisely if there is a particular X-tree in $\mathcal{P}(X)$ that refines each of the X-trees in $\mathcal{P}$. Considering the remark at the end of the last section, for biologists, this graphical interpretation implies that the X-trees in $\mathcal{P}$ are telling a 'consistent' evolutionary story.

Lemma 3.3.2. *A collection $\mathcal{P}$ of X-trees is compatible if and only if $\bigcup_{\mathcal{T}\in\mathcal{P}} \Sigma(\mathcal{T})$ is a pairwise compatible set of X-splits.*

Proof. By definition, $\mathcal{P}$ is compatible precisely if there is an X-tree, $\mathcal{T}'$ say, in $\mathcal{P}(X)$ such that $\mathcal{T} \leq \mathcal{T}'$ for all $\mathcal{T} \in \mathcal{P}$, which in turn holds if and only if $\Sigma(\mathcal{T}) \subseteq \Sigma(\mathcal{T}')$ for all $\mathcal{T} \in \mathcal{P}$. Since the latter holds if and only if $\bigcup_{\mathcal{T}\in\mathcal{P}} \Sigma(\mathcal{T}) \subseteq \Sigma(\mathcal{T}')$, it follows by the Splits-Equivalence Theorem that $\mathcal{P}$ is compatible if and only if $\bigcup_{\mathcal{T}\in\mathcal{P}} \Sigma(\mathcal{T})$ is a pairwise compatible set of X-splits. □

Theorem 3.3.3 is the main result of this section. Its short proof highlights the effectiveness of the Splits-Equivalence Theorem.

Theorem 3.3.3. *Let $\mathcal{P}$ be a collection of X-trees. Then, $\mathcal{P}$ is compatible if and only if each pair of X-trees in $\mathcal{P}$ is compatible. Moreover, if $\mathcal{P}$ is compatible, then there is a unique least upper bound in $\mathcal{P}(X)$ under $\leq$ for $\mathcal{P}$.*

Proof. The 'only if' direction is clear. For the converse, suppose that each pair of X-trees in $\mathcal{P}$ is compatible. We claim that $\bigcup_{\mathcal{T}\in\mathcal{P}} \Sigma(\mathcal{T})$ is a set of pairwise compatible X-splits. To establish this claim, let σ_1 and σ_2 be a pair of X-splits in $\bigcup_{\mathcal{T}\in\mathcal{P}} \Sigma(\mathcal{T})$ such that $\sigma_1 \in \Sigma(\mathcal{T}_1)$ and $\sigma_2 \in \Sigma(\mathcal{T}_2)$ for some X-trees $\mathcal{T}_1$ and $\mathcal{T}_2$ in $\mathcal{P}$, respectively. If $\mathcal{T}_1$ and $\mathcal{T}_2$ are equivalent, then, by the Splits-Equivalence Theorem, σ_1 and σ_2 are compatible. If not, then, by applying Lemma 3.3.2 to the pair $\mathcal{T}_1$ and $\mathcal{T}_2$, we again deduce that σ_1 and σ_2 are compatible. Consequently, $\bigcup_{\mathcal{T}\in\mathcal{P}} \Sigma(\mathcal{T})$ is a set of pairwise compatible X-splits. By invoking Lemma 3.3.2 again, we deduce that $\mathcal{P}$ is compatible. Furthermore, by the Splits-Equivalence Theorem, the least upper bound for $\mathcal{P}$ in $\mathcal{P}(X)$ under $\leq$ is the unique X-tree whose induced set of X-splits is equal to $\bigcup_{\mathcal{T}\in\mathcal{P}} \Sigma(\mathcal{T})$. □

3.4 The splits metric

In this section, we describe a way of comparing X-trees based on their induced sets of X-splits.

The *symmetric difference* of sets A and B, denoted $A\Delta B$, is the set $(A-B)\cup(B-A)$. For each pair $\mathcal{T}$ and $\mathcal{T}'$ of X-trees in $\mathcal{P}(X)$, define $d(\mathcal{T},\mathcal{T}')$ to be $|\Sigma(\mathcal{T})\Delta\Sigma(\mathcal{T}')|$. Using the Splits-Equivalence Theorem, it is easily checked that d is a metric on $\mathcal{P}(X)$. This metric on $\mathcal{P}(X)$ is called the *splits metric* on $\mathcal{P}(X)$. We now describe some of the main properties of this metric.

Suppose that $\mathcal{T}=(T;\phi)$ is an X-tree. Let e be an edge of T with end vertices u and v, and let v_e be the vertex of T/e that identifies u and v. The *contraction of e from $\mathcal{T}$* is the X-tree $(T/e;\phi_e)$, where ϕ_e is the map from X to the vertex set of T/e defined by

$$\phi_e(x)=\begin{cases}\phi(x), & \text{if } \phi(x)\notin\{u,v\},\\ v_e, & \text{if } \phi(x)\in\{u,v\}.\end{cases}$$

We denote $(T/e;\phi_e)$ by $\mathcal{T}/e$. It is easy to check that, for all edges e_1 and e_2 of T, $(\mathcal{T}/e_1)/e_2=(\mathcal{T}/e_2)/e_1$.

If an X-tree $\mathcal{T}$ is obtained from an X-tree $\mathcal{T}'$ by contracting an edge e' of $\mathcal{T}'$, then $\mathcal{T}'$ is obtained from $\mathcal{T}$ by an *expansion of e'*. Theorem 3.4.1 is due to Robinson and Foulds (1981).

Theorem 3.4.1. *Let $\mathcal{T}$ and $\mathcal{T}'$ be X-trees. Then*

(i) $d(\mathcal{T},\mathcal{T}')$ is equal to the smallest k for which there is a sequence $\mathcal{T}_0,\mathcal{T}_1,\ldots,\mathcal{T}_k$ of X-trees such that $\mathcal{T}_0=\mathcal{T}$, $\mathcal{T}_k=\mathcal{T}'$, and, for all $i\in\{1,2,\ldots,k\}$, $\mathcal{T}_i$ is obtained from $\mathcal{T}_{i-1}$ by either contracting or expanding an edge of $\mathcal{T}_i$; and

(ii) $d(\mathcal{T},\mathcal{T}')=|E(\mathcal{T})|+|E(\mathcal{T}')|-2\nu(\mathcal{T},\mathcal{T}')$, where $\nu(\mathcal{T},\mathcal{T}')$ is the number of X-splits that are induced by both $\mathcal{T}$ and $\mathcal{T}'$.

Proof. We first prove (i). Let $n(\mathcal{T},\mathcal{T}')$ denote the minimum number of edge contractions and edge expansions that are required to transform $\mathcal{T}$ into $\mathcal{T}'$. We need to show that $n(\mathcal{T},\mathcal{T}')=d(\mathcal{T},\mathcal{T}')$.

If $n(\mathcal{T},\mathcal{T}')=k$, then there is a sequence $\mathcal{T}_0,\mathcal{T}_1,\ldots,\mathcal{T}_k$ of X-trees such that $\mathcal{T}_0=\mathcal{T}$, $\mathcal{T}_k=\mathcal{T}'$, and, for all $i\in\{1,2,\ldots,k\}$, $\mathcal{T}_i$ is obtained from $\mathcal{T}_{i-1}$ by either a single-edge contraction or a single-edge expansion. Clearly, for all i, $d(\mathcal{T}_{i-1},\mathcal{T}_i)=1$. Therefore, by repeated use of the triangle inequality of metrics,

$$d(\mathcal{T},\mathcal{T}')\le n(\mathcal{T},\mathcal{T}'). \tag{3.1}$$

Now, let $E'(\mathcal{T})$ and $E'(\mathcal{T}')$ denote the sets whose elements are precisely the edges corresponding to the X-splits in $\Sigma(\mathcal{T})-\Sigma(\mathcal{T}')$ and $\Sigma(\mathcal{T}')-\Sigma(\mathcal{T})$, respectively. Let $\mathcal{T}/E'(\mathcal{T})$ and $\mathcal{T}'/E'(\mathcal{T}')$ denote the X-trees obtained from $\mathcal{T}$ and $\mathcal{T}'$ by contracting each of the edges in $E'(\mathcal{T})$ and $E'(\mathcal{T}')$, respectively. It is easily seen that the set of X-splits induced by the edges of $\mathcal{T}/E'(\mathcal{T})$ and $\mathcal{T}'/E'(\mathcal{T}')$ are equal. Therefore, by the Splits-Equivalence Theorem, these two X-trees are identical. Thus, we can transform $\mathcal{T}$ into $\mathcal{T}'$ by $|E'(\mathcal{T})|$ edge contractions and $|E'(\mathcal{T}')|$ edge expansions. Hence,

$$n(\mathcal{T},\mathcal{T}')\le|E'(\mathcal{T})|+|E'(\mathcal{T}')|=|\Sigma(\mathcal{T})\Delta\Sigma(\mathcal{T}')|. \tag{3.2}$$

Combining (3.1) and (3.2) establishes (i).

To prove (ii), note that for finite sets A and B, $|A \Delta B| = |A| + |B| - 2|A \cap B|$. Therefore, as $|\Sigma(\mathcal{T})| = |E(\mathcal{T})|$ and $|\Sigma(\mathcal{T}')| = |E(\mathcal{T}')|$, setting $A = \Sigma(\mathcal{T})$ and $B = \Sigma(\mathcal{T}')$ gives the desired result. □

Example 3.4.2. Referring to the semi-labelled trees shown in Fig. 3.3, let us calculate $d(\mathcal{T}_1, \mathcal{T}_2)$. Now, $\Sigma(\mathcal{T}) = \Sigma(\mathcal{T}_1) \cap \Sigma(\mathcal{T}_2)$. Therefore, as $|\Sigma(\mathcal{T})| = 6$, $|E(\mathcal{T}_1)| = 8$, and $|E(\mathcal{T}_2)| = 8$, it follows by Theorem 3.4.1(ii) that $d(\mathcal{T}_1, \mathcal{T}_2) = 4$. Alternatively, by Theorem 3.4.1(i), we can evaluate $d(\mathcal{T}_1, \mathcal{T}_2)$ by determining the smallest number of single-edge contractions and single-edge expansions to obtain $\mathcal{T}_2$ from $\mathcal{T}_1$. We leave it to the reader to provide a corresponding sequence of such edge operations.

The splits metric on $\mathcal{P}(X)$ was originally defined as in part (i) of Theorem 3.4.1. However, because of the emphasis on splits in this chapter, we have rephrased the definition. Using Theorem 3.4.1(ii), it is easily seen that on either $B(n)$ or $P(n)$ the splits metric has maximal value $2(n-3)$. Furthermore, on $B(n)$, the splits metric takes only even integer values and, for each tree $\mathcal{T} \in B(n)$, there are exactly $2(n-3)$ trees at distance two from $\mathcal{T}$. It is straightforward to compute the splits metric on $\mathcal{P}(X)$ in polynomial time in $|X|$.

For statistical analysis, it is useful to determine the distribution of the splits metric between pairs of trees chosen randomly according to some null model. For example, one may wish to determine whether different trees reconstructed from different data sets are any more similar to each other than would be expected if one (or more) data set was completely random. An example of such an application was provided by Penny *et al.* (1982).

For binary phylogenetic trees, two natural distributions were described in Section 2.5 by the Yule–Harding model and the uniform model. For either of these distributions and for all $n \geq 1$, let D_n denote the splits metric distance between two trees randomly selected from $B(n)$ and let $I_n = n - 3 - D_n/2$. By Proposition 1.2.3(ii) and Theorem 3.4.1(ii), I_n is the number of non-trivial splits shared by the two randomly selected binary phylogenetic trees. The following result is due to Steel and Penny (1993).

Theorem 3.4.3. *The distribution of I_n is asymptotically Poisson. In particular,*

$$\lim_{n \to \infty} \mathbb{P}[I_n = k] = \frac{e^{-\lambda}\lambda^k}{k!},$$

where $\lambda = \frac{2}{9}$ for the Yule–Harding model and $\lambda = \frac{1}{8}$ for the uniform model.

3.5 Hierarchies, clusters, and splits

The Splits-Equivalence Theorem establishes an equivalence between sets of pairwise compatible X-splits and X-trees. In this section, we introduce the concept of a hierarchy on X and show that an analogous equivalence holds between hierarchies on X and rooted X-trees.

Definition 3.5.1. Let M be a finite set. A collection $\mathcal{H}$ of non-empty subsets of M is called a *hierarchy* on M if it satisfies the property that, for all $A, B \in \mathcal{H}$,

$$A \cap B \in \{\emptyset, A, B\}.$$

A rooted X-tree $\mathcal{T} = (T; \phi)$ induces a hierarchy on X in the following way. For a vertex u of $V(T) - \{\rho\}$, let

$$C(u) = \bigcup_{v : u \leq_T v} \phi^{-1}(v),$$

where we recall that $\leq_T$ is the partial order on $V(T)$ induced by the root of T. For each vertex u of $V(T) - \{\rho\}$, the set $C(u)$ is called a *cluster* of $\mathcal{T}$ and we shall denote the set of clusters of $\mathcal{T}$ by $\mathcal{H}(\mathcal{T})$. For example, the set of clusters of the rooted semi-labelled trees shown in Fig. 3.4(a) and (b) are

$$\{\{a\}, \{b\}, \{c\}, \{d\}, \{e\}, \{f\}, \{a, b, g\}, \{a, b, c, g\}, \{d, e, f\}\}$$

and

$$\{\{a\}, \{b\}, \{c\}, \{d\}, \{e\}, \{f\}, \{a, b, g\}, \{a, b, c, g\}, \{d, e, f\}, \{a, b, c, d, e, f, g\}\},$$

respectively.

The set $\mathcal{H}(\mathcal{T})$ of clusters of a rooted X-tree $\mathcal{T}$ is a hierarchy on X. To see this, let T denote the underlying tree of $\mathcal{T}$, and let $C(u_1)$ and $C(u_2)$ be elements of $\mathcal{H}(\mathcal{T})$. If $u_1 \leq_T u_2$, then $C(u_2) \subseteq C(u_1)$, while if $u_2 \leq_T u_1$, then $C(u_1) \subseteq C(u_2)$. Furthermore, if neither holds, then $C(u_1)$ and $C(u_2)$ are disjoint, and so $C(u_1) \cap C(u_2) = \emptyset$.

One can easily reconstruct a rooted X-tree $\mathcal{T}$ from $\mathcal{H}(\mathcal{T})$ as follows. First, construct the cover digraph of $\mathcal{H}(\mathcal{T}) \cup \{X\}$ under the partial order of set inclusion. Clearly, each vertex in this digraph has in-degree one, except for X which has in-degree zero. Thus, this digraph is a directed tree with all vertices directed away from X. If $X \notin \mathcal{H}(\mathcal{T})$, label vertex X as the root vertex ρ of the tree, while if $X \in \mathcal{H}(\mathcal{T})$, adjoin a pendant edge to X and label the new leaf as the root vertex ρ. Let T denote the resulting rooted tree, whose

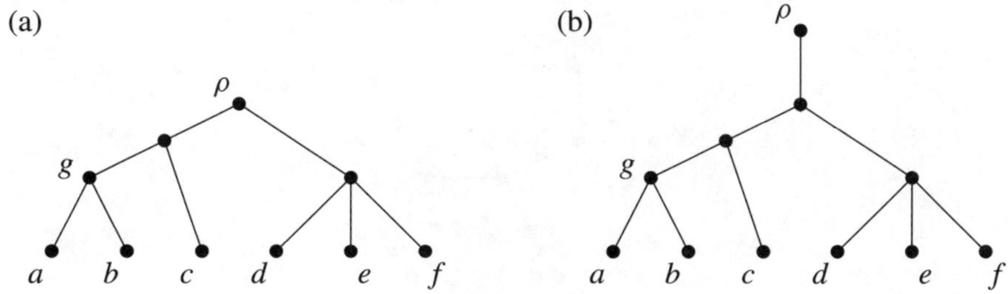

Figure 3.4

vertex set $V(T)$ consists of the elements of $\mathcal{H}(\mathcal{T}) \cup \{X\} \cup \{\rho\}$. Now, let $\phi : X \to V(T)$ be the labelling map defined by setting, for all $x \in X$, $\phi(x) = A$, where A is the unique element of $V(T)$ that has the property

$$x \in \left(A - \bigcup_{B:A<_T B} B \right).$$

The resulting pair $(T; \phi)$ is isomorphic to $\mathcal{T}$.

The next theorem describes the fundamental equivalence between hierarchies on X and rooted X-trees. It can be viewed as the rooted analogue of the Splits-Equivalence Theorem.

Theorem 3.5.2. *Let $\mathcal{H}$ be a collection of non-empty subsets of X. Then, there is a rooted X-tree $\mathcal{T}$ such that $\mathcal{H} = \mathcal{H}(\mathcal{T})$ if and only if $\mathcal{H}$ is a hierarchy on X. Moreover, if there is such a rooted X-tree, then, up to isomorphism, $\mathcal{T}$ is unique.*

Theorem 3.5.2 can be proved directly. However, we can also derive it as a consequence of the Splits-Equivalence Theorem. In order to do so, we need to describe equivalences between the sets of rooted $(X - \{x\})$-trees and X-trees, and corresponding equivalences between hierarchies on $X - \{x\}$ and sets of pairwise compatible X-splits.

Regarding the first of these two equivalences, select any element x of X, and to each X-tree $(T; \phi)$ associate the rooted $(X - \{x\})$-tree $(T'; \phi')$ where T' is the rooted tree obtained from T by distinguishing $\phi(x)$ as the root vertex, and $\phi' = \phi|(X - \{x\})$. This gives a bijection between the set of X-trees and the set of rooted $(X - \{x\})$-trees.

Example 3.5.3. Consider the X-tree $\mathcal{T}$ shown in Fig. 3.5, where $X = \{a, b, c, d, e, f, x\}$. The associated hierarchy on $X - \{x\}$ obtained by rooting T relative to the vertex $\phi(x)$ is the set of clusters

$$\{\{a\}, \{b\}, \{c, d, e, f\}, \{c\}, \{d, e\}, \{d\}, \{e\}\}.$$

We now describe the equivalence between hierarchies on $X - \{x\}$ and sets of pairwise compatible X-splits. Let Σ be a collection of X-splits and select an element x from X. Let $\mathcal{H}$ be a collection of non-empty subsets of $X - \{x\}$. We now describe two maps. Let ψ_x^- denote the map from Σ into the set of non-empty subsets of $X - \{x\}$ defined, for all

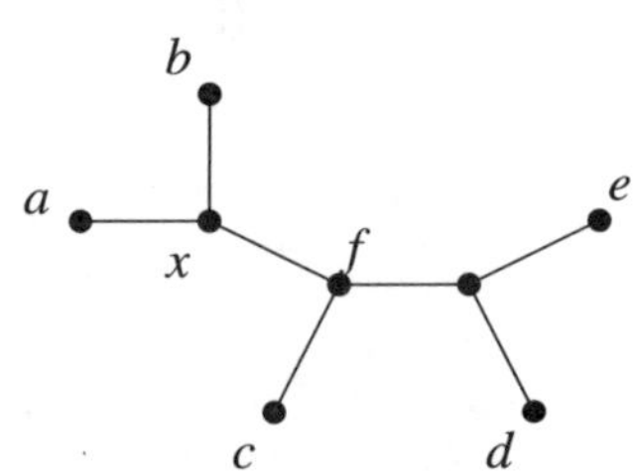

Figure 3.5

$A|B \in \Sigma$, by

$$\psi_x^-(A|B) = \begin{cases} A, & \text{if } x \in B, \\ B, & \text{if } x \in A. \end{cases}$$

Conversely, let ψ_x^+ denote the map from $\mathcal{H}$ into the complete set of splits on X defined, for all $A \in \mathcal{H}$, by $\psi_x^+(A) = A|(X - A)$. Observe that the maps ψ_x^- and ψ_x^+ are inverses of each other in the sense that

$$\psi_x^+ \circ \psi_x^-(A|B) = A|B \quad \text{and} \quad \psi_x^- \circ \psi_x^+(A) = A$$

for all $A|B$ of Σ and for all A of $\mathcal{H}$.

The following proposition describes the promised equivalence between hierarchies on $X - \{x\}$ and sets of pairwise compatible X-splits. We leave its proof as well as its application to derive Theorem 3.5.2 from the Splits-Equivalence Theorem as exercises.

Proposition 3.5.4. *Let x be an element of X.*

(i) A set Σ of X-splits is pairwise compatible if and only if

$$\{\psi_x^-(A|B): A|B \in \Sigma\}$$

is a hierarchy on $X - \{x\}$.

(ii) A set $\mathcal{H}$ of non-empty subsets of $X - \{x\}$ is a hierarchy on $X - \{x\}$ if and only if

$$\{\psi_x^+(A): A \in \mathcal{H}\}$$

is a set of pairwise compatible X-splits.

(iii) For every hierarchy $\mathcal{H}$ on $X - \{x\}$, there exists a unique collection Σ of pairwise compatible X-splits such that $\mathcal{H} = \{\psi_x^-(A|B): A|B \in \Sigma\}$.

3.6 Consensus trees

Given a collection $\mathcal{P}$ of X-trees, a *consensus tree* for $\mathcal{P}$ is an X-tree that reflects some shared or common features of the X-trees in $\mathcal{P}$. The simplest example consists of choosing our consensus tree to be the X-tree in $\mathcal{P}(X)$ that is the greatest lower bound of $\mathcal{P}$ under the partial order $\leq$; that is, the X-tree whose induced set of X-splits is the set of X-splits shared by all of the X-trees in $\mathcal{P}$. This X-tree is called the *strict consensus tree* for $\mathcal{P}$.

The problem with the strict consensus tree in applications is that it tends to be highly unrefined. A more practical consensus method—and one that is also implementable in polynomial time—is 'majority rule', which we describe next.

For a collection $\mathcal{P}$ of X-trees, let $\Sigma = \bigcup_{\mathcal{T} \in \mathcal{P}} \Sigma(\mathcal{T})$ and, for $\sigma \in \Sigma$, let $n(\sigma)$ denote the number of X-trees in $\mathcal{P}$ that induce σ. For all $0 \leq q < 1$, let

$$\Sigma_q = \{\sigma \in \Sigma: n(\sigma)/|\mathcal{P}| > q\}.$$

Theorem 3.6.1. *If $q \geq \frac{1}{2}$, then Σ_q is the set of X-splits induced by a (unique) X-tree.*

Proof. By the Splits-Equivalence Theorem, it suffices to show that the X-splits in Σ_q are pairwise compatible. Let σ and σ' be elements of Σ_q. Then, as $q \geq \frac{1}{2}$, there exists an X-tree $\mathcal{T}$ such that $\sigma, \sigma' \in \Sigma(\mathcal{T})$. Hence, σ and σ' are pairwise compatible. It follows that the X-splits in Σ_q are pairwise compatible as required. □

When $q = \frac{1}{2}$, the X-tree whose induced set of X-splits is Σ_q is called the *majority rule consensus tree* for $\mathcal{P}$. Observe that for any collection $\mathcal{P}$ of X-trees, the majority rule consensus tree for $\mathcal{P}$ refines the strict consensus tree for $\mathcal{P}$.

For rooted X-trees, there is a further canonical consensus tree known as the *Adams consensus* tree. Adams (1986) showed that this tree can be characterized axiomatically as the tree that satisfies certain natural requirements (see Exercise 6). However, we will describe the Adams consensus tree recursively as this illustrates the usefulness of the equivalence between hierarchies on X and rooted X-trees, as well as providing a constructive algorithm. In particular, because of this equivalence, it suffices to describe a procedure that takes a collection $\mathcal{P} = \{\mathcal{H}_1, \mathcal{H}_2, \ldots, \mathcal{H}_k\}$ of hierarchies on a set X and returns a hierarchy $Ad(\mathcal{P})$ on X corresponding to the set of clusters of the Adams consensus tree for $\mathcal{P}$. For each $i \in \{1, 2, \ldots, k\}$, let $\mathcal{H}_i^\circ$ denote the set of maximal sets in $\mathcal{H}_i$, and let

$$\mathcal{C}^\circ = \left\{\bigcap_{i=1}^{k} A_i \colon A_i \in \mathcal{H}_i^\circ\right\} - \{\emptyset\}.$$

It is easily checked that $\mathcal{C}^\circ$ is a partition of X. The elements of $\mathcal{C}^\circ$ make up the maximal sets of $Ad(\mathcal{P})$. The remaining sets in $Ad(\mathcal{P})$ are obtained by repeating this process recursively on the restriction of the hierarchies in $\mathcal{P}$ to the sets so far obtained. To illustrate, for each set $B \in \mathcal{C}^\circ$, let $\mathcal{P}|B = \{\mathcal{H}_1|B, \mathcal{H}_2|B, \ldots, \mathcal{H}_k|B\}$, where $\mathcal{H}_i|B = \{B \cap C \colon C \in \mathcal{H}_i \text{ and } B \cap C \neq \emptyset\}$. For all i, $\mathcal{H}_i|B$ is a hierarchy on B. For each set B, repeat the above procedure on $\mathcal{P}|B$ to obtain a partition of B. The blocks of this partition make up the maximal proper subsets of B in $Ad(\mathcal{P})$. This process continues in this way until no new sets can be formed.

Although there are exponentially (in k) many intersections involved in the definition of $\mathcal{C}^\circ$, all but $O(|X|)$ of these intersections are empty since each element of X can lie in at most one element of $\mathcal{C}^\circ$. Thus, it is straightforward to devise a fast algorithm to construct $\mathcal{C}^\circ$, and $Ad(\mathcal{P})$ can thereby be constructed in polynomial time (indeed in $O(kn^2)$ time where $n = |X|$). For details of such an algorithm, see Margush and McMorris (1981).

3.7 Cyclic split systems and weak hierarchies

A natural way to generate X-splits is to arrange the elements of X on a circle in the plane and then to draw lines to divide the points into two subsets. In this section, we will see that a collection of X-splits generated by this geometric procedure has a structure that is related to, though slightly more general than, the splits of an X-tree.

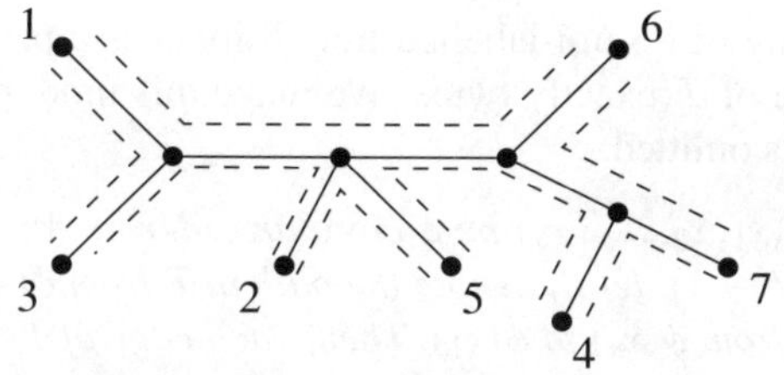

Figure 3.6 A phylogenetic tree with $(1,3,2,5,4,7,6)$ as a circular ordering.

Let $\pi = (x_1, x_2, \ldots, x_n)$ be a cyclic permutation of X. For all $1 \leq i \leq j \leq n$, let $A_{ij} = \{x_k : i \leq k \leq j\}$ and let $\Sigma^\circ(\pi)$ denote the set

$$\Sigma^\circ(\pi) = \{A_{ij}|(X - A_{ij}) : 1 \leq i \leq j \leq n\}$$

of X-splits. Arranging the elements $x_1, x_2, \ldots, x_n$ clockwise in a circle in the plane, we may view $\Sigma^\circ(\pi)$ as the set of splits that can be obtained by separating these elements according to which side of a line segment in the plane they lie on. Consequently, $|\Sigma^\circ(\pi)| = \binom{n}{2}$. A collection Σ of X-splits is said to be *circular* if $\Sigma \subseteq \Sigma^\circ(\pi)$ for some cyclic permutation π of X. In case $\Sigma(\mathcal{T}) \subseteq \Sigma^\circ(\pi)$ for some X-tree $\mathcal{T}$, we say that π provides a *circular ordering* for $\mathcal{T}$. For example, $(1,3,2,5,4,7,6)$ is a circular ordering for the phylogenetic tree (solid lines) shown in Fig. 3.6.

Theorem 3.7.1.

(i) Let $\mathcal{T} = (T; \phi)$ be a semi-labelled tree. Then, $\mathcal{T}$ has exactly

$$\prod_{v \in V(T)} (d(v) + l(v) - 1)!$$

distinct circular orderings, where $d(v)$ denotes the degree of vertex v, and $l(v) = |\phi^{-1}(v)|$.

(ii) Let $|X| = n$, where $n \geq 3$, and let π be a cyclic permutation on X. Then, the number of binary phylogenetic X-trees for which π is a circular ordering equals the (Catalan) number

$$\frac{1}{n-1}\binom{2n-4}{n-2}.$$

Proof. We leave the proof of (i) as an exercise. To prove (ii), for all n, let $C(n)$ denote the set of pairs $(\mathcal{T}, \pi)$, where $\mathcal{T}$ is a binary phylogenetic X-tree and π is a circular ordering for $\mathcal{T}$. We will count $C(n)$ in two ways. By Proposition 2.1.4, the number of choices for $\mathcal{T}$ is $(2n-4)!/[(n-2)!2^{n-2}]$ and, by (i), for each $\mathcal{T}$ the number of choices for π is 2^{n-2}. Alternatively, we may count $C(n)$ by noting that the number of distinct cyclic permutations on X is exactly $(n-1)!$ and, for each such cyclic permutation π, the number of X-trees $\mathcal{T}$ for which $(\mathcal{T}, \pi) \in C(n)$ is precisely the number we want. Equating these two counts of $C(n)$ and then rearranging gives the desired result. □

A circular ordering of a semi-labelled tree $\mathcal{T}$ induces a particular system of paths that covers each edge of $\mathcal{T}$ exactly twice. We make this more precise in the following lemma whose proof is omitted.

Lemma 3.7.2. *Let $(x_1, x_2, \ldots, x_n)$ be a circular ordering for an X-tree $\mathcal{T} = (T; \phi)$. For all $i \in \{1, 2, \ldots, n-1\}$, let P_i denote the path in T from $\phi(x_i)$ to $\phi(x_{i+1})$ and let P_n denote the path in T from $\phi(x_n)$ to $\phi(x_1)$. Then, each edge of T occurs in exactly two of the paths $P_1, P_2, \ldots, P_n$.*

To illustrate Lemma 3.7.2, consider the circular ordering $(1, 3, 2, 5, 4, 7, 6)$ for the semi-labelled tree shown in Fig. 3.6. The system of paths given by the assumptions of Lemma 3.7.2 is the collection of dashed paths shown in Fig. 3.6.

Let Σ be a set of circular splits on X and let $x \in X$. Recall from Section 3.5 that the map ψ_x^- from Σ into the set of non-empty subsets of $X - \{x\}$ defined by

$$\psi_x^-(A|B) = \begin{cases} A, & \text{if } x \in B, \\ B, & \text{if } x \in A. \end{cases}$$

It is easily seen that the set $\{\psi_x^-(A|B) : A|B \in \Sigma\}$ satisfies the following (weak Helly) property: for any three sets A_1, A_2, and A_3 in $\{\psi_x^-(A|B) : A|B \in \Sigma\}$,

$$A_1 \cap A_2 \cap A_3 \in \{A_1 \cap A_2, A_1 \cap A_3, A_2 \cap A_3\}. \tag{3.3}$$

In general, a collection of subsets of Y satisfying (3.3) is called a *weak hierarchy* on Y. Bandelt and Dress (1989) have shown that the number of elements in a weak hierarchy on a set of size $n-1$ is bounded above by $\binom{n}{2}$. Clearly, this bound is the best possible as it is realized by the hierarchy that corresponds to $\Sigma^\circ(\pi)$ for any cyclic permutation π of a set of size n.

3.8 Specialist topic: the Buneman graph

In his original proof of the Splits-Equivalence Theorem, Buneman associated a tree with a non-empty set Σ of pairwise compatible X-splits by defining the vertex set of the tree in a particular way. This tree together with a certain labelling map on X defines an X-tree $\mathcal{T}$ for which $\Sigma = \Sigma(\mathcal{T})$. Barthélemy (1989) generalized this construction, associating a graph with an arbitrary non-empty set of X-splits. In this section, we describe this general construction. Unless otherwise stated, the results in this section are due to Barthélemy (1989).

Let Σ be a non-empty set of X-splits. The *Buneman graph* on Σ, denoted $G(\Sigma)$, is defined as follows. The vertex set of $G(\Sigma)$ consists of those sets that contain exactly one block of each X-split of Σ and have the property that every pair of elements has a non-empty intersection. To describe the edge set of $G(\Sigma)$, let $\sigma = A|B$ be an element of Σ. Two vertices M and N of $G(\Sigma)$ *agree on* σ if either A is an element of both M and N, or B is an element of both M and N; otherwise, M and N *disagree on* σ. The edge

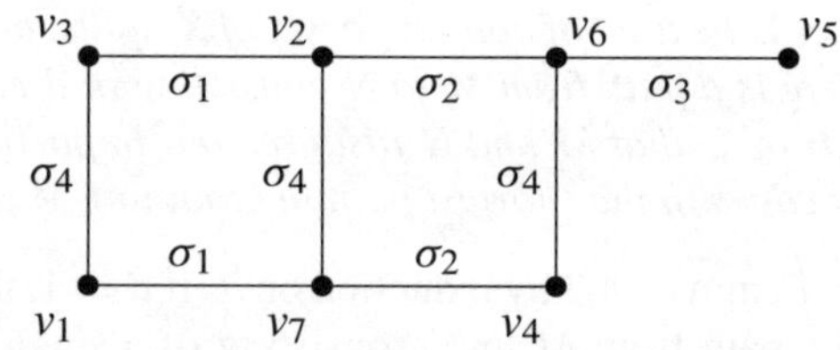

$$v_1 = \{\{a,c,d\},\{a,b,c,d\},\{a,b,c,d,e\},\{a,d,e\}\}$$
$$v_2 = \{\{b,e,f\},\{a,b,c,d\},\{a,b,c,d,e\},\{b,c,f\}\}$$
$$v_3 = \{\{a,c,d\},\{a,b,c,d\},\{a,b,c,d,e\},\{b,c,f\}\}$$
$$v_4 = \{\{b,e,f\},\{e,f\},\{a,b,c,d,e\},\{a,d,e\}\}$$
$$v_5 = \{\{b,e,f\},\{e,f\},\{f\},\{b,c,f\}\}$$
$$v_6 = \{\{b,e,f\},\{e,f\},\{a,b,c,d,e\},\{b,c,f\}\}$$
$$v_7 = \{\{b,e,f\},\{a,b,c,d\},\{a,b,c,d,e\},\{a,d,e\}\}$$

Figure 3.7 The Buneman graph on a set of X-splits.

set of $G(\Sigma)$ consists of all pairs of vertices that disagree on exactly one element of Σ. Observe that if A is a block of an element of Σ, then there is a vertex of $G(\Sigma)$ containing A. We denote the vertex set and edge set of $G(\Sigma)$ by $V(\Sigma)$ and $E(\Sigma)$, respectively.

Example 3.8.1. Let $X = \{a, b, \ldots, f\}$ and let $\Sigma = \{\sigma_1, \sigma_2, \sigma_3, \sigma_4\}$ be the set of X-splits, where $\sigma_1 = \{a,c,d\}|\{b,e,f\}$, $\sigma_2 = \{a,b,c,d\}|\{e,f\}$, $\sigma_3 = \{f\}|\{a,b,c,d,e\}$, and $\sigma_4 = \{b,c,f\}|\{a,d,e\}$. The Buneman graph on Σ is shown in Fig. 3.7. Each edge of $G(\Sigma)$ is labelled with the X-split in Σ on which its end vertices disagree.

For a non-empty set Σ of X-splits, the next sequence of lemmas establishes some basic properties of $G(\Sigma)$.

Lemma 3.8.2. *Let Σ be a non-empty set of X-splits, let N be a vertex of $G(\Sigma)$, and let A be an element of N. Then*

(i) the set obtained from N by replacing A with $X - A$ is a vertex of $G(\Sigma)$ if and only if A is a minimal element of N under set inclusion; and
(ii) the degree of N is equal to the number of minimal elements of N under set inclusion.

Proof. To prove (i), let M be the set obtained from N by replacing A with $X - A$. If A is a minimal element of N, then $X - A$ has a non-empty intersection with each of the elements in M. Thus, M is a vertex of $G(\Sigma)$. Now suppose that M is a vertex of $G(\Sigma)$ and assume that A is not a minimal element of N. Then, there is an element A' of N such that A' is a proper subset of A. But then $(X - A) \cap A'$ is empty, contradicting the assumption that M is a vertex of $G(\Sigma)$. Thus, A is a minimal element of N, completing the proof of (i). Since the end vertices of each edge of $G(\Sigma)$ disagree on exactly one element of Σ, (ii) is an immediate consequence of (i). □

Lemma 3.8.3. *Let Σ be a set of non-empty set of X-splits, and let M and N be vertices of $G(\Sigma)$. Then, there is a path from M to N consisting of d edges, where d denotes the number of elements of Σ that M and N disagree on. In particular, $G(\Sigma)$ is connected, and the number of edges in the shortest path in G joining M and N is equal to d.*

Proof. We prove Lemma 3.8.3 by induction on d. If $d = 1$, then M and N are adjacent, so there is indeed a path from M to N consisting of a single edge. Now suppose that $d = k$ and that, for each pair of vertices of $G(\Sigma)$ that disagree on $k - 1$ elements of Σ, there is a path consisting of $k - 1$ edges between the two vertices. Let A be a minimal element of $N - M$. Then, $X - A \in M$. Since M is a vertex of $G(\Sigma)$, no element of $M \cap N$ is a proper subset of A. Therefore, A is a minimal element of N, and so, by Lemma 3.8.2, $(N - \{A\}) \cup \{X - A\}$ is a vertex of $G(\Sigma)$. Moreover, M and $(N - \{A\}) \cup \{X - A\}$ disagree on $k - 1$ elements of Σ. Therefore, by the induction assumption, there is a path from M to $(N - \{A\}) \cup \{X - A\}$ consisting of $k - 1$ edges. Hence, there is a path from M to N in $G(\Sigma)$ consisting of k edges as required. □

Let σ be an element of a non-empty set Σ of X-splits. If M and N are adjacent vertices of $G(\Sigma)$ that disagree on σ, then the edge $\{M, N\}$ of $G(\Sigma)$ is said to *represent* σ.

Lemma 3.8.4. *Let σ be an element of a non-empty set Σ of X-splits. Then, there is a pair of adjacent vertices of $G(\Sigma)$ such that the edge between them represents σ.*

Proof. Let A be a block of σ. By Lemma 3.8.2, it suffices to show that A is a minimal element of some vertex of $G(\Sigma)$. Let N be a vertex of $G(\Sigma)$ containing A. If A is a minimal element of N, then the result holds. Therefore, we may assume that there is a minimal element A' of N such that $A' \subset A$. Let N' be the set obtained from N by replacing A' with $X - A'$. By Lemma 3.8.2, N' is a vertex of $G(\Sigma)$. Moreover, the number of elements of N' that are properly contained in A is strictly less than the number of elements of N that are properly contained in A. By applying this process to N' and repeating as necessary, we eventually obtained a vertex of $G(\Sigma)$ in which A is a minimal element. This completes the proof of Lemma 3.8.4. □

Lemma 3.8.5. *Let Σ be a non-empty set of X-splits. Then, $|V(\Sigma)| \geq |\Sigma| + 1$.*

Proof. The proof is by induction on the cardinality of Σ. Evidently, the result holds if $|\Sigma| = 1$. Suppose that $|\Sigma| = k$ and that the result holds for $|\Sigma| = k - 1$, where $k \geq 2$. Let σ be an element of Σ such that one block A of σ is a minimal element of the set of all blocks of Σ. Then, $X - A$ has a non-empty intersection with every element of the set of blocks of Σ. By the induction assumption, $|V(\Sigma - \{\sigma\})| \geq k$, and so, by adding $X - A$ to each of the vertices in $V(\Sigma - \{\sigma\})$, we obtain $|V(\Sigma - \{\sigma\})|$ vertices of $G(\Sigma)$. By Lemma 3.8.4, there is an edge of $G(\Sigma)$ representing σ. Therefore, there is a vertex of $G(\Sigma)$ that contains A, which implies that $|V(\Sigma)| \geq k + 1$. This completes the proof of Lemma 3.8.5. □

A set of pairwise compatible splits can be characterized in a variety of ways. Three such characterizations are stated in the next theorem. Its proof uses the previous sequence of lemmas.

For an element σ in Σ, let $E(\sigma)$ denote the set of edges of $E(\Sigma)$ that represent σ. Note that if e and f are adjacent edges of $G(\Sigma)$, then e and f represent different elements of Σ. We freely use this last fact in the proof of Theorem 3.8.6.

Theorem 3.8.6. *Let Σ be a non-empty set of X-splits. The following statements are equivalent:*

(i) Σ is pairwise compatible;
(ii) $G(\Sigma)$ is a tree;
(iii) $|V(\Sigma)| = |\Sigma| + 1$; and
(iv) $|E(\Sigma)| = |\Sigma|$.

Proof. The proof of Theorem 3.8.6 is completed by proving the following implications: (i) implies (iv); (iv) implies (iii); (iii) implies (ii); and (ii) implies (i).

To prove that (i) implies (iv), it suffices to show by Lemma 3.8.4 that $|E(\Sigma)|$ is not strictly greater than $|\Sigma|$. Suppose that $|E(\Sigma)| > |\Sigma|$. Then, there is an element $\sigma = A|(X - A)$ of Σ such that $|E(\sigma)| \geq 2$. This, in turn, means that there are distinct vertices M, M', N, and N' of $G(\Sigma)$ such that M and M' disagree on exactly σ with $A \in M$, and N and N' disagree on exactly σ with $A \in N$. Since these vertices are distinct, there is an element $\sigma' = A'|(X - A')$ of Σ such that $A' \in M$, $A' \in N$, $(X - A') \in M'$, and $(X - A') \in N'$. Moreover, each of the four intersections $A \cap A'$, $A \cap (X - A')$, $(X - A) \cap A'$, and $(X - A) \cap (X - A')$ is non-empty, contradicting that Σ is pairwise compatible. Hence, $|E(\Sigma)| = |\Sigma|$.

We next prove that (iv) implies (iii). By Lemma 3.8.5, it suffices to show that $|V(\Sigma)|$ is not strictly greater than $|\Sigma| + 1$. Suppose that this is not the case. Then, as $|E(\Sigma)| = |\Sigma|$, we have $|V(\Sigma)| > |E(\Sigma)| + 1$. However, by Lemma 3.8.3, $G(\Sigma)$ is connected; a contradiction. Hence, $|V(\Sigma)| = |\Sigma| + 1$ as required.

To prove that (iii) implies (ii), we first establish an equation and an inequality. Let σ be an element of Σ. Let V' denote the set of vertices of $G(\Sigma)$ that are not incident with an edge of $G(\Sigma)$ representing σ. Then,

$$|V'| + 2|E(\sigma)| = |V(\Sigma)|$$

and so, as $|V(\Sigma)| = |\Sigma| + 1$, we get

$$|V'| + 2|E(\sigma)| = |\Sigma| + 1. \tag{3.4}$$

Furthermore,

$$|V'| + |E(\sigma)| = |V(\Sigma - \{\sigma\})|.$$

By Lemma 3.8.5, $|V(\Sigma - \{\sigma\})| \geq |\Sigma|$, and so this last equation implies

$$|V'| + |E(\sigma)| \geq |\Sigma|. \tag{3.5}$$

Combining (3.4) and (3.5), we deduce that

$$|\Sigma| + 1 = |V'| + 2|E(\sigma)| \geq |\Sigma| + |E(\sigma)|,$$

which in turn implies that $|E(\sigma)| \leq 1$. By Lemma 3.8.4, $|E(\sigma)| \geq 1$ and so $|E(\sigma)| = 1$. It follows that $|E(\Sigma)| = |\Sigma|$ holds. Therefore, as $|V(\Sigma)| = |\Sigma| + 1$, we have $|V(\Sigma)| = |E(\Sigma)| + 1$. Since $G(\Sigma)$ is connected, $G(\Sigma)$ is a tree.

Lastly, we prove that (ii) implies (i). Suppose that Σ contains a pair of X-splits $A_1|A_2$ and $B_1|B_2$ in Σ that are not compatible. For all $i, j \in \{1, 2\}$, let N_{ij} denote a vertex of $G(\Sigma)$ that contains both A_i and B_j. Note that, by choosing the remaining elements of N_{ij} so that each one contains a particular element of $A_i \cap B_j$, such vertices exist for all $i, j \in \{1, 2\}$. By Lemma 3.8.3, there is a path P of $G(\Sigma)$ from N_{11} to N_{12} whose edge set has cardinality equal to the number of elements of Σ that N_{11} and N_{12} disagree on. Since the end vertices of each edge in P disagree on exactly one element of Σ, it follows that neither N_{22} nor N_{21} is on this path. Similarly, there are paths from N_{12} to N_{22}, from N_{22} to N_{21}, and from N_{21} to N_{11} that do not contain N_{11} and N_{21}, N_{11} and N_{12}, and N_{12} and N_{22}, respectively. Thus, $G(\Sigma)$ contains a cycle and so $G(\Sigma)$ is not a tree; a contradiction. This completes the proof of the theorem. □

Example 3.8.7. Let $X = \{a, b, c, d, e, f\}$ and let $\Sigma = \{\sigma_1, \sigma_2, \sigma_3, \sigma_4\}$ where $\sigma_1 = \{a, b\}|\{c, d, e, f\}$, $\sigma_2 = \{a, b, d\}|\{c, e, f\}$, $\sigma_3 = \{a, b, c, d, f\}|\{e\}$, and $\sigma_4 = \{a, b, d, e\}|\{c, f\}$. The Buneman graph on Σ is shown in Fig. 3.8(a), where each edge of $G(\Sigma)$ is labelled with the element of Σ on which its end vertices disagree. Observe that for each element in Σ, there is exactly one such edge. Since $G(\Sigma)$ is a tree, it follows by Theorem 3.8.6 that Σ is pairwise compatible.

For a non-empty set Σ of X-splits, we next describe a map from X into the vertex set of $G(\Sigma)$ that turns out to be an extension of the labelling map of a semi-labelled tree. For each $x \in X$, let N_x denote the necessarily unique vertex of $G(\Sigma)$ for which every element contains x. Let $\phi_\Sigma : X \to V(\Sigma)$ be the mapping defined, for all $x \in X$, by $\phi_\Sigma(x) = N_x$. We denote the Buneman graph $G(\Sigma)$ together with this mapping by $(G(\Sigma); \phi_\Sigma)$. As an

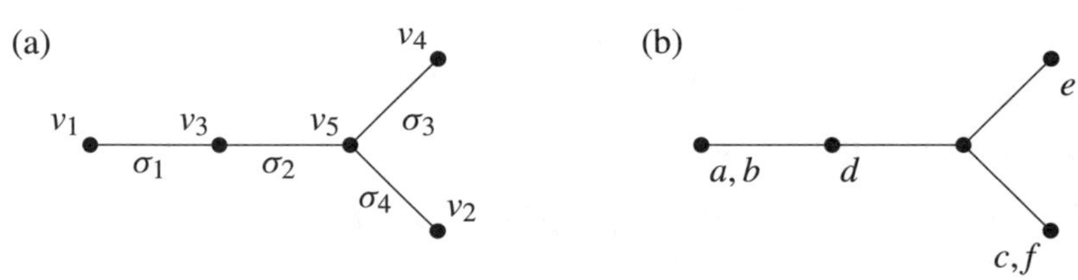

$$v_1 = \{\{a, b\}, \{a, b, d\}, \{a, b, c, d, f\}, \{a, b, d, e\}\}$$
$$v_2 = \{\{c, d, e, f\}, \{c, e, f\}, \{a, b, c, d, f\}, \{c, f\}\}$$
$$v_3 = \{\{c, d, e, f\}, \{a, b, d\}, \{a, b, c, d, f\}, \{a, b, d, e\}\}$$
$$v_4 = \{\{c, d, e, f\}, \{c, e, f\}, \{e\}, \{a, b, d, e\}\}$$
$$v_5 = \{\{c, d, e, f\}, \{c, e, f\}, \{a, b, c, d, f\}, \{a, b, d, e\}\}$$

Figure 3.8 (a) $G(\Sigma)$. (b) $(G(\Sigma); \phi_\Sigma)$.

example, Fig. 3.8(b) shows the graph $(G(\Sigma); \phi_\Sigma)$, where Σ is the set of splits given in Example 3.8.7.

Lemma 3.8.8. *Let Σ be a non-empty set of X-splits and let N be a vertex of $G(\Sigma)$ with degree at most two. Then, N is an element of $\phi_\Sigma(X)$.*

Proof. If N has degree one, then, by Lemma 3.8.2, there is a unique minimal element of N, and therefore there is an element x of X that is contained in every element of N. Thus in this case, N is an element of $\phi(X)$.

If N has degree two, then, again by Lemma 3.8.2, there are exactly two minimal elements, A and A' say, of N. Since N is a vertex of $G(\Sigma)$, the intersection $A \cap A'$ is non-empty. Let $y \in A \cap A'$. Then, y is contained in every element of N, and so N is an element of $\phi_\Sigma(X)$. □

It follows from Theorem 3.8.6 and Lemma 3.8.8 that if Σ is a pairwise compatible set of X-splits, then $(G(\Sigma); \phi_\Sigma)$ is an X-tree $\mathcal{T}$, in which case ϕ_Σ is the labelling map of $\mathcal{T}$.

Theorem 3.8.9. *Let Σ be a non-empty set of X-splits. If Σ is pairwise compatible, then the set of X-splits induced by the X-tree $(G(\Sigma); \phi_\Sigma)$ is equal to Σ.*

Proof. Let $\sigma = A|B$ be an element of Σ. By Lemma 3.8.4 and Theorem 3.8.6, there is a unique edge e of $G(\Sigma)$ that represents σ. Consider the graph $G(\Sigma)\backslash e$ and suppose that x is an element of X. Since e is the only edge of $G(\Sigma)$ that represents σ, it follows that the set of vertices of one component, G_A say, of $G(\Sigma)\backslash e$ is the set of vertices of $G(\Sigma)$ containing A. Therefore, x is contained in A precisely if N_x is a vertex of G_A. Hence, σ is an element of the set of X-splits induced by $(G(\Sigma); \phi_\Sigma)$. Furthermore, by the equivalence of (i) and (iv) in the statement of Theorem 3.8.6, the cardinality of Σ is equal to the cardinality of the set of X-splits induced by $(G(\Sigma); \phi_\Sigma)$. Theorem 3.8.9 now readily follows. □

We now present a result of Dress *et al.* (1997*a*) on the number of vertices and edges of the Buneman graph on a non-empty set Σ of X-splits. To this end, let $\mathcal{I}(\Sigma)$ denote the collection of subsets of Σ that are either pairwise incompatible or have cardinality at most one. Note that the latter implies $\emptyset \in \mathcal{I}(\Sigma)$.

Theorem 3.8.10. *Let Σ be a non-empty set of X-splits. Then*

(i) $|V(\Sigma)| = |\mathcal{I}(\Sigma)|$; *and*
(ii) $|E(\Sigma)| = \sum_{I \in \mathcal{I}(\Sigma)} |I|$.

To establish this theorem, we require some preliminaries. Let Σ be a non-empty set of X-splits and let x be an element of X. If N is a vertex of $G(\Sigma)$, denote by $N_{\min}$ the set of minimal elements of N under set inclusion. We now define a map on which the proof of Theorem 3.8.10 is based. Let θ_x be the map from $V(\Sigma)$ into $\mathcal{I}(\Sigma)$ defined by

$$\theta_x(N) = \{A|(X - A) \colon A \in N_{\min} \text{ and } x \notin A\}.$$

To see that $\theta_x(N)$ is indeed an element of $\mathcal{I}(\Sigma)$, suppose that $A_1|B_1$ and $A_2|B_2$ are distinct elements of $\theta_x(N)$ for some $N \in V(\Sigma)$, where A_1 and A_2 are elements of N. Since A_1 and A_2 are elements of N, $A_1 \cap A_2$ is non-empty. Furthermore, as A_1 and A_2 are minimal, it follows that $A_1 \cap B_2$ and $A_2 \cap B_1$ are both non-empty. Lastly, as x is neither an element of A_1 nor A_2, we deduce that $x \in B_1 \cap B_2$, and so $B_1 \cap B_2$ is also non-empty. Hence, $A_1|B_1$ and $A_2|B_2$ are incompatible, so $\theta_x(N)$ is indeed an element of $\mathcal{I}(\Sigma)$.

The next lemma shows that θ_x is a bijection. If σ is an X-split and x is an element of X, we denote the block of σ containing x by $\sigma(x)$ and the block of σ not containing x by $\sigma(\overline{x})$.

Lemma 3.8.11. *Let Σ be a non-empty set of X-splits and let x be an element of X. Then, the map θ_x:$V(\Sigma) \to \mathcal{I}(\Sigma)$ is a bijection.*

Proof. We first show that θ_x is one-to-one. To do this, it suffices to show that if N is an element of $V(\Sigma)$, then the set $\{A\colon A \in N_{\min} \text{ and } x \notin A\}$ uniquely determines N.

Let σ be an element of Σ. There are two cases to consider. In the first case, suppose that, for some element A' in $\{A\colon A \in N_{\min} \text{ and } x \notin A\}$, either $\sigma(x)$ or $\sigma(\overline{x})$ contains A'. If $\sigma(x)$ contains A', then $\sigma(\overline{x}) \cap A'$ is empty, so $\sigma(x)$ has to be an element of N. Similarly, if $\sigma(\overline{x})$ contains A', then $\sigma(\overline{x})$ has to be an element of N. For the second case, suppose that neither $\sigma(x)$ nor $\sigma(\overline{x})$ contains an element of $\{A\colon A \in N_{\min} \text{ and } x \notin A\}$. In this case, $\sigma(x)$ has to be an element of N. To see this, observe that if $\sigma(\overline{x})$ is an element of N, then, as $x \notin \sigma(\overline{x})$, the block $\sigma(\overline{x})$ must contain an element of $\{A\colon A \in N_{\min} \text{ and } x \notin A\}$. Thus, θ_x is one-to-one.

Now let I be an element of $\mathcal{I}(\Sigma)$ and let N be the set

$$\{h(\sigma)\colon \sigma \in \Sigma\},$$

where

$$h(\sigma) = \begin{cases} \sigma(\overline{x}), & \text{if } \sigma(\overline{x}) \text{ contains a block of an element of } I, \\ \sigma(x), & \text{otherwise.} \end{cases}$$

We now show that (i) $N \in V(\Sigma)$ and (ii) $\theta_x(N) = I$, thus showing that θ_x is onto. We divide the proof of (i) into three cases. Let A and B be elements of N. Firstly, if neither A nor B contain x, then both A and B contain some block of an element of I. Therefore, as I is pairwise incompatible, it follows that in this case $A \cap B$ is non-empty. Secondly, if both A and B contain x, then $A \cap B$ is non-empty. Lastly, suppose exactly one of A and B, A say, contains x. Since A is an element of N, $X - A$ does not contain a block of an element of I, and so A has a non-empty intersection with both blocks of each element of I. Therefore, as B contains a block of an element of I, it follows that $A \cap B$ is non-empty. Thus, N is an element of $V(\Sigma)$. By the construction of N it is clear that $\theta_x(N) = I$ so that (ii) holds. Hence, θ_x is onto, and therefore θ_x is a bijection as required. □

Proof of Theorem 3.8.10. Part (i) immediately follows from Lemma 3.8.11. For the proof of (ii), it follows by Lemma 3.8.2 that the degree of a vertex N of $G(\Sigma)$ is equal to the number of minimal elements of N under set inclusion. Furthermore, if $\{M, N\}$ is an edge of $G(\Sigma)$, then M and N disagree on exactly one element σ of Σ with one end vertex containing $\sigma(\overline{x})$ and the other end vertex containing $\sigma(x)$. It follows that the number of edges of $G(\Sigma)$ is equal to

$$\sum_{N \in V(\Sigma)} |\{A\colon A \in N_{\min} \text{ and } x \notin A\}|. \tag{3.6}$$

However, by the definition of θ_x and Lemma 3.8.11, (3.6) is equal to $\sum_{I \in \mathcal{I}(\Sigma)} |I|$. This completes the proof of (ii). □

3.9 Exercises

1. Let A be a non-empty proper subset of X. Show that the number of binary phylogenetic X-trees that induce the split $A|(X - A)$ is $|RB(a)| \times |RB(n - a)|$, where $a = |A|$ and $n = |X|$.
2. For a set X of size n, find the maximum possible size for any pairwise compatible set of X-splits.
3. Let $\mathcal{T}$ be a binary phylogenetic X-tree. Show that the following three sets of X-splits are equal:
   (a) $\Sigma(\mathcal{T}) \cup \{(A - C)|(B \cup C) : A|B, C|D \in \Sigma(\mathcal{T}) \text{ and } C \subset A\}$;
   (b) $\{A|B\colon A|B \in \Sigma(\mathcal{T}'), \mathcal{T}'$ is obtained from $\mathcal{T}$ by one TBR operation$\}$; and
   (c) $\{A|B\colon A|B \in \Sigma(\mathcal{T}'), \mathcal{T}'$ is obtained from $\mathcal{T}$ by one SPR operation$\}$.
4. Let $n \geq 2$, and let d be the splits metric on $P(n)$. Let $\mathcal{P} = \{\mathcal{T}_1, \mathcal{T}_2, \ldots, \mathcal{T}_k\}$ be a subset of $P(n)$. A phylogenetic tree $\mathcal{T}$ of $P(n)$ is a *median tree* for $\mathcal{P}$ if $\mathcal{T}$ minimizes the sum
$$\sum_{i=1}^{k} d(\mathcal{T}, \mathcal{T}_i).$$
   (i) (Barthélemy and McMorris 1986) Prove that if k is odd, the median tree for $\mathcal{P}$ is unique and is the majority rule consensus tree for $\mathcal{P}$. Classify the median tree(s) for $\mathcal{P}$ when k is even.
   (ii) Suppose that $\mathcal{P}$ consists of binary phylogenetic trees and has the additional property that, for all distinct i and j, $d(\mathcal{T}_i, \mathcal{T}_j)$ takes the maximum possible value of $2(n - 3)$. Prove that $k \leq \frac{1}{2}\binom{n}{2}$. Construct a set of binary phylogenetic trees that realizes this upper bound.
5. Let $X = \{x_1, x_2, \ldots, x_n\}$ and let Σ be a pairwise compatible set of X-splits. Let $\mathcal{T}$ be the X-tree such that $\Sigma = \Sigma(\mathcal{T})$. Describe a polynomial–time algorithm that constructs $\mathcal{T}$ from Σ by generating a sequence $\mathcal{T}_1, \mathcal{T}_2, \ldots, \mathcal{T}_n$ of semi-labelled trees such that $\mathcal{T}_n = \mathcal{T}$ and, for all i, $\mathcal{T}_i$ is an X_i–tree where $X_i = \{x_1, x_2, \ldots, x_i\}$.
6. (Adams 1986) Let $\mathcal{T} = (T; \phi)$ be a rooted phylogenetic X-tree, and let A and B be subsets of X. The set A *nests in* B in $\mathcal{T}$ if the most recent common ancestor of A is a proper descendant of the most recent common ancestor of B. Now, let $\mathcal{P}$ be a

collection of rooted phylogenetic X-trees and let $Ad(\mathcal{P})$ denote the Adams consensus tree for $\mathcal{P}$. Prove that $Ad(\mathcal{P})$ is the unique rooted phylogenetic X-tree that satisfies the following two properties:

(Ad1) If A and B are subsets of X with the property that A nests in B in every phylogenetic tree in $\mathcal{P}$, then A nests in B in $Ad(\mathcal{P})$.

(Ad2) If C and D are clusters of $Ad(\mathcal{P})$ such that C nests in D in $Ad(\mathcal{P})$, then C nests in D in every phylogenetic tree in $\mathcal{P}$.

7. (Margush and McMorris 1981) Describe a polynomial–time algorithm that determines the set of maximal clusters of the Adams consensus tree for a set of rooted X-trees.
8. (Bandelt and Dress 1989) Let $\mathcal{P}$ be a collection of rooted phylogenetic X-trees. Prove that the collection of subsets of X that occur as clusters in more than $|\mathcal{P}|/3$ of the phylogenetic trees in $\mathcal{P}$ is a weak hierarchy on X.
9. (Bandelt and Dress 1989) Let $\mathcal{H}$ be a collection of subsets of X. Prove that $\mathcal{H}$ is a weak hierarchy on X if and only if, for all non-empty subsets A of X, there exists a pair of elements $a, b \in A$ such that the intersection of all subsets of $\mathcal{H}$ containing A is equal to the intersection of all subsets of $\mathcal{H}$ containing $\{a, b\}$. Deduce that, if $\mathcal{H}$ is a weak hierarchy on X, then

$$|\mathcal{H}| \leq \binom{n+1}{2},$$

where $n = |X|$.
10. Prove Theorem 3.7.1(i).
11. A *cut-set* of a graph G is a set E' of edges of G such that $G \backslash E'$ has more components than G. An edge e for which $\{e\}$ is a cut-set is called a *cut-edge*. Let Σ be a set of X-splits and let $\sigma = A|B$ be an element of Σ.
    (i) Show that σ is compatible with every element of Σ if and only if $E(\sigma)$ is a cut-edge of $G(\Sigma)$.
    (ii) Show that $E(\sigma)$ is a minimal cut-set of $G(\Sigma)$. Furthermore, show that $\phi_\Sigma(A)$ and $\phi_\Sigma(B)$ are subsets of the vertex sets of different components of the graph obtained from $(G(\Sigma); \phi_\Sigma)$ by deleting all of the edges in $E(\sigma)$.
12. (Barthélemy and Guénoche 1991) Let X' be a subset of X. For a set Σ of X-splits, let $\Sigma|X'$ denote the set

$$\Sigma|X' = \{(X' \cap A)|(X' \cap B) \colon X' \cap A, X' \cap B \neq \emptyset \text{ and } A|B \in \Sigma\}.$$

Suppose that $|X| \geq 4$, and let Σ_1 and Σ_2 be sets of pairwise compatible X-splits. Prove that $\Sigma_1 = \Sigma_2$ if and only if $\Sigma_1|X' = \Sigma_2|X'$ for every subset X' of X with $|X'| = 4$.

4

Compatibility of characters

In biology, 'characters' describe attributes of the species under consideration and are the data that biologists typically use to reconstruct phylogenetic trees. Mathematically, characters are functions. A natural biological constraint is that the reconstructed phylogenetic trees have the property that each of the characters could have 'evolved' without 'reverse' or 'convergent transitions'. In this chapter, we formalize this notion and explore the underlying mathematical properties of characters that allow us to construct such phylogenetic trees.

4.1 Characters and convexity

A (qualitative or discrete) *character on* X is a function χ from a non-empty subset X' of X into a set C of *character states*. We sometimes refer to C as the *state set* of χ. The character χ is said to be *trivial* if there is at most one element $\alpha \in C$ for which $|\chi^{-1}(\alpha)| \geq 2$; otherwise, χ is *non-trivial*. Furthermore, if $X' = X$, we say χ is a *full character*, and if $|\chi(X')| = r$, we say χ is an *r-state character*. A character χ on X is a *binary character* if χ is a two-state full character.

In biology, characters can be morphological (e.g. wings versus no-wings), biochemical, physiological, behavioural, embryological, or genetic (e.g. the nucleotide at a particular DNA sequence position or the order of certain genes on a chromosome). Allowing X' to be a strict subset of X recognizes that a character state may not be defined or determined for some species.

Definition 4.1.1. Let χ be a character on X from X' into a set C of character states. Then, χ is *convex* on an X-tree $(T; \phi)$ with $T = (V, E)$ if there is a function $\bar{\chi} : V \to C$ satisfying the following two properties:

(C1) $\bar{\chi} \circ (\phi|X') = \chi$; and

(C2) for each $\alpha \in C$, the subgraph of T induced by $\{v \in V : \bar{\chi}(v) = \alpha\}$ is connected.

Convexity has a fundamental biological interpretation that we will describe later in this section.

It immediately follows from the definition of convexity that a binary character χ on X is convex on an X-tree $\mathcal{T}$ precisely if the bipartition of X induced by χ is an X-split of $\mathcal{T}$.

Example 4.1.2. Let X be the set $\{1, 2, \ldots, 7\}$ and let C be the set $\{\alpha, \beta, \gamma, \delta, \epsilon\}$ of character states. Let $\chi : X \to \{\alpha, \beta, \gamma, \delta, \epsilon\}$ be the full character on X defined by

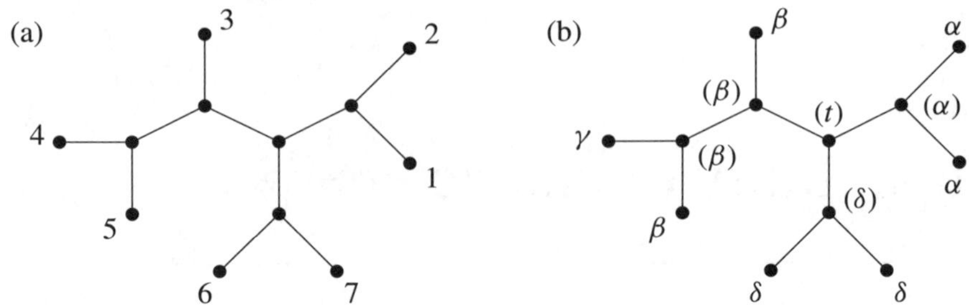

Figure 4.1 An X-tree and a mapping $\bar{\chi}_t$.

$\chi(1) = \chi(2) = \alpha$, $\chi(3) = \chi(5) = \beta$, $\chi(4) = \gamma$, and $\chi(6) = \chi(7) = \delta$. Thus, χ is a four-state character. Consider the X-tree $\mathcal{T} = (T; \phi)$ with $T = (V, E)$ shown in Fig. 4.1(a). For each $t \in \{\alpha, \beta, \delta\}$, let $\bar{\chi}_t$ be the map from V into C indicated in Fig. 4.1(b). Clearly, if $t = \alpha$, then $\bar{\chi}_t$ satisfies (C1) and (C2). Therefore, χ is convex on $\mathcal{T}$. Observe that for each $t \in \{\beta, \delta\}$, the function $\bar{\chi}_t$ also satisfies (C1) and (C2). In fact, it is easily seen that $\bar{\chi_\alpha}$, $\bar{\chi_\beta}$, and $\bar{\chi_\delta}$ are the only such functions satisfying both (C1) and (C2).

Let $\mathcal{T} = (T; \phi)$ be an X-tree with $T = (V, E)$ and let $\chi : X' \to C$ be a character on X. For each state α in $\chi(X')$, let $T(\alpha)$ denote the minimal subtree of T containing $\phi(\chi^{-1}(\alpha))$. Proposition 4.1.3 provides alternative descriptions of convexity.

Proposition 4.1.3. *Let $\mathcal{T} = (T; \phi)$ be an X-tree, and let $\chi : X' \to C$ be a character on X. The following statements are equivalent:*

(i) χ is convex on $\mathcal{T}$;
(ii) the members of $\{T(\alpha) : \alpha \in C\}$ are pairwise vertex disjoint; and
(iii) for all distinct $\alpha, \beta \in \chi(X')$, there exists an X-split $A|B$ of $\mathcal{T}$ such that $\chi^{-1}(\alpha) \subseteq A$ and $\chi^{-1}(\beta) \subseteq B$.

Proof. Here, we prove the equivalence of (i) and (ii), and leave their equivalence with (iii) to the reader. Suppose that χ is convex on $\mathcal{T}$. Then, there is a function $\bar{\chi} : V(T) \to C$ satisfying (C1) and (C2). Let α_1 and α_2 be elements of C. It follows by (C2) that, for each $i \in \{1, 2\}$, $T(\alpha_i)$ is a subtree of the subgraph of T induced by $\{v \in V(T) : \bar{\chi}(v) = \alpha_i\}$, and so $T(\alpha_1)$ and $T(\alpha_2)$ are vertex disjoint.

Conversely, suppose that the members of $\{T(\alpha) : \alpha \in C\}$ are pairwise vertex disjoint. Define $\bar{\chi} : V(T) \to C$ to be any function such that the following hold:

(I) For each $\alpha \in C$, if v is a vertex of $T(\alpha)$, then $\bar{\chi}(v) = \alpha$.
(II) Let F denote the subgraph obtained from T by deleting the elements in the set $\{v \in T : v \in T(\alpha)$ for some $\alpha \in C\}$. If u and u' are elements of the vertex set of the same component of F, then $\bar{\chi}(u) = \bar{\chi}(u')$ and the subgraph of T induced by $\{v \in V(T) : \bar{\chi}(v) = \bar{\chi}(u)\}$ is connected.

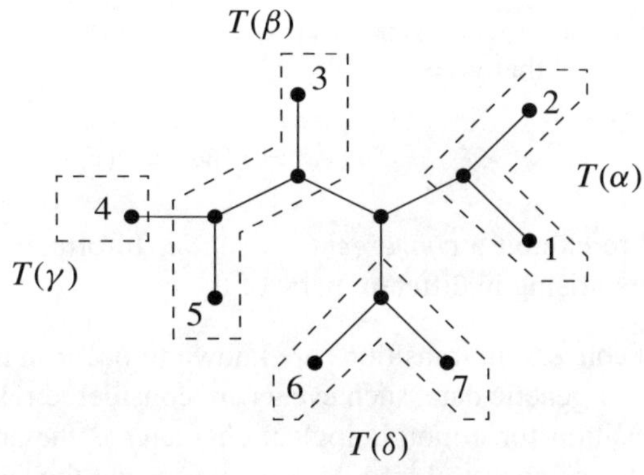

Figure 4.2 Vertex disjoint subtrees.

Since the members of $\{T(\alpha) : \alpha \in C\}$ are pairwise vertex disjoint, it is easily seen that there is such a function $\bar{\chi}$ satisfying (I) and (II). Moreover, $\bar{\chi}$ satisfies (C1) and one can easily check that it also satisfies (C2). Thus, χ is convex on $\mathcal{T}$ as required. □

To illustrate Proposition 4.1.3, consider the X-tree $\mathcal{T}$ and character χ on X in Example 4.1.2. The vertex disjoint subtrees $T(\alpha)$, $T(\beta)$, $T(\gamma)$, and $T(\delta)$ are shown in Fig. 4.2.

Biological justification for convexity. There is a simple biological rationale for regarding convexity as fundamental. Viewing a rooted phylogenetic X-tree $\mathcal{T} = (T; \phi)$ as describing the evolution of the set X of extant species from an ancestral species at the root ρ of T. The remaining interior vertices of $\mathcal{T}$ correspond to other hypothetical ancestral species that are descended from the ancestral species at ρ. Now, suppose each extant and ancestral species has an associated character state lying in some set of character states. In this way, we can regard the character state as also 'evolving' from ρ towards the elements of X on $\mathcal{T}$. This leads to a concept of evolutionary 'innovation', namely that each time a species changes its character state, the new state it acquires is arising for the first time in the tree. There are several equivalent ways to formalize this concept. One way is as follows. For all $v \in V(T)$, let $c(v)$ denote the character state assigned to vertex v. Then, the 'innovation' concept corresponds to the requirement that neither of the following two events occur, in which case we will say that c is *homoplasy-free*:

(i) Suppose that $v_1, v_2, \ldots, v_k$ is a path in $\mathcal{T}$ directed away from the root ρ. If, for some $i \in \{2, 3, \ldots, k-1\}$,

$$c(v_1) = c(v_k) \neq c(v_i),$$

c is said to exhibit a *reverse transition*. Informally, this corresponds to a new state arising but then reverting to an earlier state.

(ii) Suppose that $v_1, v_2, \ldots, v_k$ and $w_1, w_2, \ldots, w_l$ are paths in $\mathcal{T}$ directed away from the root ρ, and that $v_1 = w_1$. If

$$c(v_k) = c(w_l) \neq c(v_1),$$

c is said to exhibit a *convergent transition*. Informally, this corresponds to the same state arising in different parts of the tree.

Reverse and convergent transitions are known to occur in biology. However, except for certain types of genetic data, such events are considered relatively rare. An example of a reverse transition for a morphological character is the acquisition of teeth in the lineage leading to the ancestral bird *Archaeopteryx* and the subsequent loss of teeth in modern birds. An example of a convergent transition is the acquisition of some similar morphological features by the Australian marsupial wolf *Thylacinus cynocephalus* and by distantly related South American marsupials.

We now explain the connection between these biology concepts and convexity. Let $\mathcal{T} = (T; \phi)$ be a rooted phylogenetic X-tree with $T = (V, E)$ and suppose that each vertex v of T has an associated character state $c(v)$ that is an element of a set C of character states. Consider the associated phylogenetic X-tree $\mathcal{T}^{-\rho}$. If we restrict our attention to the values that c takes at the leaves of T, we obtain an induced full character χ on X by setting $\chi(x) = c(\phi(x))$ for all $x \in X$. This character is precisely the character describing the states assigned to the set of present-day species X. Now if c is homoplasy-free, then χ is convex on $\mathcal{T}^{-\rho}$ since $\bar{\chi} : V \to C$ defined by $\bar{\chi}(u) = c(u)$ for all $u \in V$, satisfies conditions (C1) and (C2).

Conversely, if a character χ_1 is convex on a phylogenetic X-tree $\mathcal{T}_1$ with $T_1 = (V_1, E_1)$ and a corresponding function $\bar{\chi}_1 : V_1 \to C$ that satisfies conditions (C1) and (C2), then, for all choices of a root ρ, we can extend $\bar{\chi}_1$ to a map from $V_1 \cup \{\rho\}$ to C that is homoplasy-free.

It is important to note, however, that even if c is not homoplasy-free on a rooted phylogenetic tree $\mathcal{T}$, it is still entirely possible that the associated character χ may be convex on $\mathcal{T}^{-\rho}$.

We end this section by presenting an exact expression for the number of full r-state characters $\chi : X \to C$ on X that are convex on a given binary phylogenetic X-tree. Surprisingly, this expression depends only on r, $|C|$, and $|X|$, and not on the shape of the tree (Steel 1992).

Proposition 4.1.4. *Let $\mathcal{T}$ be a binary phylogenetic X-tree and let C be a set of $c \geq r$ states. Then, the number of full r-state characters $\chi : X \to C$ that are convex on $\mathcal{T}$ is*

$$\frac{c!}{(c-r)!}\binom{2n-r-1}{r-1},$$

where $n = |X|$ and $c = |C|$.

4.2 Character compatibility (perfect phylogeny)

Definition 4.2.1. A collection of characters on X is said to be *compatible* if there exists an X-tree on which all the characters in the collection are convex.

Informally, a collection of characters is compatible if they could all have evolved on some tree without any reverse or convergent transitions. Note that the condition of compatibility is unaltered if one substitutes 'phylogenetic X-tree' or 'binary phylogenetic X-tree' for 'X-tree' in the definition. Also, a trivial character on X is convex on every phylogenetic X-tree.

Determining whether a collection of characters is compatible and, if so, constructing a tree on which they are all convex is a fundamental problem in classification. This problem was initially called the *character compatibility problem* but, more recently, has become known in computer science as the *perfect phylogeny problem.*

The Splits-Equivalence Theorem shows that a collection of binary characters is compatible if and only if the characters are pairwise compatible, and in that case, under $\leq$, there is a unique minimal tree on which the characters are convex. However, neither of these claims apply for collections of characters that contain non-binary characters. For non-binary characters, an appropriate framework for characterizing character compatibility is via chordal graphs, which we now describe.

For a character $\chi : X' \to C$ on X, let $\pi(\chi)$ denote the partition of X' corresponding to $\{\chi^{-1}(\alpha) : \alpha \in C\}$. In particular, binary characters on X produce X-splits in this way.

Let $\mathcal{C}$ be a collection of characters on X and let $\mathcal{T} = (T; \phi)$ be an X-tree. We next define two graphs each of which has vertex set

$$\bigcup_{\chi \in \mathcal{C}} \{(\chi, A) : A \in \pi(\chi)\}. \tag{4.1}$$

(i) The *partition intersection graph of* $\mathcal{C}$ is the graph that has vertex set (4.1) and an edge joining two vertices precisely if the intersection of the second coordinates is non-empty. We denote this graph by $\mathrm{int}(\mathcal{C})$.

(ii) The *subtree intersection graph of* $\mathcal{T}$ *induced by* $\mathcal{C}$ is the graph that has vertex set (4.1) and an edge $\{(\chi, A), (\chi', B)\}$ if the intersection of the vertex sets of $T(\phi(A))$ and $T(\phi(B))$ is non-empty. This graph is denoted by $\mathrm{int}(\mathcal{C}, \mathcal{T})$.

Lastly, recall that a chordalization of a graph $G = (V, E)$ is a graph $G' = (V, E')$ in which G' is chordal and $E \subseteq E'$. A graph G is a *restricted chordal completion* of $\mathrm{int}(\mathcal{C})$ if G is a chordalization of $\mathrm{int}(\mathcal{C})$ and, for all edges $\{(\chi, A), (\chi', B)\}$ of G, $\chi \neq \chi'$.

McMorris and Meacham (1983) characterized when a graph can be realized as a partition intersection graph (see Exercise 5). Our main concern is to characterize when a set of characters is compatible. Theorem 4.2.2 (indicated by Buneman 1974*a* and Meacham 1983, and formally proved by Steel 1992) is the main result of this section. Originally proved for a collection of full characters, the extension to a collection of (not necessarily full) characters is straightforward.

Theorem 4.2.2. *Let $\mathcal{C}$ be a collection of characters on X. Then, $\mathcal{C}$ is compatible if and only if there exists a restricted chordal completion of* $\mathrm{int}(\mathcal{C})$.

x	$\chi_1(x)$	$\chi_2(x)$	$\chi_3(x)$
1	α	β	γ
2	α	β'	γ'
3	α'	β	γ'
4	α'	β''	γ''
5	α''	β''	γ

Figure 4.3 Characters χ_1, χ_2, and χ_3.

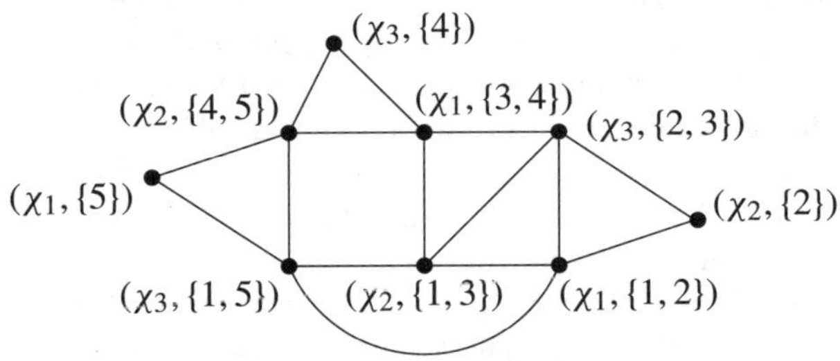

Figure 4.4 A partition intersection graph of a set of characters.

Before proving Theorem 4.2.2, we illustrate it with an example.

Example 4.2.3. Suppose that $X = \{1,2,3,4,5\}$ and let χ_1, χ_2, and χ_3 be the characters on X as defined in Fig. 4.3. Then, $\pi(\chi_1) = \{\{1,2\},\{3,4\},\{5\}\}$, $\pi(\chi_2) = \{\{1,3\},\{2\},\{4,5\}\}$, and $\pi(\chi_3) = \{\{1,5\},\{2,3\},\{4\}\}$. Let $\mathcal{C} = \{\chi_1,\chi_2,\chi_3\}$, and consider the graph $\text{int}(\mathcal{C})$ shown in Fig. 4.4. We now make direct use of Theorem 4.2.2 to show that $\mathcal{C}$ is not compatible.

Suppose that there is a restricted chordal completion G of $\text{int}(\mathcal{C})$. Then, as the vertices in

$$\{(\chi_1,\{3,4\}),(\chi_2,\{4,5\}),(\chi_3,\{1,5\}),(\chi_2,\{1,3\})\}$$

induce a four-cycle of $\text{int}(\mathcal{C})$ and G is a restricted chordal completion of $\text{int}(\mathcal{C})$, it follows that $\{(\chi_1,\{3,4\}),(\chi_3,\{1,5\})\}$ is an edge of G. This, in turn, implies that G must have an edge joining either $(\chi_1,\{1,2\})$ and $(\chi_1,\{3,4\})$, or $(\chi_3,\{1,5\})$ and $(\chi_3,\{2,3\})$, for otherwise the vertices in

$$\{(\chi_1,\{1,2\}),(\chi_3,\{2,3\}),(\chi_1,\{3,4\}),(\chi_3,\{1,5\})\}$$

induce a four-cycle of G. But then G is not a restricted chordal completion of $\text{int}(\mathcal{C})$; a contradiction. Hence, by Theorem 4.2.2, $\mathcal{C}$ is not compatible.

Proof of Theorem 4.2.2. Suppose that $\mathcal{C}$ is compatible. Then, there exists an X-tree $\mathcal{T}$ on which every character in $\mathcal{C}$ is convex. By Theorem 1.3.2, $\text{int}(\mathcal{C},\mathcal{T})$ is chordal.

Therefore, as the edge set of int($\mathcal{C}$) is a subset of the edge set of int($\mathcal{C}, \mathcal{T}$) and every character in $\mathcal{C}$ is convex on $\mathcal{T}$, it follows that int($\mathcal{C}, \mathcal{T}$) is a restricted chordal completion of int($\mathcal{C}$).

To prove the converse, suppose that G is a restricted chordal completion of int($\mathcal{C}$). By the equivalence of (i) and (iv) in Theorem 1.3.2, there exists a tree T' whose vertex set $\mathcal{K}$ is the set of maximal cliques of G and, for each vertex (χ, A), the subgraph of T' induced by the elements of $\mathcal{K}$ containing (χ, A) is a subtree of T'. We complete the proof of the converse by defining an X-tree via T' on which every character in $\mathcal{C}$ is convex.

Define $\phi : X \to \mathcal{K}$ to be a map such that, for each element x in X, $\phi(x)$ contains the vertices of the maximum-sized clique in G in which x is an element of the second coordinate of every vertex. Observe that, as int($\mathcal{C}$) is a subgraph of G, a vertex of G is in this clique precisely if the second coordinate of this vertex contains x. Note that such a map may not be unique. Now, define T to be the tree obtained from T' by suppressing all vertices of degree two that are not identified by an element of X. It is easily checked that all degree-one vertices of T' are identified by an element of X, and so $\mathcal{T} = (T; \phi)$ is an X-tree.

To show that every character in $\mathcal{C}$ is convex on $\mathcal{T}$, let A_1 and A_2 be members of $\pi(\chi)$ for some $\chi \in \mathcal{C}$. Then, the subtrees T_1' and T_2' of T' induced by the elements of $\mathcal{K}$ containing (χ, A_1) and (χ, A_2), respectively, do not intersect. Since the elements of A_i can only be identified with vertices in T_i', for each $i \in \{1, 2\}$, it follows that the intersection of the vertex sets of $T(A_1)$ and $T(A_2)$ is empty. Thus, every element of $\mathcal{C}$ is convex on $\mathcal{T}$, and therefore, by definition, $\mathcal{C}$ is compatible. □

For a collection $\mathcal{C}$ of compatible characters on X, the next example illustrates the construction of an X-tree on which each of the characters in $\mathcal{C}$ is convex. This construction uses the method outlined in the proof of Theorem 4.2.2. Before giving this example, we first need to explain how to construct a maximal clique tree representation of a chordal graph.

Suppose that $G = (V, E)$ is a chordal graph and let $v_1, v_2, \ldots, v_k$ be a perfect elimination ordering of the vertices of G, where $k = |V|$. Since every chordal graph has at least one simplicial vertex (Dirac 1961) and every vertex-induced subgraph of a chordal graph is chordal, obtaining such an ordering is elementary.

From this ordering, we next construct a maximal clique tree representation of G. This construction follows Golumbic (1980). Let $i \in \{1, 2, \ldots, k-1\}$ and let K_i denote the vertex set of the maximal clique of $G[\{v_i, v_{i+1}, \ldots, v_k\}]$ that contains v_i. Define T_k to be the tree consisting of the single vertex v_k. Thus, T_k is a maximal clique tree representation of $G[\{v_k\}]$. In general, for all i, define T_i to be the tree obtained from T_{i+1} as follows:

(i) if $K_i - v_i$ is a vertex of T_{i+1}, then replace $K_i - v_i$ with K_i to get T_i;

(ii) otherwise, join a new vertex K_i to a vertex of T_{i+1} containing $K_i - v_i$ to get T_i.

It is easily checked that, for all i, T_i is a maximal clique tree representation of $G[\{v_i, v_{i+1}, \ldots, v_k\}]$. Hence, T_1 is a maximal clique tree representation of G. Note that such a tree is not necessarily unique.

Example 4.2.4. Let $X = \{1,2,3,4,5,6\}$ and let $\mathcal{C} = \{\chi_1, \chi_2, \chi_3\}$ be a set of characters on X such that $\pi(\chi_1) = \{\{1,2\},\{3,5\}\}$, $\pi(\chi_2) = \{\{2,6\},\{3,4\}\}$, and $\pi(\chi_3) = \{\{1,4\},\{5,6\}\}$. Let G denote the graph shown in Fig. 4.5(a). Then, by observing that the first coordinates of the end vertices of each of the dashed lines are distinct, G is a restricted chordal completion of int($\mathcal{C}$), where int($\mathcal{C}$) is the graph induced by the solid lines of this graph. Therefore, by Theorem 4.2.2, $\mathcal{C}$ is compatible. Note that G is not the only restricted chordal completion of int($\mathcal{C}$).

We next describe the construction of an X-tree on which all of the characters in $\mathcal{C}$ are convex. The following sequence is a perfect elimination ordering of G:

$$(\chi_3, \{1,4\}), (\chi_2, \{2,6\}), (\chi_1, \{3,5\}), (\chi_1, \{1,2\}), (\chi_2, \{3,4\}), (\chi_3, \{5,6\}).$$

Using this ordering and by the process described immediately prior to this example, one can construct the maximal clique tree representation $T = (\mathcal{K}, E)$ of G shown in Fig. 4.5(b), where $\mathcal{K}$ is the collection of maximal cliques of G and, for each vertex v in G, the subgraph induced by the elements of $\mathcal{K}$ containing v is a subtree of T. In this case, this is the only maximal clique tree representation of G. Lastly, to obtained our desired X-tree, we define a map $\phi : X \to \mathcal{K}$ so that, for each element x in X, $\phi(x)$ contains the vertices of the maximum-sized clique in G in which x is an element of the second coordinate of every vertex. The X-tree $(T;\phi)$ shown in Fig. 4.5(c) has the property that all of the characters in $\mathcal{C}$ is convex on it.

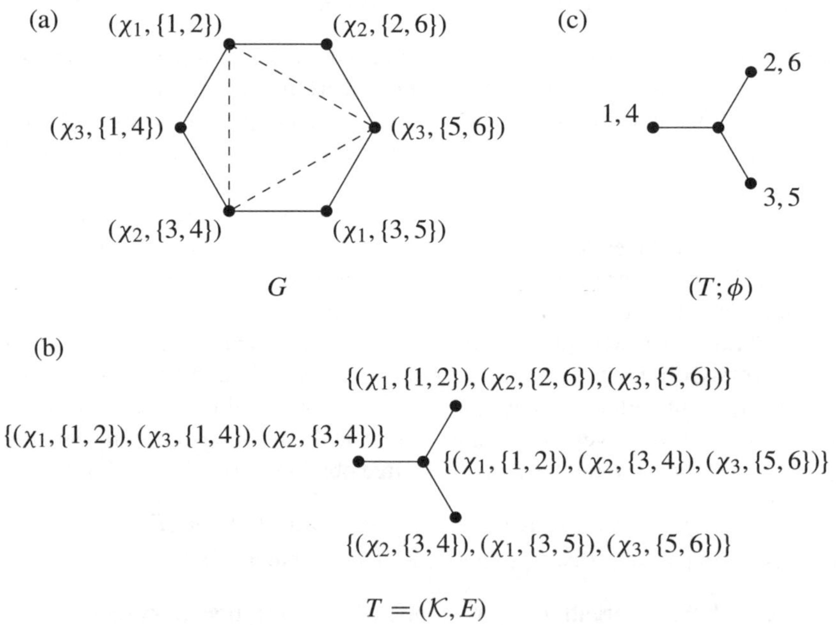

Figure 4.5 (a) A restricted chordal completion of int($\mathcal{C}$). (b) A maximal clique tree representation of G. (c) An X-tree on which each character in $\mathcal{C}$ is convex.

The following two corollaries are attractive consequences of Theorem 4.2.2. The first is given by Estabrook and McMorris (1977).

Corollary 4.2.5. *Two characters χ and χ' on X are compatible if and only if* $\text{int}(\{\chi, \chi'\})$ *is acyclic.*

Proof. Suppose that $\text{int}(\{\chi, \chi'\})$ is acyclic. Then, $\text{int}(\{\chi, \chi'\})$ is chordal, and so, by Theorem 4.2.2, χ and χ' are compatible. Now, suppose that $\text{int}(\{\chi, \chi'\})$ contains a cycle. Let G be a chordalization of $\text{int}(\{\chi, \chi'\})$. Then, G must contain a 3-cycle, which implies that G is not a restricted chordal completion of $\text{int}(\{\chi, \chi'\})$. Therefore, by Theorem 4.2.2, χ and χ' are not compatible. □

Corollary 4.2.6. *Let $\mathcal{C}$ be a collection of binary characters on X. Then, $\mathcal{C}$ is compatible if and only if* $\text{int}(\mathcal{C})$ *is chordal.*

Proof. If $\text{int}(\mathcal{C})$ is chordal, then $\text{int}(\mathcal{C})$ is a restricted chordal completion of itself. Therefore, by Theorem 4.2.2, $\mathcal{C}$ is compatible.

Now suppose that $\mathcal{C}$ is compatible. Then, by the Splits-Equivalence Theorem, there is a unique X-tree $\mathcal{T}$ such that $\Sigma(\mathcal{T})$ is equal to the set of X-splits induced by the elements in $\mathcal{C}$. We next show that $\text{int}(\mathcal{C})$ is equal to $\text{int}(\mathcal{C}, \mathcal{T})$ by verifying the claim that the edge sets of $\text{int}(\mathcal{C})$ and $\text{int}(\mathcal{C}, \mathcal{T})$ are equal.

Let (χ, A) and (χ', B) be distinct vertices of $\text{int}(\mathcal{C})$. If $\chi = \chi'$, then, as $A|B$ is an X-split induced by $\mathcal{T}$, the claim clearly holds. Therefore, we may assume that $\chi \neq \chi'$. If $A \cap B$ is non-empty, then trivially the claim holds. Assume that $A \cap B$ is empty. Now, $A|(X - A)$ and $B|(X - B)$ are distinct X-splits induced by $\mathcal{T}$. Therefore, by Lemma 3.1.5, there is a partitioning of X into three sets X_1, X_2, and X_3 so that $X_1 \in \{A, X - A\}$, $X_3 \in \{B, X - B\}$, and the intersection of the vertex sets of $T(\phi(X_1))$ and $T(\phi(X_3))$ is empty. The only possible choice for X_1 and X_3 is A and B, respectively. The claim now readily follows for this case, completing the proof of the corollary. □

Corollary 4.2.6 does not extend to collections of two-state characters on X. For example, the collection $\mathcal{C}$ in Example 4.2.4 is a collection of compatible two-state characters on X but $\text{int}(\mathcal{C})$, which is the graph induced by the solid lines in Fig. 4.5(a), is not chordal.

We end this section by noting that there is a natural way to convert any set $\mathcal{C}$ of characters into a set $\mathcal{C}'$ of full characters, so that $\mathcal{C}$ is compatible if and only if $\mathcal{C}'$ is compatible. Furthermore, the set of phylogenetic X-trees on which each character in $\mathcal{C}$ is convex is identical to the set of phylogenetic X-trees on which each character in $\mathcal{C}'$ is convex. For a subset X' of X, one natural way to extend a character $\chi : X' \to C$ to a full character χ' on X so as to achieve these objectives is as follows. Let C' be any set that contains C and has $|X - X'|$ additional elements, and let $h : (X - X') \to (C' - C)$ denote any one-to-one map. Let $\chi' : X \to C'$ be defined by

$$\chi'(x) = \begin{cases} \chi(x), & \text{if } x \in X', \\ h(x), & \text{otherwise.} \end{cases}$$

Then, for all phylogenetic X-trees $\mathcal{T}$, it is easily verified that χ is convex on $\mathcal{T}$ if and only if χ' is convex on $\mathcal{T}$.

4.3 Strongly compatible characters

A pair of characters χ and χ' on X is said to be *strongly compatible* if there exists an element of $\pi(\chi)$ and an element of $\pi(\chi')$ whose union is X. Strong compatibility was introduced by Dress *et al.* (1997*b*). Viewing binary characters as splits, it is easily checked that a pair of binary characters is compatible precisely if the pair is strongly compatible. More generally, it is easily seen that if a pair of characters χ and χ' on X is strongly compatible, then $\text{int}(\{\chi, \chi'\})$ is acyclic. Consequently, by Corollary 4.2.5, χ and χ' are compatible. The following theorem (Dress *et al.* 1997*b*) extends this last observation by showing that pairwise strong compatibility suffices to guarantee compatibility of any set of characters.

Theorem 4.3.1. *Suppose that $\mathcal{C}$ is a collection of pairwise strongly compatible characters on X. Then, $\mathcal{C}$ is compatible. Furthermore, for such a collection $\mathcal{C}$, an X-tree on which each character in $\mathcal{C}$ is convex can be reconstructed from $\mathcal{C}$ in polynomial time.*

Proof. Without loss of generality, we may assume that $|\pi(\chi)| > 1$ for all $\chi \in \mathcal{C}$. Construct the set of X-splits

$$\Sigma(\mathcal{C}) = \bigcup_{\chi \in \mathcal{C}} \{A|A' : A \in \pi(\chi)\},$$

where we adopt the notation $A' = X - A$ throughout this proof. We first show that $\Sigma(\mathcal{C})$ is a pairwise compatible set of X-splits. To this end, suppose that $A \in \pi(\chi)$ and $B \in \pi(\chi')$, and consider the X-splits $A|A'$ and $B|B'$. We will show that $\emptyset \in \{A \cap B, A \cap B', A' \cap B, A' \cap B'\}$ for all choices of A and B. Firstly, if $\chi = \chi'$, then $A \cap B = \emptyset$. Thus, we may assume that $\chi \neq \chi'$. Since χ and χ' are strongly compatible, there exists $A_1 \in \pi(\chi)$ and $B_1 \in \pi(\chi')$ with $A_1 \cup B_1 = X$. If $A = A_1$ and $B = B_1$, then $A' \cap B' = X - (A_1 \cup B_1) = \emptyset$. If $A \neq A_1$ and $B \neq B_1$, then $A_1 \subseteq A'$ and $B_1 \subseteq B'$, and so $A' \cup B' = X$, which implies that $A \cap B = X - (A' \cup B') = \emptyset$. For the case $A = A_1$ and $B \neq B_1$, we have $B_1 \subseteq B'$ and so $A_1 \cup B' = X$, which implies that $A' \cap B = X - (A \cup B') = X - (A_1 \cup B') = \emptyset$. Similarly, if $A \neq A_1$ and $B = B_1$, we have $A \cap B' = \emptyset$. This covers all possible cases and so justifies our claim that each pair of X-splits in $\Sigma(\mathcal{C})$ is compatible. It now follows from the Splits-Equivalence Theorem that there is an X-tree $\mathcal{T} = (T; \phi)$ with $\Sigma(\mathcal{C}) = \Sigma(\mathcal{T})$. We now show that each character χ in $\mathcal{C}$ is convex on $\mathcal{T}$. Take $C, D \in \pi(\chi)$. Since $C|C' \in \Sigma(\mathcal{T})$ and $D \subseteq C'$, there is an edge of T whose deletion results in the vertices in $\phi(C)$ being in a separate component from the vertices in $\phi(D)$. Since this holds for all pairs $C, D \in \pi(\chi)$, it follows that χ is indeed convex on $\mathcal{T}$. Finally, the construction of $\Sigma(\mathcal{C})$, and thereby of $\mathcal{T}$, can be carried out in polynomial time using the techniques described in Section 3.1. □

4.4 Complexity of determining character compatibility

As we have seen, it is easy to determine whether a collection of binary characters is compatible, and therefore to construct the unique minimal tree on which each character is convex. However, for an arbitrary collection $\mathcal{C}$ of characters, determining if $\mathcal{C}$ is compatible is NP-complete even if all of the characters in $\mathcal{C}$ are two-state, and we will demonstrate this in Chapter 6. Moreover, in further contrast to the binary setting, pairwise compatibility is not sufficient for compatibility of the entire collection of characters (as Example 4.2.3 demonstrates) and there may exist (exponentially) many minimal trees on which each character in $\mathcal{C}$ is convex (Böcker 1999).

Despite the NP-completeness result stated in the previous paragraph, for collections of full characters on X, there are some positive results. Given a bound r on the number of distinct states that each character can take, the compatibility problem can be solved by a polynomial–time algorithm. This remarkable discovery by Agarwala and Fernández-Baca (1994) followed a number of polynomial–time algorithms by various authors for the cases $r = 3$ and $r = 4$. Their algorithm runs in time $O(2^{3r}(nk^3 + k^4))$, where k is the number of characters and n is the cardinality of X. This algorithm was improved to an $O(2^{2r}k^2n)$ algorithm by Kannan and Warnow (1997). Alternatively, given a bound k on the number of full characters on X, the compatibility problem can also be solved in polynomial time. McMorris *et al.* (1994) obtained an $O(r^{k+1}k^{k+1} + nk^2)$ algorithm for this problem.

4.5 Maximum compatibility

In applications, it is unusual to encounter situations where an entire collection of given characters is compatible. In that case, two options present themselves:

(i) to adopt a more refined measure to describe how well the evolution of the characters can be described by a tree; or

(ii) to retain compatibility as the desired criterion and search for maximum-sized subsets of characters that form a compatible set.

Option (i) will be explored in the next chapter; here, we explore option (ii), called the *maximum compatibility* approach. We will describe how, although being computationally intractable in general, the approach (and an extension of it to allow weighted characters) has a polynomial–time solution under certain restrictions.

Given a collection $\mathcal{C}$ of characters on X the associated *compatibility graph* is the graph that has vertex set $\mathcal{C}$ and that has an edge joining two characters in $\mathcal{C}$ precisely if they are compatible. Suppose that $\mathcal{C}$ is a collection of binary characters on X. Then, the Splits-Equivalence Theorem implies that the compatible subsets of $\mathcal{C}$ correspond to the cliques of the compatibility graph of $\mathcal{C}$. Hence, in this case, the problem of determining a maximum compatible subset of $\mathcal{C}$ is equivalent to finding a maximum-sized clique in the compatibility graph of $\mathcal{C}$. This problem is NP-hard (Day and Sankoff 1986). However, we may also ask whether the collection $\mathcal{C}$ is 'almost' compatible, in the sense that there is a subset of $\mathcal{C}$ of size t (small) whose removal from $\mathcal{C}$ results in a compatible set of characters on X. The smallest such value of t is called the *imperfection* of $\mathcal{C}$. To determine the imperfection of a set $\mathcal{C}$ of binary characters, one can invoke an algorithm that applies to a more general graph-theoretic problem, as we now describe.

A *vertex cover* of a graph $G = (V, E)$ is a subset V' of V with the property that every edge of G is incident with at least one element of V'. If t is the size of a minimum vertex cover of G, it is possible to find a minimum-sized vertex cover of G by an algorithm having complexity $O(|V|t + (4/3)^t t^2)$ (Balasubramanian *et al.* 1998). In our setting, we take G to be the *incompatibility graph* of $\mathcal{C}$; that is, the graph that has vertex set $\mathcal{C}$ and an edge joining two characters in $\mathcal{C}$ precisely if they are incompatible. In this case, a vertex cover corresponds to a subset of $\mathcal{C}$ whose removal from $\mathcal{C}$ results in a pairwise compatible set of characters on X. Thus, removing from $\mathcal{C}$ the elements in a minimum-sized vertex cover of the incompatibility graph of $\mathcal{C}$ results in a maximum-sized compatible subset of $\mathcal{C}$ if $\mathcal{C}$ consists of binary characters. In this way, we obtain an $O(k^2n + kt + (4/3)^t t^2)$ algorithm for the maximum compatibility problem on k binary characters on a set X of size n and having imperfection t (Bonet *et al.* 1998).

In applications, a set $\mathcal{C}$ of characters often comes with an associated weight function $w : \mathcal{C} \to \mathbb{R}^{>0}$. This weight function may record, for example, how often each character appears in a sequence (e.g. with DNA sequence data). In this more general setting, the maximum compatibility approach seeks compatible subsets $\mathcal{C}_0$ of $\mathcal{C}$ that maximize $\sum_{\chi \in \mathcal{C}_0} w(\chi)$.

Suppose that we are given a collection $\mathcal{C}$ of binary characters on X and a weight function $w : \mathcal{C} \to \mathbb{R}^{>0}$. In this setting, it follows by the Splits-Equivalence Theorem that the maximum compatibility approach seeks a subset $\mathcal{C}_0$ of $\mathcal{C}$ that (i) maximizes $\sum_{\chi \in \mathcal{C}_0} w(\chi)$ and (ii) has the property that $\{\pi(\chi) : \chi \in \mathcal{C}_0\} = \Sigma(\mathcal{T})$ for some X-tree $\mathcal{T}$. Bryant (1996) found a tractable variation on this problem by requiring that the X-tree $\mathcal{T}$ in (ii) to be a phylogenetic X-tree in which the degree of any vertex is at most d for some prechosen integer $d \geq 3$. In fact, there may not exist any such subset set $\mathcal{C}_0$ satisfying this modification of (ii) for a fixed value of d. Nevertheless, there is an $O(nk + ndk^{d-1})$ algorithm to determine whether such a set $\mathcal{C}_0$ exists, where $k = |\mathcal{C}|$ and $n = |X|$. Moreover, if this set does exist, then one can efficiently construct such a phylogenetic X-tree that satisfies the modified form of (ii). See Bryant (1996) for details.

The maximum compatibility approach has been used only to a limited extent as a technique for reconstructing phylogenies from character data (see, e.g. LeQuesne 1982). Reasons for this include the computational intractability of the problem of determining a maximum-weighted compatible subset of a set of characters, and the fact that compatibility is based on convexity, which is an 'all-or-nothing' condition, and so it is somewhat coarse compared to measures described in the next chapter. Nevertheless, the method has been quite useful for data that is close to being compatible. One such example from linguistics is the analysis of the Indo-European family of languages described by Warnow (1997).

4.6 Generalized characters and directional constraints

In this section, we extend the notion of characters and convexity in two ways. First, we allow for the character state of some species to be ambiguous. Second, under the viewpoint of a character state evolving on a rooted semi-labelled tree, we allow for directional constraints on the evolution of a character state. This section is based on Benham *et al.* (1995) and Pe'er *et al.* (2000).

A *generalized character* g on X with state set C is a map $g : X \to (2^C - \{\emptyset\})$. Generalized characters arise in biology when the character state assigned to one or more species is either unknown or ambiguous.

The earlier concepts of convexity and compatibility for characters extend naturally to generalized characters. Specifically, a generalized character $g : X \to (2^C - \{\emptyset\})$ is said to be *convex* on an X-tree $\mathcal{T}$ if there exists a full character $\chi : X \to C$ such that $\chi(x) \in g(x)$ for all $x \in X$, and χ is convex on $\mathcal{T}$. A collection $\mathcal{C}$ of generalized characters on X is *compatible* if there is an X-tree on which all of the generalized characters in $\mathcal{C}$ are convex.

Although the definition of a generalized character on X requires the character to have domain X, it still captures the concept of convexity of an ordinary character on X whose domain is some subset of X. More precisely, let $X' \subseteq X$ and suppose that $\chi : X' \to C$ is a character on X. Then, χ is convex on $\mathcal{T}$ if and only if the associated generalized character $g_\chi : X \to C \cup \{C\}$ defined, for all $x \in X$, by

$$g_\chi(x) = \begin{cases} \{\chi(x)\}, & \text{if } x \in X', \\ C, & \text{if } x \in X - X', \end{cases}$$

is convex on $\mathcal{T}$.

We now describe a further extension of the concept of convexity. In certain applications, character states evolve with directional constraints; e.g., state α may only be able to 'evolve' into state β but not conversely. Such constraints can be conveniently summarized by means of a directed graph D that has vertex set C and that has an arc directed from α to β precisely if the transition from α to β is allowed. In this setting, suppose we are given a generalized character $g : X \to (2^C - \{\emptyset\})$ on X, a rooted tree $T = (V, E)$, and a labelling map $\phi : X \to V$. If necessary, we may allow T to have vertices of degree two. We say that g is *convex on* (T, ϕ) *relative to* D if there exists a map $\bar{\chi} : V \to C$ satisfying the following properties:

(i) $\bar{\chi}(\phi(x)) \in g(x)$ for all $x \in X$;

(ii) for all $\alpha \in C$, the subgraph of T induced by $\{v \in V : \bar{\chi}(v) = \alpha\}$ is connected; and

(iii) if $u, v \in V$ and v is an immediate descendant of u in T, then either $\bar{\chi}(u) = \bar{\chi}(v)$ or there is an arc in D directed from $\bar{\chi}(u)$ to $\bar{\chi}(v)$.

As a special case, this definition of convexity includes the convexity of generalized characters (and thereby of ordinary characters) provided we take D to be the directed graph that has an arc directed from every element of C to every other element of C.

Biological examples. A simple example of a generalized character with directional constraints arises when the set of character states is $\{0, 1\}$ and $0 \to 1$ is the only allowable transition. This is useful for modelling situations in which 0 represents some 'ancestral' state, 1 represents a 'derived' state, and where transitions from 1 to 0 are regarded as highly unlikely. The generalized character value $\{0, 1\}$ is used to denote uncertainty as to whether an observation is ancestral or derived. This set-up has been applied, for example, to certain molecular genetic data known as SINEs ('short interspersed nuclear elements'), where the states 0 and 1 denote the absence or presence, respectively, of a particular sequence inserted into a particular region in a genome (see Pe'er *et al.* 2000). Biological settings for a generalized character with character state set $\{0, 1, 2\}$ and directional constraints indicated by the directed graph $0 \rightarrow 1 \rightarrow 2$ are discussed by Benham *et al.* (1995).

The notion of compatibility of a collection of generalized characters relative to a collection of directional constraints is defined in the obvious way. In particular, suppose that $\mathcal{C} = \{g_1, g_2, \ldots, g_k\}$ is a collection of generalized characters on X and $\mathcal{D} = \{D_1, D_2, \ldots, D_k\}$ is a collection of directed graphs such that, for all $i \in \{1, 2, \ldots, k\}$, the directed graph D_i summarizes the directional constraints associated with the generalized character g_i. The collection $\mathcal{C}$ is *compatible relative to* $\mathcal{D}$ if there is a pair (T, ϕ), where $T = (V, E)$ is a rooted tree and $\phi : X \rightarrow V$ a labelling map, such that g_i is convex on (T, ϕ) relative to D_i for all $i \in \{1, 2, \ldots, k\}$.

The compatibility of a collection $\mathcal{C}$ of generalized characters on X relative to a collection $\mathcal{D}$ of directional constraints is computationally interesting. Benham *et al.* (1995) showed that deciding the compatibility of $\mathcal{C}$ relative to $\mathcal{D}$ is NP-complete, even if D_i is the directed graph $0 \rightarrow 1 \rightarrow 2$ for all $i \in \{1, 2, \ldots, k\}$. However, in the same paper Benham *et al.* identified a setting where a polynomial–time algorithm applies. Suppose that, for all $i \in \{1, 2, \ldots, k\}$, the directed graph D_i is a rooted tree, and g_i satisfies the condition that, for each $x \in X$, $g_i(x)$ is either a single-element subset of C or a subset of C whose elements induce a directed path in the rooted tree D_i. Under these conditions, there is an $O(n^2k)$ algorithm for deciding the compatibility of $\mathcal{C}$ subject to the directional constraints $\mathcal{D}$, where $n = |X|$. Furthermore, when compatibility applies, the algorithm allows for the construction of a pair (T, ϕ), where $T = (V, E)$ is a rooted tree and $\phi : X \rightarrow V$ is a labelling map, on which g_i is convex relative to D_i for all $i \in \{1, 2, \ldots, k\}$.

Although the requirement that D_i is a rooted tree may seem restrictive, it nevertheless includes the example described above where $\{g_1, g_2, \ldots, g_k\}$ is a collection of generalized characters on X with state set $\{0, 1\}$ and $0 \rightarrow 1$ is the only allowable transition for all $i \in \{1, 2, \ldots, k\}$. Thus, for this example, compatibility can be determined in $O(n^2k)$ time, where $n = |X|$; a faster algorithm was subsequently found by Pe'er *et al.* (2000). We will also describe an algorithm in Chapter 6 that can also be used to solve the compatibility problem for this example (see Exercise 4, Chapter 6).

4.7 Specialist topic: defining an X-tree by characters

In Section 4.2, we gave a graph-theoretic characterization for when a collection of characters on X is compatible. In this section, we present a graph-theoretic characterization for when, up to isomorphism, there exists a unique X-tree on which a collection of characters on X is convex.

Suppose that $\mathcal{C}$ is a collection of characters on X. The set $\mathcal{C}$ is *convex* on an X-tree $\mathcal{T}$ if every character in $\mathcal{C}$ is convex on $\mathcal{T}$. We say that $\mathcal{C}$ *defines* an X-tree $\mathcal{T}$ if $\mathcal{C}$ is convex on $\mathcal{T}$ and all other X-trees on which $\mathcal{C}$ is convex are isomorphic to $\mathcal{T}$. For example, the set of characters given in Example 4.2.4 does not define a semi-labelled tree, as the semi-labelled tree shown in Fig. 4.6 also has the property that this set is convex on it. Analogous to Theorem 4.2.2, the main result of this section (Theorem 4.7.1) gives a graph-theoretic characterization for when a set of characters on X defines an X-tree. To state this theorem, we need some definitions in addition to those given in Section 4.2.

Let $\mathcal{C}$ be a collection of compatible characters on X and let $\mathcal{T}$ be an X-tree on which $\mathcal{C}$ is convex. A restricted chordal completion $G = (V, E)$ of $\text{int}(\mathcal{C})$ is *minimal* if, for every non-empty subset F of edges of $E - E(\text{int}(\mathcal{C}))$, $(V, E - F)$ is not chordal. An edge $e = \{u, v\}$ of $\mathcal{T}$ is said to be *distinguished by a character* χ of $\mathcal{C}$ if, for every mapping $\bar{\chi}$ satisfying (C1) and (C2), $\bar{\chi}(u) \neq \bar{\chi}(v)$. If each edge of $\mathcal{T}$ is distinguished by a character in $\mathcal{C}$, then $\mathcal{T}$ is *distinguished by* $\mathcal{C}$. The following theorem is due to Semple and Steel (2002*a*).

Theorem 4.7.1. *Let $\mathcal{C}$ be a collection of characters on X. Then, $\mathcal{C}$ defines an X-tree if and only if the following two conditions are satisfied:*

(i) there is a binary phylogenetic X-tree on which $\mathcal{C}$ is convex and which is distinguished by $\mathcal{C}$; and

(ii) there is a unique minimal restricted chordal completion of $\text{int}(\mathcal{C})$.

We present an outline rather than a formal proof of Theorem 4.7.1 (for full details and proofs of the lemmas used in this outline, see Semple and Steel 2002*a*).

The 'only if' part of the proof of Theorem 4.7.1 follows from Lemmas 4.7.2 and 4.7.3.

Lemma 4.7.2. *Let $\mathcal{T}$ be an X-tree and let $\mathcal{C}$ be a set of characters on X. Then,* $\text{int}(\mathcal{C}, \mathcal{T})$ *is a restricted chordal completion of* $\text{int}(\mathcal{C})$ *if and only if $\mathcal{C}$ is convex on $\mathcal{T}$.*

Lemma 4.7.3. *Let $\mathcal{C}$ be a set of characters on X and let G be a minimal restricted chordal completion of* $\text{int}(\mathcal{C})$*. Then, there exists a phylogenetic X-tree $\mathcal{T}$ such that* $\text{int}(\mathcal{C}, \mathcal{T}) = G$.

Note that the assumption on 'G' in Lemma 4.7.3 cannot be weakened by removing 'minimal'.

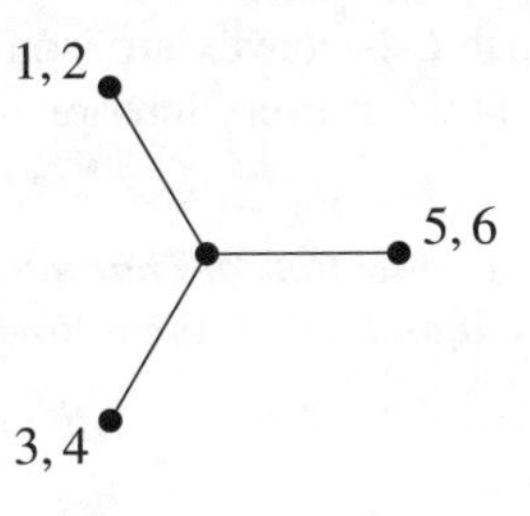

Figure 4.6

The 'if' part of the proof of Theorem 4.7.1 essentially reduces to proving a special case, which is the content of Lemma 4.7.4.

Lemma 4.7.4. *Let $\mathcal{C}$ be a collection of two-state characters on X. Suppose that the following two conditions hold:*

(i) there is a binary phylogenetic X-tree on which $\mathcal{C}$ is convex and which is distinguished by $\mathcal{C}$; and

(ii) there is a unique minimal restricted chordal completion of $\text{int}(\mathcal{C})$.

Then, $\mathcal{C}$ defines an X-tree.

We complete the outline of the proof of Theorem 4.7.1 by showing how Lemma 4.7.4 is used.

Let $\mathcal{C}$ be a collection of characters on X. For a character χ in $\mathcal{C}$, let X_χ denote the domain of χ and C_χ denote the codomain of χ. For each character χ in $\mathcal{C}$, define, for all distinct $\alpha, \beta \in C_\chi$,

$$X_\chi^{\alpha,\beta} = \{x : x \in X_\chi \text{ and } \chi(x) \in \{\alpha, \beta\}\}.$$

The collection of 'two-state characters on X induced by $\mathcal{C}$' is the set

$$\mathcal{C}_2 = \bigcup_{\chi \in \mathcal{C}} \{\chi | X_\chi^{\alpha,\beta} : \alpha, \beta \in C_\chi, \alpha \neq \beta, \alpha \in \chi(X_\chi), \text{ and } \beta \in \chi(X_\chi)\}.$$

Lemma 4.7.5. *Let $\mathcal{C}$ be a collection of characters on X and let $\mathcal{T}$ be an X-tree.*

(i) $\mathcal{C}$ is convex on $\mathcal{T}$ if and only if $\mathcal{C}_2$ is convex on $\mathcal{T}$.

(ii) Suppose that $\mathcal{C}$ is convex on $\mathcal{T}$. If $\text{int}(\mathcal{C}_2, \mathcal{T})$ *is a minimal restricted chordal completion of* $\text{int}(\mathcal{C}_2)$, *then* $\text{int}(\mathcal{C}, \mathcal{T})$ *is a minimal restricted chordal completion of* $\text{int}(\mathcal{C})$.

The 'if' part of the proof of Theorem 4.7.1 is completed by first using Lemma 4.7.5 to show that (i) and (ii) of Lemma 4.7.4 hold with '$\mathcal{C}_2$' replacing '$\mathcal{C}$', and then, as $\mathcal{C}_2$ defines an X-tree, it follows by (i) of Lemma 4.7.5 that $\mathcal{C}$ defines an X-tree. This completes the outline of the proof.

Next, we provide an application of Theorem 4.7.1. A set $\mathcal{C}$ of characters on X *phylogenetically defines* a phylogenetic X-tree $\mathcal{T}$ if $\mathcal{C}$ is convex on $\mathcal{T}$ and all other phylogenetic X-trees on which $\mathcal{C}$ is convex are isomorphic to $\mathcal{T}$. An X-tree $\mathcal{T}'$ is said to be *internally* distinguished by $\mathcal{C}$ if every interior edge of $\mathcal{T}'$ is distinguished by some character in $\mathcal{C}$.

Corollary 4.7.6. *Let $\mathcal{C}$ be a collection of characters on X. Then, $\mathcal{C}$ phylogenetically defines a phylogenetic X-tree if and only if the following two conditions are satisfied:*

(i) there is a binary phylogenetic X-tree on which $\mathcal{C}$ is convex and which is internally distinguished by $\mathcal{C}$; and

(ii) there is unique minimal restricted chordal completion of $\text{int}(\mathcal{C})$.

Proof. Suppose that $\mathcal{C}$ phylogenetically defines a phylogenetic X-tree $\mathcal{T}$. Then, it is clear that $\mathcal{T}$ satisfies the properties in (i). Let G be a minimal restricted chordal completion of $\text{int}(\mathcal{C})$. To show that (ii) holds, it suffices to show that $G = \text{int}(\mathcal{C}, \mathcal{T})$. By Lemma 4.7.3, there is a phylogenetic X-tree $\mathcal{T}'$ such that $\text{int}(\mathcal{C}, \mathcal{T}') = G$ and, by Lemma 4.7.2, $\mathcal{C}$ is convex on $\mathcal{T}'$. Since $\mathcal{C}$ phylogenetically defines $\mathcal{T}$, it follows that $\mathcal{T}' \cong \mathcal{T}$. Hence, $G = \text{int}(\mathcal{C}, \mathcal{T})$.

To prove the converse, suppose that (i) and (ii) hold. Let $\mathcal{T}$ be an X-tree satisfying all of the properties in (i) and let $\mathcal{C}^+$ denote the set $\mathcal{C} \cup \mathcal{C}_{\text{triv}}$ of characters on X, where $\mathcal{C}_{\text{triv}}$ is a collection of characters on X such that

$$\{\pi(\chi) : \chi \in \mathcal{C}_{\text{triv}}\} = \{\{\{x\}, X - \{x\}\} : x \in X\}.$$

Since $\mathcal{T}$ satisfies the properties of (i), $\mathcal{C}^+$ is convex on $\mathcal{T}$ and $\mathcal{C}^+$ distinguishes $\mathcal{T}$. We next show that there is a unique minimal restricted chordal completion of $\text{int}(\mathcal{C}^+)$.

Let G be the unique minimal restricted chordal completion of $\text{int}(\mathcal{C})$, and let G^+ be the graph with vertex set $V(\text{int}(\mathcal{C}^+))$ and edge set $E(\text{int}(\mathcal{C}^+)) \cup E(G)$. We claim that G^+ is the unique minimal restricted chordal completion of $\text{int}(\mathcal{C}^+)$. To justify this claim, first note that a routine case analysis gives $\text{int}(\mathcal{C}^+, \mathcal{T}) = G^+$. Therefore, as $\mathcal{C}^+$ is convex on $\mathcal{T}$, it follows by Lemma 4.7.2 that G^+ is a restricted chordal completion of $\text{int}(\mathcal{C}^+)$.

To complete the proof of the claim, let G_1^+ be a minimal restricted chordal completion of $\text{int}(\mathcal{C}^+)$. Then, by Lemma 4.7.3, there is a phylogenetic X-tree $\mathcal{T}_1^+$ such that $\text{int}(\mathcal{C}^+, \mathcal{T}_1^+) = G_1^+$. By Lemma 4.7.2, this implies that $\mathcal{C}^+$ and, in particular $\mathcal{C}$, is convex on $\mathcal{T}_1^+$. Therefore, using Lemma 4.7.3 again, $\text{int}(\mathcal{C}, \mathcal{T}_1^+)$ is a restricted chordal completion of $\mathcal{C}$. Since G is the unique minimal restricted chordal completion of $\mathcal{C}$, we deduce that

$$E(G) \subseteq E(\text{int}(\mathcal{C}, \mathcal{T}_1^+)) \subseteq E(\text{int}(\mathcal{C}^+, \mathcal{T}_1^+)).$$

Consequently, by the construction of G^+, $E(G^+) \subseteq E(\text{int}(\mathcal{C}^+, \mathcal{T}_1^+))$, which implies that $E(G^+) = E(\text{int}(\mathcal{C}^+, \mathcal{T}_1^+))$ as G_1^+ is minimal. Hence, G^+ is the unique minimal restricted chordal completion of $\text{int}(\mathcal{C}^+)$ as claimed.

It follows by Theorem 4.7.1 that $\mathcal{C}^+$ defines $\mathcal{T}$. Now, $\mathcal{T}$ is a phylogenetic X-tree on which $\mathcal{C}$ is convex. Moreover, every other phylogenetic X-tree on which $\mathcal{C}$ is convex is isomorphic to $\mathcal{T}$, for otherwise, there is a phylogenetic X-tree not isomorphic to $\mathcal{T}$, and on which $\mathcal{C}^+$ is convex. This completes the proof of the corollary. □

Example 4.7.7. To illustrate Corollary 4.7.6, let $X = \{1, 2, \ldots, 6\}$ and let $\mathcal{C} = \{\chi_1, \chi_2, \chi_3\}$ be a set of characters on X such that $\pi(\chi_1) = \{\{1, 4\}, \{3, 5, 6\}\}$, $\pi(\chi_2) = \{\{1, 2\}, \{3, 5\}\}$, and $\pi(\chi_3) = \{\{2, 3\}, \{4, 6\}\}$. Let G denote the graph (including dashed lines) shown in Fig. 4.7(a). One can easily check that G is the unique minimal restricted chordal completion of $\text{int}(\mathcal{C})$, where $\text{int}(\mathcal{C})$ is the subgraph of G induced by the solid edges. Furthermore, the binary phylogenetic X-tree $\mathcal{T}$ shown in Fig. 4.7(b) has the properties that $\mathcal{C}$ is convex on $\mathcal{T}$ and is internally distinguished by $\mathcal{C}$. It now follows by Corollary 4.7.6 that $\mathcal{C}$ defines a phylogenetic X-tree, in particular, $\mathcal{C}$ defines $\mathcal{T}$.

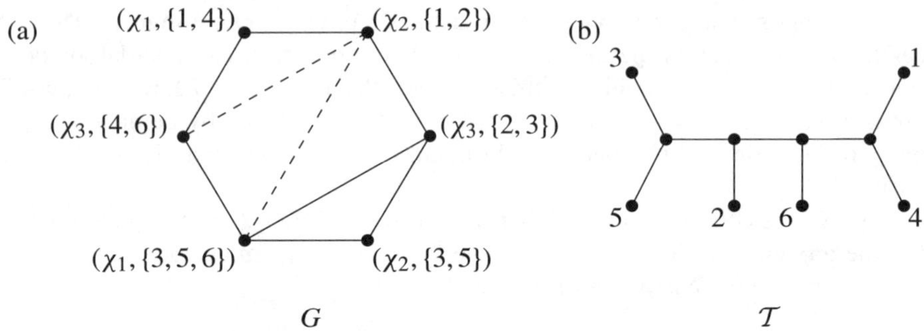

Figure 4.7 (a) A minimal restricted chordal completion of int($\mathcal{C}$). (b) An X-tree on which $\mathcal{C}$ is convex.

4.8 Exercises

1. Suppose that a character $\chi : X \to C$ is convex on an X-tree $\mathcal{T}$. Let C' be a set of character states and let $h : C \to C'$ be a function. Is it necessarily the case that the character $h \circ \chi$ is convex on $\mathcal{T}$?
2. Select an element x_0 in X and let $\mathcal{C}_0$ be the set of all full characters $\chi : X \to \{0, 1\}$ such that $\chi(x_0) = 0$. Show that every maximal clique of the compatibility graph of $\mathcal{C}_0$ has size $2n-3$, where $n = |X|$. Furthermore, show that there are $b(n) = (2n-5)!!$ such cliques. Each maximal clique of the compatibility graph of $\mathcal{C}_0$ corresponds to which sort of X-tree?
3. Show that a collection $\mathcal{C} = \{\chi_1, \chi_2, \ldots, \chi_k\}$ of full characters on X is compatible if and only if there exists a compatible collection $\mathcal{P}$ of X-trees

$$\{(T_1; \phi_1), (T_2; \phi_2), \ldots, (T_k; \phi_k)\}$$

such that $\{\phi_i^{-1}(v) : v \in V(T_i)\} - \{\emptyset\} = \pi(\chi_i)$ for all $i \in \{1, 2, \ldots, k\}$. Describe the X-trees in $\mathcal{P}$ when $\mathcal{C}$ is *strongly* compatible.
4. Let $\mathcal{C}$ be the collection of non-trivial binary characters on a set X of size five. Show that the compatibility graph of $\mathcal{C}$ is isomorphic to the Petersen graph shown in Fig. 4.8.
5. (McMorris and Meacham 1983) For some $k > 1$, let G be a graph that is properly k-colourable but not properly $(k-1)$-colourable. Show that G is isomorphic to a partition intersection graph if and only if G has no isolated vertices, and there is a set of cliques of G with each clique consisting of k vertices and with every edge of G in some clique.
6. Given a phylogenetic X-tree $\mathcal{T}$ and a generalized character $g : X \to (2^C - \{\emptyset\})$, describe a polynomial–time algorithm for determining if g is convex on $\mathcal{T}$.

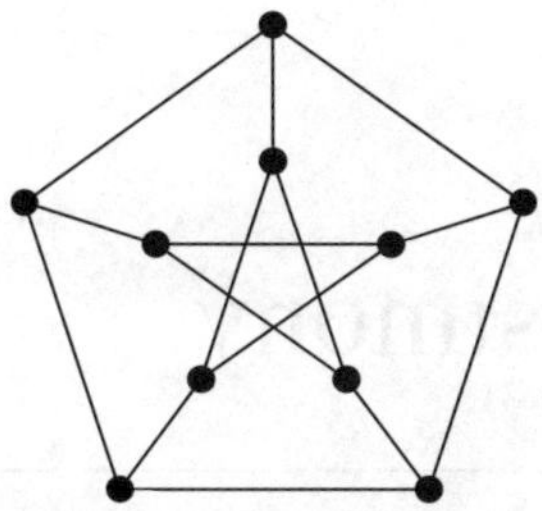

Figure 4.8 The Petersen graph.

7. Suppose that $g : X \rightarrow (2^{\{0,1\}} - \{\emptyset\})$ is a generalized character on X. Let $X' = \{x \in X : g(x) \neq \{0,1\}\}$. Select an element $y \notin X$ and consider the following two-state character $\chi : X' \cup \{y\} \rightarrow \{0,1\}$ defined by

$$\chi(x) = \begin{cases} g(x), & \text{if } x \in X'; \text{ and} \\ 0, & \text{if } x = y. \end{cases}$$

Show that g is convex on a rooted X-tree $\mathcal{T}$ relative to the directed graph $0 \rightarrow 1$ if and only if χ is convex on the $(X \cup \{y\})$-tree obtained from $\mathcal{T}$ by attaching a new leaf to the root of $\mathcal{T}$ and labelling this leaf y.

8. (McMorris *et al.* 1994) Using Exercise 11 in Chapter 1, prove that if $\mathcal{C}$ is a compatible set of k characters, then the tree-width of the intersection graph of $\mathcal{C}$ satisfies

$$tw(\text{int}(\mathcal{C})) \leq k - 1.$$

9. Prove Lemma 4.7.2.
10. Using Corollary 4.7.6 or otherwise, show that every caterpillar tree can be phylogenetically defined by just two characters.
11. (Semple and Steel 2002*b*) Prove that every binary phylogenetic tree can be defined by at most five characters.

5
Maximum parsimony

The *maximum parsimony method* is a popular technique for reconstructing phylogenetic trees from characters. It can be viewed as a refinement of the maximum compatibility approach described in Chapter 4.

The central idea behind maximum parsimony is to fit character data to a semi-labelled tree so as to minimize the number of reverse and convergent transitions. One justification that has been offered for this approach is 'Ockham's Razor', the principle that one should always prefer a simpler explanation over a more complex one. In this setting, complexity is being measured by the number of reverse and convergent transitions. A related viewpoint is that such transitions are typically rare, and so the most 'likely' explanation for the data is one requiring the fewest transitions.

In this chapter, we describe the classical parsimony problem as well as some extensions. The classical problem has direct connections with graph theory and enumerative combinatorics. Throughout the chapter, the following convention will hold. When we consider an X-tree $\mathcal{T}$ together with a character χ on X, we will suppose that if $\phi(x) = \phi(y)$, then $\chi(x) = \chi(y)$. This convention is redundant if $\mathcal{T}$ is a phylogenetic X-tree.

Except for the next section, we will mostly deal with *full* characters in this chapter (this is for convenience and many of the results described are not confined to full characters). Also, we will mostly deal with sequences of characters rather than sets of characters. This is because, in applications, a character may appear multiple times in some data set, and we can handle this repetition by describing the data as an ordered sequence of characters.

5.1 Classical parsimony

Definition 5.1.1. For a graph $G = (V, E)$ and a function f on V, the *changing set* of f is the subset

$$\mathrm{Ch}(f) = \{\{u, v\} \in E : f(u) \neq f(v)\}$$

of edges of G. The *changing number* of f, denoted $\mathrm{ch}(f)$, is the cardinality of $\mathrm{Ch}(f)$.

Let $\chi : X' \to C$ be a character on X and let $\mathcal{T} = (T; \phi)$ be an X-tree. An *extension of* χ to $\mathcal{T}$ is a function $\overline{\chi} : V(T) \to C$ for which $\overline{\chi} \circ (\phi|X') = \chi$. The *parsimony score* of χ on $\mathcal{T}$ is the minimum value of $\mathrm{ch}(\overline{\chi})$ over all extensions $\overline{\chi}$ of χ to $\mathcal{T}$. We denote this score by $l(\chi, \mathcal{T})$. Furthermore, if $\overline{\chi}$ is an extension of χ to $\mathcal{T}$ and $\mathrm{ch}(\overline{\chi}) = l(\chi, \mathcal{T})$,

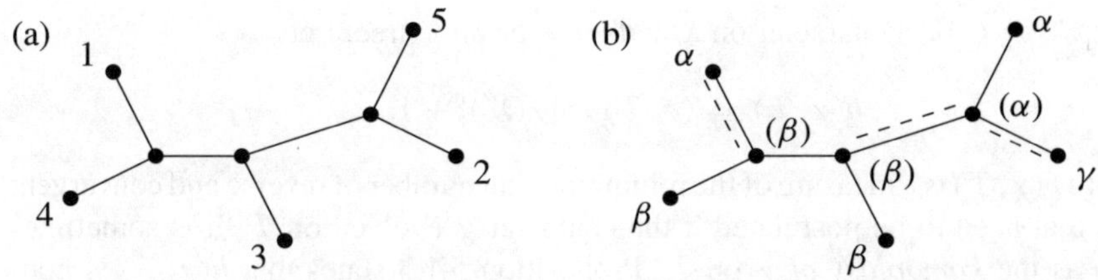

Figure 5.1 A minimum extension of a character.

then $\overline{\chi}$ is called a *minimum extension* of χ to $\mathcal{T}$. All of the above concepts extend to rooted X-trees in the obvious way.

Let $\mathcal{C} = (\chi_1, \ldots, \chi_k)$ be a sequence of characters on X. The *parsimony score of* $\mathcal{C}$ on an X-tree $\mathcal{T}$, denoted $l(\mathcal{C}, \mathcal{T})$, is the sum of the individual parsimony scores of the characters on $\mathcal{T}$; thus,

$$l(\mathcal{C}, \mathcal{T}) = \sum_{i=1}^{k} l(\chi_i, \mathcal{T}).$$

An X-tree $\mathcal{T}'$ that minimizes $l(\mathcal{C}, \mathcal{T}')$ is called a *maximum parsimony tree* for $\mathcal{C}$ and the corresponding value of $l(\mathcal{C}, \mathcal{T}')$ is denoted by $l(\mathcal{C})$.

Example 5.1.2. With $X = \{1, 2, 3, 4, 5\}$, let $\mathcal{T}$ be the X-tree shown in Fig. 5.1(a). Let $\chi : X \to \{\alpha, \beta, \gamma\}$ be the character defined by $\chi(1) = \chi(5) = \alpha$, $\chi(2) = \gamma$, and $\chi(3) = \chi(4) = \beta$. One can easily check that $l(\chi, \mathcal{T}) = 3$. A minimum extension $\overline{\chi}$ of χ to $\mathcal{T}$ is indicated in Fig. 5.1(b) along with the three corresponding edges in $\mathrm{Ch}(\overline{\chi})$. Note that there are other minimum extensions of χ to $\mathcal{T}$.

We describe the setting detailed in Definition 5.1.1 as *classical parsimony*. Later, we will discuss several variations and extensions of this concept.

The parsimony score of a sequence of characters captures and generalizes the concept of convexity as Proposition 5.1.3 makes clear.

Proposition 5.1.3. *Let χ be an r-state character on X and let $\mathcal{T}$ be an X-tree. Then,*

$$l(\chi, \mathcal{T}) \geq r - 1. \tag{5.1}$$

Moreover, equality holds in (5.1) *if and only if χ is convex on $\mathcal{T}$.*

Proof. Let T denote the underlying tree of $\mathcal{T}$ and let $\overline{\chi}$ be a minimum extension of χ to $\mathcal{T}$. Let $T_{\overline{\chi}}$ denote the tree obtained from T by contracting every edge in $E(T) - \mathrm{Ch}(\overline{\chi})$, and consider the mapping on $V(T_{\overline{\chi}})$ induced by $\overline{\chi}$. Since the cardinality of the image of this mapping is r, we have $|V(T_{\overline{\chi}})| \geq r$. Therefore, as $|E(T_{\overline{\chi}})| = \mathrm{ch}(\overline{\chi})$ and $T_{\overline{\chi}}$ is a tree,

$$l(\chi, \mathcal{T}) = \mathrm{ch}(\overline{\chi}) \geq r - 1.$$

Furthermore, if χ is convex on $\mathcal{T}$, then $|V(T_{\overline{\chi}})| = r$. Hence, equality holds in (5.1) if and only if χ is convex on $\mathcal{T}$. □

Let $\chi : X' \to C$ be a character on X and let $\mathcal{T}$ be an X-tree. Let

$$h(\chi, \mathcal{T}) = l(\chi, \mathcal{T}) - |\chi(X')| + 1.$$

The quantity $h(\chi, \mathcal{T})$ is a measure of the minimum total number of reverse and convergent transitions that need to be postulated if the character χ evolved on $\mathcal{T}$. It is sometimes referred to as the *homoplasy of* χ on $\mathcal{T}$. Proposition 5.1.3 states that $h(\chi, \mathcal{T})$ is non-negative, and is equal to zero precisely if χ is convex on $\mathcal{T}$. Now, let $\mathcal{C}$ be a sequence $(\chi_1, \chi_2, \ldots, \chi_k)$ of characters on X and let $\mathcal{T}$ be a maximum parsimony tree for $\mathcal{C}$. The quantity

$$h(\mathcal{C}) = \sum_{i=1}^{k} h(\chi_i, \mathcal{T})$$

is the *total homoplasy* of $\mathcal{C}$ and is a measure of the total number of reverse and convergent transitions that need to be postulated if all the characters in $\mathcal{C}$ evolved on a common X-tree. The following corollary is an immediate consequence of Proposition 5.1.3.

Corollary 5.1.4. *Suppose that* $\mathcal{C} = (\chi_1, \ldots, \chi_k)$ *is a sequence of characters on* X. *Then,* $h(\mathcal{C}) \geq 0$ *with equality precisely if* $\mathcal{C}$ *is compatible.*

Two further ways to view the parsimony score of a character on a semi-labelled tree are as follows:

(i) in terms of sets of edges that separate vertices assigned different character states; and

(ii) in terms of a maximal system of paths under certain restrictions.

We describe these now, beginning with (i).

Let $\chi : X' \to C$ be a character on X and let $\mathcal{T} = (T; \phi)$ be an X-tree with $T = (V, E)$. A subset E_1 of E is a *cut-set for* χ on $\mathcal{T}$ if, for each pair $x, y \in X'$ with $\chi(x) \neq \chi(y)$, the vertices $\phi(x)$ and $\phi(y)$ lie in different components of the forest $T \backslash E_1$. A *minimum cut-set* for χ on $\mathcal{T}$ is a cut-set for χ of minimum size and the size of such a cut-set is denoted by $\mathrm{cut}(\chi, \mathcal{T})$. The set of all minimum cut-sets for χ on $\mathcal{T}$ is denoted by $\mathrm{Cut}(\chi, \mathcal{T})$.

An arbitrary cut-set E_1 for χ does not necessarily correspond to $\mathrm{Ch}(\overline{\chi})$ for some extension $\overline{\chi}$ of χ. To illustrate this, consider the full character χ on X and the X-tree $\mathcal{T}$ described in Example 5.1.2. The set $\{e_1, e_2, e_3, e_4\}$ of edges of $\mathcal{T}$ as shown in Fig. 5.2 is

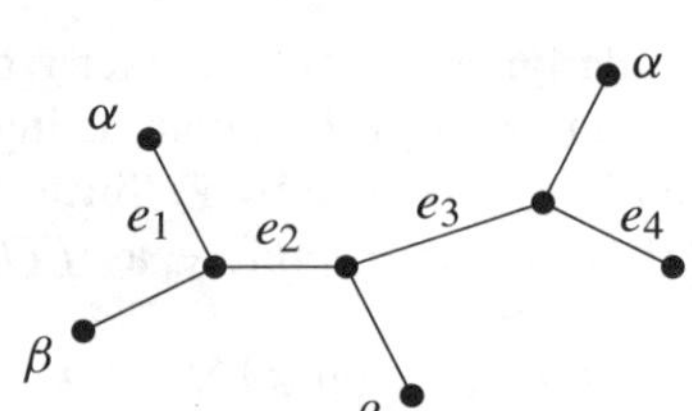

Figure 5.2 Edges forming a cut-set for χ.

a cut-set for χ, but it is easily seen that it is not the changing set of any extension of χ. However, Lemma 5.1.5 shows that this is not the case if E_1 is a minimum cut-set for χ.

Lemma 5.1.5. *Let χ be a character on X, let $\mathcal{T} = (T; \phi)$ be an X-tree with $T = (V, E)$, and let E_1 be a subset of E. If E_1 is a minimum cut-set for χ, then $E_1 = \text{Ch}(\overline{\chi}_1)$ for a unique extension $\overline{\chi}_1$ of χ.*

Proof. Suppose that E_1 is a minimum cut-set for χ. If V' is the set of vertices of a component of the forest $T \backslash E_1$, then, as E_1 is a cut-set for χ, we must have $|\chi(\phi^{-1}(V'))| \leq 1$. Furthermore, $|\chi(\phi^{-1}(V'))| \neq 0$, for otherwise, by selecting an edge e in E_1 that is incident with a vertex in V', the set $E_1 - \{e\}$ is a cut-set for χ, contradicting the minimality of E_1. Thus, $|\chi(\phi^{-1}(V'))| = 1$.

We define an extension $\overline{\chi}_1$ of χ as follows. Each vertex v of V lies in exactly one of the components of $T \backslash E_1$. If the vertex set of that component is V', set $\overline{\chi}_1(v)$ equal to the unique element in $\chi(\phi^{-1}(V'))$. Now, each edge $\{u, v\}$ in E_1 satisfies $\overline{\chi}_1(u) \neq \overline{\chi}_1(v)$, for otherwise E_1 would not be minimal, and so $\text{Ch}(\overline{\chi}_1) = E_1$. Furthermore, it is easily checked that every extension $\overline{\chi}$ of χ that satisfies $\text{Ch}(\overline{\chi}) = E_1$ must agree with $\overline{\chi}_1$ at all vertices in V. This completes the proof of the lemma. □

Proposition 5.1.6 shows that the parsimony score of a character χ on X on an X-tree $\mathcal{T}$ is equal to the size of a minimum cut-set for χ on $\mathcal{T}$. Furthermore, any two minimum extensions that induce the same changing set are equal.

Proposition 5.1.6. *Let χ be a character on X and let $\mathcal{T}$ be an X-tree. Then,* $\text{cut}(\chi, \mathcal{T}) = l(\chi, \mathcal{T})$. *Furthermore, the map ψ from the set of minimum extensions of χ on $\mathcal{T}$ into* $\text{Cut}(\chi, \mathcal{T})$ *defined by* $\psi(\overline{\chi}) = \text{Ch}(\overline{\chi})$, *for all such extensions, is a bijection.*

Proof. Let $\overline{\chi}$ be a minimum extension of χ on $\mathcal{T}$. Then, $\text{ch}(\overline{\chi}) = l(\chi, \mathcal{T})$, and since $\text{Ch}(\overline{\chi})$ is a cut-set for χ on $\mathcal{T}$, $\text{cut}(\chi, \mathcal{T}) \leq l(\chi, \mathcal{T})$. Now, suppose that $E_1 \in \text{Cut}(\chi, \mathcal{T})$. Then, by Lemma 5.1.5,

$$\text{cut}(\chi, \mathcal{T}) = |E_1| \geq \min\{\text{ch}(\overline{\chi}_1) : \overline{\chi}_1 \text{ is an extension of } \chi\} = l(\chi, \mathcal{T}),$$

establishing $\text{cut}(\chi, \mathcal{T}) \geq l(\chi, \mathcal{T})$, and thereby the first part of the proposition.

We now prove the second part of the proposition. Since $\text{Ch}(\overline{\chi})$ is a cut-set for χ on $\mathcal{T}$ and since $\text{ch}(\overline{\chi}) = l(\chi, \mathcal{T})$, it follows from the first part of the proposition that $\text{Ch}(\overline{\chi}) \in \text{Cut}(\chi, \mathcal{T})$. Moreover, by Lemma 5.1.5, the map ψ is a bijection. This completes the proof. □

We next describe the parsimony score of characters in terms of a maximal system of paths. We first do this for two-state characters by invoking the following classical graph theory result (Menger 1927).

Lemma 5.1.7 (Menger's Theorem). *Let $G = (V, E)$ be a graph, and let V_1 and V_2 be disjoint subsets of V. Then, the maximum number of edge-disjoint paths in G with the property that each path has one endpoint in V_1 and the other endpoint in V_2 is equal to the minimum number of edges whose removal from G leaves the vertices in V_1 in different components from the vertices in V_2.*

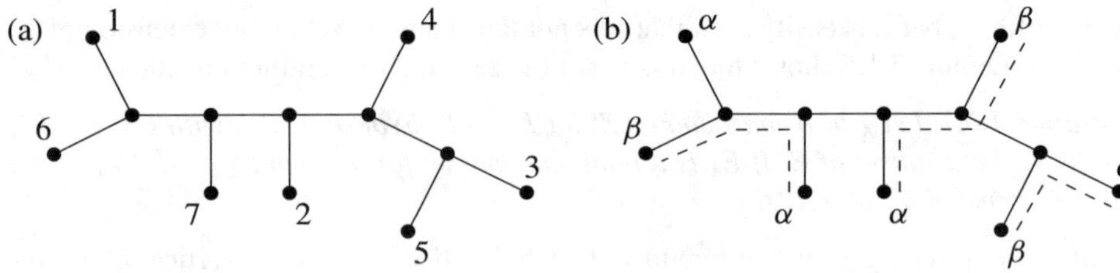

Figure 5.3 A maximum-sized set of edge-disjoint proper paths.

Menger's theorem has direct bearing on the parsimony score of binary characters. For an X-tree $\mathcal{T} = (T;\phi)$, a path P in T is a *proper path* relative to a character χ on X if, for some $x, y \in X$, P connects $\phi(x)$ and $\phi(y)$, and $\chi(x) \neq \chi(y)$. Combining Lemma 5.1.7 with Proposition 5.1.6, we obtain the following result.

Corollary 5.1.8. *Let χ be a two-state character on X and let $\mathcal{T} = (T;\phi)$ be an X-tree. Then, $l(\chi, \mathcal{T})$ is equal to the maximum number of edge-disjoint proper paths of T relative to χ.*

Example 5.1.9. To illustrate Corollary 5.1.8, consider the phylogenetic X-tree $\mathcal{T}$ shown in Fig. 5.3(a) and the binary character $\chi : X \rightarrow \{\alpha, \beta\}$, where $\chi(1) = \chi(2) = \chi(3) = \chi(7) = \alpha$ and $\chi(4) = \chi(5) = \chi(6) = \beta$. Each of the dashed paths shown in Fig. 5.3(b) is proper relative to χ. Moreover, as these paths are edge disjoint and as $|\chi^{-1}(\beta)| = 3$, the set consisting of these paths is a maximum-sized set of edge-disjoint proper paths of $\mathcal{T}$ relative to χ. Hence, by Corollary 5.1.8, $l(\chi, \mathcal{T}) = 3$.

Although Corollary 5.1.8 only applies to two-state characters, Erdös and Székely (1992) developed an extension to arbitrary characters that differs by permitting paths to intersect provided certain conditions are met.

Suppose that we have a phylogenetic X-tree $\mathcal{T} = (T;\phi)$ and a character χ on X. A collection $\mathcal{D}$ of directed paths in T is an *Erdös–Székely path system* for χ on $\mathcal{T}$ if it satisfies the following two conditions:

(i) If $P \in \mathcal{D}$, then P connects two leaves $\phi(x)$ and $\phi(y)$ of T for which $\chi(x) \neq \chi(y)$.
(ii) Let P and P' be paths in $\mathcal{D}$ that share some edge. Then, P and P' traverse this edge in the same direction and, if $\phi(x)$ and $\phi(y)$ denote the terminal vertices of P and P', $\chi(x) \neq \chi(y)$.

Example 5.1.10. Consider the phylogenetic X-tree $\mathcal{T}$ shown in Fig. 5.4(a) and the full character $\chi : X \rightarrow \{\alpha, \beta, \gamma\}$, where $\chi(1) = \chi(2) = \beta$, $\chi(3) = \chi(4) = \alpha$, and $\chi(5) = \chi(6) = \gamma$. An Erdös–Székely path system for χ on $\mathcal{T}$ is shown in Fig. 5.4(b).

Theorem 5.1.11. *Let χ be a character on X and let $\mathcal{T}$ be a phylogenetic X-tree. Then, $l(\chi, \mathcal{T})$ is equal to the maximum size of an Erdös–Székely path system for χ on $\mathcal{T}$.*

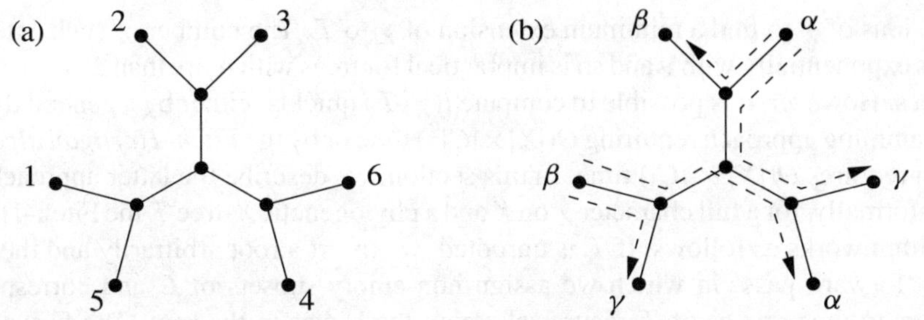

Figure 5.4 (a) A phylogenetic tree $\mathcal{T}$. (b) An Erdös–Székely path system for $\mathcal{T}$.

Theorem 5.1.11 is due to Erdös and Székely (1992), who also provided a polynomial-time algorithm to construct an explicit Erdös–Székely path system for a given character on X and phylogenetic X-tree.

Remark. For a character χ on X, Theorem 5.1.11 describes the parsimony score of χ on a phylogenetic X-tree. However, we can extend this result, and any other such results, to X-trees by making the following observation. Let $\mathcal{T} = (T;\phi)$ be an X-tree, and suppose that, for some $x \in X$, $\phi(x)$ is an internal vertex of T. Let $V_1, V_2, \ldots, V_k$ denote the vertex sets of the components of $T\backslash\phi(x)$ and, for each $i \in \{1, 2, \ldots, k\}$, let T_i denote the subgraph of T induced by $V_i \cup \{\phi(x)\}$. For all i, let $X_i = \phi^{-1}(V(T_i))$, let $\phi_i = \phi|X_i$, and let $\mathcal{T}_i = (T_i;\phi_i)$. It is now easily seen that

$$l(\chi, \mathcal{T}) = \sum_{i=1}^{k} l(\chi_i, \mathcal{T}_i),$$

where $\chi_i = \chi|X_i$ for all i. This equality implies that by repeatedly applying this process one can obtain a collection of phylogenetic trees in which the sum of the appropriate individual parsimony scores equals the parsimony score of χ on $\mathcal{T}$.

Two basic computational questions arise concerning the maximum parsimony approach to reconstructing phylogenetic trees. First, given a phylogenetic tree, we may wish to compute the parsimony score of a sequence of characters and perhaps some (or all) minimum extensions of them. This fixed-tree problem is sometimes called the 'small parsimony problem'. The second problem is to find a maximum parsimony tree for a sequence of characters. We begin with the first (easier) problem in the next section. As mentioned earlier, for the remainder of this chapter we will mostly deal, for convenience, with full characters.

5.2 Optimization on a fixed tree

Suppose we have a character $\chi : X \to C$ and a phylogenetic X-tree $\mathcal{T}$ with ν internal vertices. One could compute $l(\chi, \mathcal{T})$ using the brute-force approach of searching all $|C|^{\nu}$

extensions of χ to find a minimum extension of χ to $\mathcal{T}$. The number of such extensions grows exponentially with ν and so is impractical for trees with more than 20 or so internal vertices. However, it is possible to compute $l(\chi, \mathcal{T})$ quickly, either by a general dynamic programming approach requiring $O(|X| \times |C|^2)$ time or by the *Fitch–Hartigan algorithm*, which requires $O(|X| \times |C|)$ time. In this section, we describe the latter approach.

Informally, for a full character χ on X and a phylogenetic X-tree $\mathcal{T}$ the Fitch–Hartigan algorithm works as follows. If $\mathcal{T}$ is unrooted, we insert a root arbitrarily and then carry out a 'forward pass' in which we assign non-empty subsets of C and corresponding integers to the vertices of $\mathcal{T}$ recursively from the leaves to the root. The forward pass allows us to compute $l(\chi, \mathcal{T})$. A subsequent 'backward pass' from the root back to the leaves constructs a minimum extension of χ to $\mathcal{T}$. Here, we consider only the first of these two passes.

Let $\chi : X \to C$ be a character and let $\mathcal{T}$ be a rooted phylogenetic X-tree with root ρ. Define two maps

$$\psi : V(\mathcal{T}) \to (2^C - \{\emptyset\})$$

and

$$l : V(\mathcal{T}) \to \{0, 1, 2, \ldots\}$$

recursively as follows. Let $v \in V(\mathcal{T})$.

(i) If v is a leaf of $\mathcal{T}$ labelled by $x \in X$, set $\psi(v) = \{\chi(x)\}$ and $l(v) = 0$.

(ii) Let $v_1, v_2, \ldots, v_k$ denote the immediate descendants of v, and suppose that, for each $i \in \{1, 2, \ldots, k\}$, $\psi(v_i)$ and $l(v_i)$ have been determined. Let

$$f(v) = \max_{\alpha \in C} |\{i : \alpha \in \psi(v_i)\}|.$$

Set $\psi(v)$ to be the set of character states of C that appear in $f(v)$ of the sets $\psi(v_1), \psi(v_2), \ldots, \psi(v_k)$ and set $l(v) = \sum_{i=1}^{k} l(v_i) + k - f(v)$.

Now associate with each vertex v of $V(\mathcal{T})$ the ordered pair $(\psi(v), l(v))$. This assignment of an ordered pair to the vertices of $\mathcal{T}$ is called a *forward pass*, and its significance is highlighted in Theorem 5.2.1 (Hartigan 1973). We leave the proof as an exercise.

Theorem 5.2.1. *Let $\chi : X \to C$ be a character and let $\mathcal{T}$ be a rooted phylogenetic X-tree with root ρ. Suppose that we have completed a forward pass on the vertices of $\mathcal{T}$. Then*

(i) $l(\rho) = l(\chi, \mathcal{T})$ and

(ii) $\psi(\rho) = \{\alpha \in C :$ there is a minimum extension $\overline{\chi}$ of χ to $\mathcal{T}$ with $\overline{\chi}(\rho) = \alpha\}$.

Example 5.2.2. To illustrate the forward pass of the Fitch–Hartigan algorithm, consider the phylogenetic X-tree shown in Fig. 5.5(a), and the full character $\chi : X \to \{\alpha, \beta, \gamma\}$ defined by $\chi(1) = \chi(6) = \beta$, $\chi(2) = \gamma$, and $\chi(3) = \chi(4) = \chi(5) = \alpha$. A forward pass for χ and $\mathcal{T}$ is shown in Fig. 5.5(b).

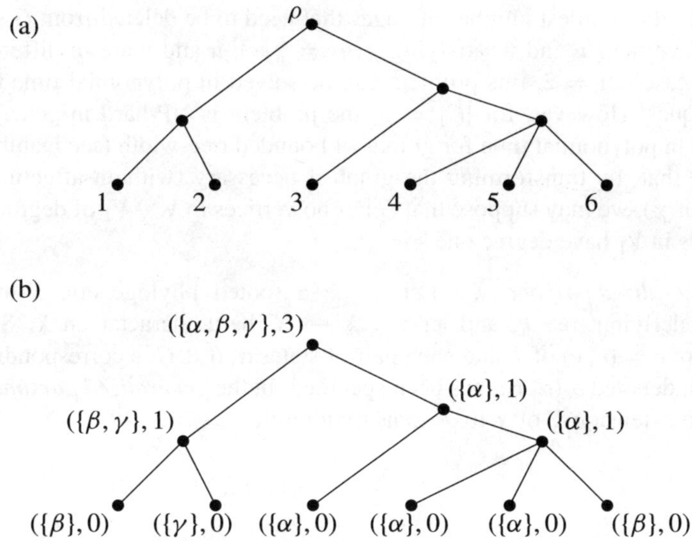

Figure 5.5 A forward pass of the Fitch–Hartigan algorithm.

Rinsma *et al.* (1990) describe an extension of Theorem 5.2.1 to count the number of minimum extensions of a character to a phylogenetic tree.

5.3 Extensions of classical parsimony

The fixed-tree problem for parsimony can be generalized in a number of different directions, some of which we discuss in this section. In each case, applications of notions such as parsimony score and minimum extension should be clear, and we will often not define these in precise detail.

Unspecified character states. In certain applications, the character state assigned by some particular character(s) and species may be ambiguous or simply unspecified. A simple remedy is to work with generalized characters as introduced in Section 4.6. Recall that a generalized character on X is a function $g : X \rightarrow (2^C - \{\emptyset\})$, where, for all $x \in X$, $g(x)$ denotes the set of possible character states that species x could be assigned. Given a phylogenetic X-tree $\mathcal{T} = (T; \phi)$, one can extend the definition of the parsimony score given in Section 5.1 to seek a function $\overline{g} : V(T) \rightarrow C$ such that $\overline{g} \circ \phi(x) \in \chi(x)$ for all $x \in X$, and which minimizes $\mathrm{ch}(\overline{g})$. It is straightforward to modify the Fitch–Hartigan algorithm to handle this extension.

Extension to graphs. One can easily formulate the parsimony problem for graphs. Given a graph $G = (V, E)$ and a function χ from a subset V_1 of V into a set C, we may wish to find an extension $\overline{\chi}$ of χ to G to minimize the number of edges $\{u, v\}$ of G for which $\overline{\chi}(u) \neq \overline{\chi}(v)$. As in the classical parsimony setting, this minimum value is equal to the smallest number of edges that need to be deleted from G so that, for every pair of vertices u and v satisfying $\chi(u) \neq \chi(v)$, u and v are in different components. In the case $|C| = 2$, this problem can be solved in polynomial time by network-flow techniques. However, for $|C| = 3$, the problem is NP-hard in general, but it can be solved in polynomial time for graphs of bounded tree-width (see Dahlhaus *et al.* 1994). Notice that, by transforming the graph if necessary (without affecting the parsimony score of χ), we may suppose that G has no vertices in $V - V_1$ of degree two and that all vertices in V_1 have degree one.

Generalized parsimony. Let $\mathcal{T}$ be a rooted phylogenetic X-tree with root ρ and underlying tree T, and let $\chi : X \rightarrow C$ be a character on X. Suppose that, for each arc $e = (u, v)$ of T and each pair of states $\alpha, \beta \in C$, a corresponding non-negative weight, denoted $\delta_e(\alpha, \beta)$, has been specified. In the *generalized parsimony* problem, we seek an extension $\overline{\chi}$ of χ to $\mathcal{T}$ so as to minimize

$$\sum_{e=(u,v)\in E(T)} \delta_e(\overline{\chi}(u), \overline{\chi}(v)).$$

In the special case where the function δ_e is a metric on C that takes the same values over all edges e of T, we write $d(\alpha, \beta)$ in place of $\delta_e(\alpha, \beta)$ for all $\alpha, \beta \in C$, and refer to this as the *metric space setting* for the parsimony problem. In this setting, we can regard $\mathcal{T}$ as an unrooted phylogenetic X-tree since the parsimony score of χ on $\mathcal{T}$ is the same regardless of where T is rooted. Generalized parsimony includes classic parsimony as a special case by taking δ_e to be the discrete metric on C.

In biology, the function δ_e allows for the differential weighting of transitions so that rare transitions can be penalized heavily. The dependence of δ_e on e may reflect prior knowledge that transitions on particular edges are rare or frequent.

Following Erdös and Székely (1994), we now describe a dynamic programming algorithm to compute $l(\chi, \mathcal{T})$ for a character $\chi : X \rightarrow C$ and a rooted phylogenetic X-tree $\mathcal{T} = (T; \phi)$. For each vertex v of T, let $\mathcal{T}_v$ denote the rooted phylogenetic subtree of $\mathcal{T}$ with root vertex v and let χ_v denote the restriction of χ to the label set of $\mathcal{T}_v$. For each $v \in V(T)$ and $\alpha \in C$, let $l(v, \alpha)$ denote the minimum value of

$$\sum_{e=(u,u')\in E(\mathcal{T}_v)} \delta_e(\overline{\chi}(u), \overline{\chi}(u'))$$

over all extensions $\overline{\chi}$ of χ_v to $\mathcal{T}_v$ that satisfy $\overline{\chi}(v) = \alpha$. We calculate $l(v, \alpha)$ recursively as follows. If $v = \phi(x)$ for some $x \in X$, then, for each $\alpha \in C$, set

$$l(v, \alpha) = \begin{cases} 0, & \text{if } \chi(x) = \alpha, \\ \infty, & \text{otherwise.} \end{cases}$$

Now, suppose that v is an interior vertex of T. Let $v_1, v_2, \ldots, v_k$ denote the immediate descendants of v and suppose that $l(v_i, \beta)$ has been determined for each $i \in \{1, 2, \ldots, k\}$

and each $\beta \in C$. Then, for all $\alpha \in C$, set

$$l(v, \alpha) = \sum_{i=1}^{k} \min_{\beta \in C}\{\delta_{(v,v_i)}(\alpha, \beta) + l(v_i, \beta)\}.$$

If ρ is the root of $\mathcal{T}$, then

$$l(\chi, \mathcal{T}) = \min_{\alpha \in C}\{l(\rho, \alpha)\}.$$

In this way we obtain an $O(|X| \times |C|^2)$ algorithm for computing $l(\chi, \mathcal{T})$. A minimum extension of χ to $\mathcal{T}$ can be constructed by carrying out a second pass, proceeding from the root to the leaves of $\mathcal{T}$.

Large state sets. The running times of the algorithms we have presented in Section 5.2 and above for the classical and generalized parsimony problems on a fixed phylogenetic X-tree grow with the size of the state set C, which can be a problem if $|C|$ is very large. This problem arises for molecular data in two settings: (i) aligning DNA sequences on a tree, sometimes called 'phylogenetic alignment', and (ii) fitting ordered genome data on a tree. An algorithm for solving a problem of type (i) was described in a classical paper by Sankoff (1975). This algorithm is exponential in $|X|$, but we cannot expect a general polynomial–time algorithm for this problem as it has been shown to be NP-hard even when the phylogenetic tree is binary (Wang and Jiang 1994). Nevertheless, polynomial–time results can be obtained for the fixed-tree problem by investigating the mathematical structure of the parsimony problem in the metric space setting, and we describe some of these results here.

Let d be a metric on a (finite or infinite) state set C and let S be a finite subset of C. The *median of* S, denoted $M(S, d)$, is the set of elements $\alpha \in C$ that minimizes the sum $\sum_{\beta \in S} d(\alpha, \beta)$. An extension $\overline{\chi}$ of a C-valued full character χ on X to a phylogenetic X-tree $\mathcal{T}$ is said to be a *median extension* if, for every interior vertex v of $\mathcal{T}$,

$$\overline{\chi}(v) \in M(N(v), d),$$

where $N(v) = \{\overline{\chi}(u): \{u, v\} \in E(\mathcal{T})\}$. It is easily seen that every minimum extension of χ to $\mathcal{T}$ is a median extension. However, the converse fails even for the discrete metric on a state set of size two. Nevertheless, a median extension of χ to $\mathcal{T}$ may provide an extension that comes close to minimizing the parsimony score of χ on $\mathcal{T}$, and heuristics have been developed from this idea (see Swofford *et al.* 1996).

A related approach is to explore the effect of restricting the state set C to the values that the character χ takes on X. In order to describe this further, we introduce some notation.

Given a character $\chi : X \to C$, a metric d on C, and a phylogenetic X-tree $\mathcal{T}$, let $l_d(\chi, \mathcal{T})$ denote the minimum value of $\sum_{\{u,v\} \in E(\mathcal{T})} d(\overline{\chi}(u), \overline{\chi}(v))$ over all extensions $\overline{\chi}$ of χ to $\mathcal{T}$. Define $l_d^*(\chi, \mathcal{T})$ in the same way, except that we impose the additional restriction that each such extension $\overline{\chi}$ of χ must take values in the set $C^* = \{\chi(x): x \in X\}$. Note that

$l_d^*(\chi, \mathcal{T})$ can be computed in polynomial time by regarding C^* as the state set and using the dynamic programming algorithm for generalized parsimony. Clearly, $l_d(\chi, \mathcal{T}) \leq l_d^*(\chi, \mathcal{T})$. Moreover, we next show that for certain metric spaces, we have equality.

We say that a metric space (C, d) has the *median intersection property* if, for all finite non-empty subsets S of C, we have

$$M(S, d) \cap S \neq \emptyset.$$

Proposition 5.3.1. *Let d be a metric on a state space C, let $\chi : X \to C$ be a character, and let $\mathcal{T}$ be a phylogenetic X-tree. If (C, d) satisfies the median intersection property, then*

$$l_d^*(\chi, \mathcal{T}) = l_d(\chi, \mathcal{T}).$$

Proof. The proof is by induction on the number of interior edges of $\mathcal{T} = (T; \phi)$. If $\mathcal{T}$ has no interior edges, then $\mathcal{T}$ is a star tree, and the proposition follows directly from the definition of the median intersection property. Now suppose that $\mathcal{T}$ has r interior edges, where $r \geq 1$, and that the result holds for all phylogenetic trees with $r-1$ interior edges. Let v be an interior vertex of T that is adjacent to exactly one other interior vertex, w say, and let $\phi(x_1), \phi(x_2), \ldots, \phi(x_s)$ denote the leaves of T adjacent to v. Let $\overline{\chi}$ be a minimum extension of χ to $\mathcal{T}$, and consider the set

$$S = \{\overline{\chi}(w), \overline{\chi}(x_1), \overline{\chi}(x_2), \ldots, \overline{\chi}(x_s)\}.$$

By the median intersection property, $M(S, d) \cap S \neq \emptyset$. Consequently, at least one of the following two cases must apply:

(i) $\overline{\chi}(w) \in M(S, d)$; or
(ii) $\overline{\chi}(x_k) \in M(S, d)$ for some $k \in \{1, 2, \ldots, s\}$.

Consider case (i). Let $\mathcal{T}'$ denote the phylogenetic X-tree $\mathcal{T}/\{v, w\}$. Since $\overline{\chi}(w) \in M(S, d)$, a (possible further) minimum extension of χ to $\mathcal{T}$ can be obtained from $\overline{\chi}$ by replacing $\overline{\chi}(v)$ with $\overline{\chi}(w)$. It follows that $l_d(\chi, \mathcal{T}') \leq l_d(\chi, \mathcal{T})$. Furthermore, by considering a minimum extension of χ to $\mathcal{T}'$, it is easily checked that $l_d(\chi, \mathcal{T}')$ is not strictly less than $l_d(\chi, \mathcal{T})$. Thus, $l_d(\chi, \mathcal{T}') = l_d(\chi, \mathcal{T})$ and so, by the induction assumption, $l_d(\chi, \mathcal{T}) = l_d^*(\chi, \mathcal{T}')$. By considering a minimum extension of χ to $\mathcal{T}'$ corresponding to the quantity $l_d^*(\chi, \mathcal{T}')$, it is straightforward to deduce that $l_d(\chi, \mathcal{T}) = l_d^*(\chi, \mathcal{T})$. This establishes the induction step for case (i).

For case (ii), let $X' = X - \{x_1, \ldots, x_{k-1}, x_{k+1}, \ldots, x_s\}$ and let T' be the tree obtained from T by deleting $\phi(x_1), \ldots, \phi(x_{k-1}), \phi(x_{k+1}), \ldots, \phi(x_s)$ and suppressing the resulting degree-two vertex. Let $\mathcal{T}''$ be the phylogenetic X'-tree $(T'; \phi|X')$. Since $\overline{\chi}(x_k) \in M(S, d)$, a minimum extension of χ to $\mathcal{T}$ can be obtained from $\overline{\chi}$ by replacing $\overline{\chi}(v)$ with $\overline{\chi}(x_k)$. It follows that $l_d(\chi, \mathcal{T}) \leq l_d(\chi', \mathcal{T}'') + \sum_{i=1}^{s} d(x_k, x_i)$, where χ' denotes the restriction of χ to X'. By considering a minimum extension of χ' to $\mathcal{T}''$, it is easily deduced that this inequality is an equality. By the induction assumption, $l_d(\chi', \mathcal{T}'') = l_d^*(\chi', \mathcal{T}'')$, and so

$$l_d(\chi, \mathcal{T}) = l_d^*(\chi', \mathcal{T}'') + \sum_{i=1}^{s} d(x_k, x_i). \tag{5.2}$$

By choosing a minimum extension of χ' to $\mathcal{T}''$ corresponding to the quantity $l_d^*(\chi', \mathcal{T}'')$ and extending this extension to a minimum extension of χ to $\mathcal{T}$, we deduce that the right-hand side of eqn (5.2) is equal to $l_d^*(\chi, \mathcal{T})$. This establishes the induction step for case (ii) and thereby completes the proof. □

The significance of Proposition 5.3.1 is that one can solve the parsimony problem in the metric space setting on a phylogenetic X-tree in polynomial time in $|X|$, regardless of the size of the state set C, when working with a metric space that satisfies the median intersection property. An example of such a metric space is any finite set with the discrete metric. Consequently, independent of $|C|$, the classical parsimony problem can be solved in $O(|X|^2)$ time by the Fitch–Hartigan algorithm. A second example of a metric space that satisfies the median intersection property is the real line with the usual metric. This example has been important in certain applications where a full character $\chi : X \to \mathbb{R}$ is used to describe some continuously variable attribute of the species such as the average body length. This parsimony problem for $C = \mathbb{R}$ was analysed by Robinson (1973) in the more general setting of graphs. Narushima and Hanazawa (1997) have developed a detailed investigation of the algorithmic and mathematical aspects of the set of solutions of this problem on phylogenetic trees.

Given a C-valued character χ on X, a phylogenetic X-tree $\mathcal{T}$, and a metric space (C, d), a natural question is how much larger can $l_d^*(\chi, \mathcal{T})$ be in comparison to $l_d(\chi, \mathcal{T})$. Proposition 5.3.2 (see Gusfield 1997) shows that the ratio of these two quantities is bounded by a constant factor.

Proposition 5.3.2. *Let χ be a C-valued character on X, let $\mathcal{T}$ be a phylogenetic X-tree, and let (C, d) be a metric space. Then,*

$$l_d^*(\chi, \mathcal{T}) \leq 2l_d(\chi, \mathcal{T}).$$

5.4 Maximum parsimony trees

Even in the setting of classical parsimony, it is an NP-hard problem to find a maximum parsimony tree for a sequence $\mathcal{C}$ of binary characters (Foulds and Graham 1982). For these tasks, branch-and-bound algorithms have been effective for modest numbers of species, while heuristics methods, based on the tree rearrangement operations in Section 2.6, have been employed for larger numbers. In this section, we establish some bounds on the size of $l(\mathcal{C})$. In fact, we will work in the more general metric setting, where quantities such as $l(\mathcal{C}, \mathcal{T})$ and $l(\mathcal{C})$ generalize in the obvious way [by replacing $l(\chi, \mathcal{T})$ with $l_d(\chi, \mathcal{T})$ throughout the earlier definitions] and we denote them by $l_d(\mathcal{C}, \mathcal{T})$ and $l_d(\mathcal{C})$, respectively.

A *spanning tree* of a connected graph G is a subgraph T of G such that T is a tree and $V(T) = V(G)$. Suppose that w is a non-negative weight function on the edges of G. A *minimum-weight spanning tree* of G is a spanning tree T of G that minimizes $\sum_{e \in E(T)} w(e)$. Kruskal (1956) gives a polynomial–time algorithm for finding a minimum-weight spanning tree of any connected graph.

Proposition 5.4.1. *Let $\mathcal{C} = (\chi_1, \ldots, \chi_k)$ be a sequence of C-valued characters on X and let d be a metric on C. Let G be the complete graph (X, E) and let w be the weight function on the edges of G defined by*

$$w(\{x, x'\}) = \sum_{i=1}^{k} d(\chi_i(x), \chi_i(x')).$$

Let $w(\mathcal{C})$ denote the weight of a minimum-weight spanning tree of G. Then,

$$\tfrac{1}{2} w(\mathcal{C}) \leq l_d(\mathcal{C}) \leq w(\mathcal{C}).$$

Proof. Clearly, $l_d(\mathcal{C}) \leq w(\mathcal{C})$. To prove the lower bound, suppose that $\mathcal{T}$ is a maximum parsimony tree for $\mathcal{C}$. Without loss of generality, we may assume that $\mathcal{T} = (T; \phi)$ is a phylogenetic X-tree. Let $(x_1, \ldots, x_n)$ denote a circular ordering of X for $\mathcal{T}$ as described in Section 3.7, and let

$$W = \sum_{j=1}^{n-1} \sum_{i=1}^{k} d(\chi_i(x_j), \chi_i(x_{j+1})). \tag{5.3}$$

Since W is the weight of the spanning tree of G consisting of the path $x_1, x_2, \ldots, x_n$,

$$w(\mathcal{C}) \leq W. \tag{5.4}$$

Consider the set of paths $P_1, \ldots, P_n$ in T where, for each $j \in \{1, \ldots, n-1\}$, P_j is the path from $\phi(x_j)$ to $\phi(x_{j+1})$ and P_n is the path from $\phi(x_n)$ to $\phi(x_1)$. For each $i \in \{1, \ldots, k\}$, let $\overline{\chi}_i$ denote a minimum extension of χ_i to $\mathcal{T}$. The triangle inequality for metrics implies that, for each $j \in \{1, \ldots, n-1\}$, and each $i \in \{1, \ldots, k\}$,

$$d(\chi_i(x_j), \chi_i(x_{j+1})) \leq \sum_{\{u,v\} \in P_j} d(\overline{\chi}_i(u), \overline{\chi}_i(v)). \tag{5.5}$$

Let $\overline{w}$ denote the edge weighting of T defined, for all $\{u, v\} \in E(T)$, by $\overline{w}(\{u, v\}) = \sum_{i=1}^{k} d(\overline{\chi}_i(u), \overline{\chi}_i(v))$. For each j, (5.5) implies

$$\sum_{i=1}^{k} d(\chi_i(x_j), \chi_i(x_{j+1})) \leq \sum_{e \in P_j} \overline{w}(e).$$

Consequently, by (5.3),

$$W \leq \sum_{j=1}^{n-1} \sum_{e \in P_j} \overline{w}(e). \tag{5.6}$$

By Lemma 3.7.2, each edge of T occurs in exactly two of the paths from the collection $P_1, \ldots, P_n$. Thus, from (5.6),

$$W \leq 2 \sum_{e \in E(T)} \overline{w}(e). \tag{5.7}$$

Observing that, by definition of $\overline{w}(e)$ and $\mathcal{T}$, we have

$$\sum_{e \in E(\mathcal{T})} \overline{w}(e) = l_d(\mathcal{C}, \mathcal{T}) = l_d(\mathcal{C}), \tag{5.8}$$

the proposition follows by combining (5.4), (5.7), and (5.8). □

By exploiting the observation that only $n - 1$ of the paths $P_1, \ldots, P_n$ are involved in inequality (5.6) one can improve Proposition 5.4.1 slightly by establishing $l_d(\mathcal{C}) \geq \frac{1}{2}[1 + (1/n)]w(\mathcal{C})$. However, Foulds (1984) showed that the factor of $\frac{1}{2}$ in this bound cannot be made any larger (see Exercise 10). Results from Prömel and Steger (2000) provide slightly better bounds for $l_d(\mathcal{C})$ using a more refined approach than the minimum-weight spanning tree construction. Proposition 5.4.2 provides an alternative type of lower bound on $l_d(\mathcal{C})$ in terms of the total homoplasy $h(\mathcal{C})$ of $\mathcal{C}$.

Proposition 5.4.2. *Let $\mathcal{C}$ be a sequence $(\chi_1, \chi_2, \ldots, \chi_k)$ of full characters on X and let $I(\mathcal{C})$ be the number of (unordered) pairs of characters in $\mathcal{C}$ that are incompatible. Then,*

$$h(\mathcal{C}) \geq \frac{I(\mathcal{C})}{(k-1)}.$$

Proof. Let $\mathcal{T}$ be a maximum parsimony tree for $\mathcal{C}$ and, for each $i \in \{1, 2, \ldots, k\}$, let $h_i = h(\chi_i, \mathcal{T})$. Then, $h(\mathcal{C}) = \sum_{i=1}^{k} h_i$. Furthermore, if χ_i and χ_j are incompatible, then $h_i + h_j \geq 1$. Thus,

$$\begin{aligned} I(\mathcal{C}) &\leq \sum_{1 \leq i < j \leq k} (h_i + h_j) = \frac{1}{2} \sum_{i=1}^{k} \sum_{1 \leq j \leq k:\, i \neq j} (h_i + h_j) \\ &= \frac{1}{2} \sum_{i=1}^{k} ((k-1)h_i + h(\mathcal{C}) - h_i) = (k-1)h(\mathcal{C}), \end{aligned}$$

as required. □

Classical parsimony and the Steiner problem. We end this section by relating the determination of a maximum parsimony tree for a sequence of characters to a more general problem in combinatorial optimization. Let $G = (V, E)$ be a connected graph, let w be a strictly positive real-valued weighting of the edges of G, and let V_1 be a subset of V. The *Steiner problem for graphs* is to find a connected subgraph $G' = (V', E')$ of G with $V_1 \subseteq V'$ that minimizes $\sum_{e \in E'} w(e)$. Evidently, such a graph G' is necessarily a tree and is called a *minimum Steiner tree* for V_1 in G.

Now consider the task of finding a maximum parsimony tree for a sequence $\mathcal{C} = (\chi_1, \chi_2, \ldots, \chi_k)$ of C-valued full characters on X. Let C^k denote the set of all sequences of length k over the alphabet set C. Consider the following edge-weighted graphs:

(i) Let G be the complete graph on vertex set C^k, and let w be the weighting on the edges of G defined by setting, for all $\{u, v\} \in E(G)$, $w(\{u, v\})$ equal to the number of co-ordinates on which u and v disagree.

(ii) Let G be the graph that has vertex set C^k and an edge $\{u, v\}$ precisely if u and v differ in exactly one coordinate, and let w to be the weight function on the edges of G defined by $w(e) = 1$ for all edges e in $E(G)$.

In both (i) and (ii), a minimum Steiner tree for the set $\{(\chi_1(x), \chi_2(x), \ldots, \chi_k(x)) : x \in X\}$ in G is a maximum parsimony tree for $\mathcal{C}$ up to degree-two vertices. Such a minimum Steiner tree describes not just a maximum parsimony tree $\mathcal{T}$ for $\mathcal{C}$, but also a minimum extension of each character in $\mathcal{C}$ to $\mathcal{T}$. This recasting of the maximum parsimony problem will be developed further in Section 5.5.

5.5 Specialist topic: networks of maximum parsimony trees

In this section, we define what it means for an X-tree to be displayed in a graph and show that, for a sequence $\mathcal{C}$ of distinct binary characters on X, every maximum parsimony tree for $\mathcal{C}$ is displayed in the Buneman graph on the set of X-splits induced by $\mathcal{C}$. This section deals exclusively with classical parsimony.

Suppose that Σ is a set $\{\sigma_1, \sigma_2, \ldots, \sigma_k\}$ of X-splits. In Section 3.8, we defined the vertex set of the Buneman graph $G(\Sigma)$ on Σ as a particular collection of sets. In this section, we fix an ordering of Σ and view a vertex of $G(\Sigma)$ as a k-tuple such that, for all $i \in \{1, 2, \ldots, k\}$, the i-th component is one block of σ_i and every pair of components has a non-empty intersection.

Let $\mathcal{T} = (T; \phi)$ be an X-tree and let G be a graph. Let ϕ' be a map from X into the vertex set of G. Then, $\mathcal{T}$ is said to be *displayed in* $(G; \phi')$ if the following two conditions hold:

(i) there exists a tree T' isomorphic to T under a map $\psi : V(T) \rightarrow V(T')$ such that T' can be obtained from a subgraph of G by suppressing degree-two vertices; and

(ii) for all $x \in X$, $\psi(\phi(x)) = \phi'(x)$.

The main result of this section, due to Bandelt *et al.*, is Theorem 5.5.1. For a sequence $\mathcal{C}$ of binary characters on X, let $\Sigma_{\mathcal{C}}$ denote the set of X-splits induced by $\mathcal{C}$.

Theorem 5.5.1. *Let $\mathcal{C}$ be a sequence of binary characters on X that induce distinct X-splits and let $\mathcal{T}$ be a maximum parsimony tree for $\mathcal{C}$. Then, $\mathcal{T}$ is displayed in $(G(\Sigma_{\mathcal{C}}); \phi_\Sigma)$.*

The proof of Theorem 5.5.1 is a consequence of several results. We begin with some preliminaries. A connected graph G is a *median graph* if, for every three vertices u_1, u_2, and u_3 of G, there is exactly one vertex v of G, denoted $m_G(u_1, u_2, u_3)$, that has the property that it simultaneously lies on shortest paths joining u_1 and u_2, u_1 and u_3, and u_2 and u_3. The vertex $m_G(u_1, u_2, u_3)$ is called the *median of* u_1, u_2, *and* u_3. Trees are examples of median graphs. Two other examples of median graphs are given in Examples 5.5.2 and 5.5.3.

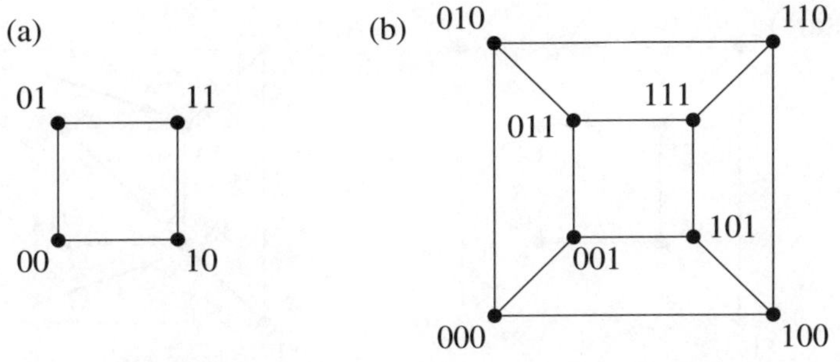

Figure 5.6 (a) The 2-cube. (b) The 3-cube.

Example 5.5.2. For all positive integers k, the *k-cube* is the graph whose vertex set consists of all 0–1 sequences of length k and where two such sequences are adjacent precisely if they differ in exactly one position. The 2- and 3-cubes are shown in Fig. 5.6. For all k, the k-cube is a median graph. The median of three vertices u, v, and w of the k-cube is the vertex where, for all $i \in \{1, 2, \ldots, k\}$, the i-th component is 0 if at least two of u_i, v_i, and w_i is 0, and 1 if at least two of u_i, v_i, and w_i is 1. Thus, in Fig. 5.6(b), the median of 010, 111, and 100 is 110. In the study of median graphs, the cubes are a fundamental class.

Example 5.5.3. The Buneman graph $G(\Sigma)$ on a set Σ of X-splits is a median graph. To see this, first note that, by Lemma 3.8.3, $G(\Sigma)$ is connected and the length of a shortest path between two vertices of $G(\Sigma)$ is equal to the number of components that the two vertices disagree on. It is now straightforward to show that if N_1, N_2, and N_3 are vertices of $G(\Sigma)$, then the vertex of $G(\Sigma)$ where, for all $i \in \{1, 2, \ldots, k\}$, the i-th component is the block of σ_i that is the i-th component of at least two of N_1, N_2, and N_3 is the unique median of N_1, N_2, and N_3.

A connected subgraph G' of a median graph G is a *median subgraph* if, for every three vertices u, v, and w of $V(G')$, $m_G(u, v, w)$ is an element of $V(G')$. Observe that if G' is a median subgraph of a median graph G, then G' is a median graph.

For a set $\Sigma = \{\sigma_1, \sigma_2, \ldots, \sigma_k\}$ of X-splits, the Buneman graph on Σ is a median subgraph of the k-cube. To see this, observe that one can relabel the vertices of the k-cube as follows. Let Σ^k denote the collection of all k-tuples in which, for all $i \in \{1, 2, \ldots, k\}$, the i-th component is one block of σ_i. Now, fix an element x of X. If v is a vertex of the k-cube, then relabel v with the element of Σ^k where the i-th component contains x precisely if the i-th component of v is 0. Note that there is nothing special about the choice of x here. It now follows from Examples 5.5.2 and 5.5.3 that $G(\Sigma)$ is a median subgraph of the k-cube.

Example 5.5.4. Let $\mathcal{C}$ be a sequence (χ_1, χ_2, χ_3) of binary characters on $X = \{a, b, c, d\}$ such that the set of X-splits induced by $\mathcal{C}$ is

$$\{\{a, b, c\}|\{d\}, \{a, d\}|\{b, c\}, \{a, b\}|\{c, d\}\}.$$

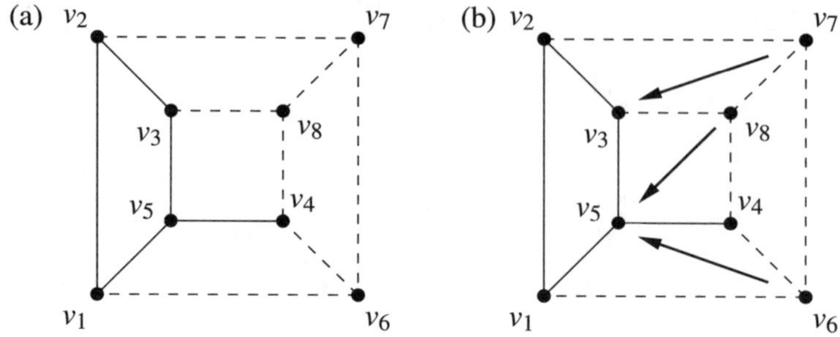

$v_1 = (\{a, b, c\}, \{a, d\}, \{a, b\})$ $\quad$ $v_6 = (\{d\}, \{a, d\}, \{a, b\})$
$v_2 = (\{a, b, c\}, \{b, c\}, \{a, b\})$ $\quad$ $v_7 = (\{d\}, \{b, c\}, \{a, b\})$
$v_3 = (\{a, b, c\}, \{b, c\}, \{c, d\})$ $\quad$ $v_8 = (\{d\}, \{b, c\}, \{c, d\})$
$v_4 = (\{d\}, \{a, d\}, \{c, d\})$
$v_5 = (\{a, b, c\}, \{a, d\}, \{c, d\})$

Figure 5.7 (a) $G(\Sigma_C)$ as a subgraph of the 3-cube. (b) $G(\Sigma_C)$ as a retract of the 3-cube.

The Buneman graph $G(\Sigma_C)$ as a subgraph of the 3-cube is shown in Fig. 5.7(a). The solid edges are the edges of $G(\Sigma_C)$ and the solid edges together with the dashed edges are the edges of the 3-cube.

The fact that the Buneman graph on a set of X-splits is a median subgraph of the k-cube, where $k = |X|$, is an immediate consequence of the following characterization of median graphs (see Mulder 1978; Mulder and Schrijver 1979).

Proposition 5.5.5. *A graph G is a median graph if and only if, for some positive integer k, G is a median subgraph of the k-cube.*

The characterization of a median graph given above is one of many such characterizations. Another due to Bandelt (1984) is essential to the proof of Theorem 5.5.1 and we present this characterization next.

Let G be a graph and let G' be a vertex induced subgraph of G. The subgraph G' is a *retract* of G if there is a map $\psi : V(G) \to V(G')$ such that the following two properties hold:

(i) if vertices u_1 and u_2 are adjacent in G; then either $\psi(u_1) = \psi(u_2)$, or $\psi(u_1)$ and $\psi(u_2)$ are adjacent in G'; that is, ψ is *edge-preserving*; and
(ii) if v is a vertex of G', then $\psi(v) = v$.

The map ψ is called a *retraction*.

Example 5.5.6. Referring back to Example 5.5.4 and Fig. 5.7(a), let ψ be the map from the vertex set of the 3-cube to the vertex set of $G(\Sigma_C)$ defined by $\psi(v_i) = v_i$ for each $i \in \{1, 2, 3, 4, 5\}$, $\psi(v_6) = v_5$, $\psi(v_7) = v_3$, and $\psi(v_8) = v_5$. This mapping is indicated in Fig. 5.7(b) and it is easily checked that ψ is a retraction. Hence, $G(\Sigma_C)$ is a retract of the 3-cube.

Example 5.5.6 is an illustration of the following general result of Bandelt (1984). We refer the reader to the original paper for the proof.

Proposition 5.5.7. *Let G' be a subgraph of a median graph G. Then, G' is a retract of G if and only if G' is a median subgraph of G.*

Propositions 5.5.5 and 5.5.7 imply that a graph G is a median graph if and only if G is a retract of a k-cube for some positive integer k. For us, an important and immediate consequence is the following corollary.

Corollary 5.5.8. *Let $\mathcal{C}$ be a sequence of k binary characters on X that induce distinct X-splits. Then, $G(\Sigma_{\mathcal{C}})$ is a retract of the k-cube.*

One last concept is needed before we can derive Theorem 5.5.1. Let $\mathcal{C}$ be a sequence of k binary characters on X that induce distinct X-splits and view $(\Sigma_{\mathcal{C}})^k$ as the vertex set of the k-cube. For all $x \in X$, let N_x denote the unique vertex of the k-cube in which every component contains x. With respect to the number of edges, let T be a minimum Steiner tree for the subset $\{N_x\colon x \in X\}$ of vertices of the k-cube. Let T' be the tree obtained from T by suppressing all vertices of degree two not in $\{N_x\colon x \in X\}$. Let ϕ be the map from X into the vertex set of T' defined by $\phi(x) = N_x$ for all x. Then, $(T'; \phi)$ is a maximum parsimony tree for $\mathcal{C}$. Thus, given a minimum Steiner tree for the set of vertices $\{N_x\colon x \in X\}$ of the k-cube, one can determine a maximum parsimony tree for $\mathcal{C}$.

Theorem 5.5.1 is a direct consequence of Proposition 5.5.9, which uses the previous corollary in its statement.

Proposition 5.5.9. *Let $\mathcal{C}$ be a sequence of k binary characters on X that induce distinct X-splits and let T be a minimum Steiner tree for the subset $\{N_x : x \in X\}$ of vertices of the k-cube. Let ψ be a retraction from $(\Sigma_{\mathcal{C}})^k$ into the vertex set of $G(\Sigma_{\mathcal{C}})$. Then, the image of T under ψ is isomorphic to T.*

Proof. No two distinct vertices v_1 and v_2 of T have the property that $\psi(v_1) = \psi(v_2)$, for otherwise there is a tree of the k-cube that spans the subset $\{N_x\colon x \in X\}$ of vertices and has fewer edges than T. Since ψ is an edge-preserving map, the result immediately follows. □

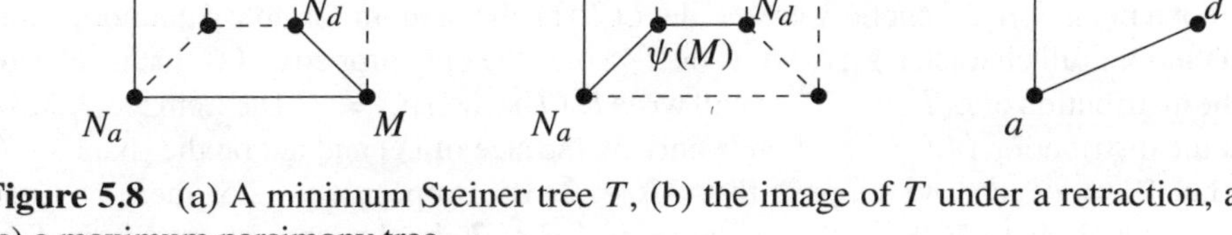

Figure 5.8 (a) A minimum Steiner tree T, (b) the image of T under a retraction, and (c) a maximum parsimony tree.

Example 5.5.10. To illustrate Proposition 5.5.9 and Theorem 5.5.1, consider the sequence $\mathcal{C}$ of binary characters given in Example 5.5.4. A minimum Steiner tree T for $\{N_x : x \in X\}$ in the 3-cube is shown in Fig. 5.8(a), where $M = (\{d\}, \{a, d\}, \{a, b\})$. Under the retraction ψ of Example 5.5.6, the image $\psi(T)$ of T is shown in Fig. 5.8(b) where $\psi(M) = (\{a, b, c\}, \{a, d\}, \{c, d\})$ and the associated maximum parsimony tree for $\mathcal{C}$ is shown in Fig. 5.8(c).

5.6 Specialist topic: combinatorial statistics of parsimony

In this section, we consider two probabilistic questions concerning the parsimony score of a character on a binary phylogenetic tree. Informally, we seek to describe the probability distribution of the parsimony score of (i) a random character on a fixed tree and (ii) a fixed character on a random tree. Although results for a single character may seem somewhat limiting, they often have implications for sequences of characters. For example, if we generate random characters independently, then the parsimony score of the resulting sequence of characters on a fixed tree is simply a sum of independent random variables (one for each character). Similarly, for any given sequence of characters on X, the average parsimony score of this sequence over all binary phylogenetic X-trees is simply the sum (over all characters in the sequence) of the average score of each character on a binary phylogenetic X-tree randomly selected with uniform probability. Statistical applications of these probability distributions are discussed by Maddison and Slatkin (1991), Archie and Felsenstein (1993), Steel *et al.* (1993), Hamel and Steel (1997), Faith and Trueman (2001).

Throughout this section, a *random binary phylogenetic X-tree* will refer to a binary phylogenetic X-tree selected with uniform probability from the set of all binary phylogenetic X-trees. A *random (C-valued) character* on X will generally refer to an independent assignment of character states from a set C to each element of X under the uniform distribution. Equivalently, we may select uniformly at random one of the $|C|^{|X|}$ possible full characters $\chi : X \to C$.

Formally, this section addresses the following two problems:

(i) for a fixed binary phylogenetic X-tree $\mathcal{T}$ and a random C-valued full character χ, describe the probability distribution of the parsimony score $l(\chi, \mathcal{T})$;

(ii) for a fixed C-valued full character χ and a random binary phylogenetic X-tree $\mathcal{T}$, describe the probability distribution of the parsimony score $l(\chi, \mathcal{T})$.

Both (i) and (ii) have closed form solutions when $|C| = 2$. We describe these next [as well as one other closed form solution of (ii)], beginning with (i).

For a binary phylogenetic X-tree $\mathcal{T}$, let $L(\mathcal{T})$ be the random variable parsimony score of a random full character $\chi : X \to C$ on $\mathcal{T}$. Since the only property of C that is relevant to the distribution of $L(\mathcal{T})$ is $|C|$, we will write $L_c(\mathcal{T})$ when $|C| = c$. Theorem 5.6.2 shows that the distribution of $L_2(\mathcal{T})$ depends only on the size of $|X|$ and not on the shape of $\mathcal{T}$.

Let $\mathcal{T} = (T; \phi)$ and suppose that $|X| \geq 3$. By Proposition 1.2.5, there is a pair $x, y \in X$ such that $\{\phi(x), \phi(y)\}$ is a cherry of T. Let T' be the tree obtained from T by

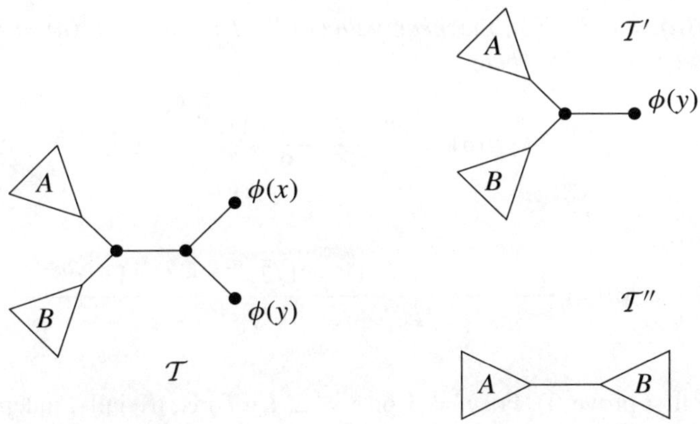

Figure 5.9 Constructing $\mathcal{T}'$ and $\mathcal{T}''$ from $\mathcal{T}$.

deleting $\phi(x)$ and suppressing the resulting degree-two vertex and let $\mathcal{T}'$ denote the binary phylogenetic $(X - \{x\})$-tree $(T'; \phi|(X - \{x\}))$. Let T'' be the tree obtained from T' by deleting $\phi(y)$ and, provided $|X| \geq 4$, suppressing the resulting degree-two vertex. Let $\mathcal{T}''$ denote the binary phylogenetic $(X - \{x, y\})$-tree $(T''; \phi|(X - \{x, y\}))$. These constructions are illustrated in Fig. 5.9 when $|X| \geq 4$.

Lemma 5.6.1. *Let $\mathcal{T}$ be a binary phylogenetic X-tree with $|X| \geq 3$, and let $\mathcal{T}'$ and $\mathcal{T}''$ be the binary phylogenetic trees obtained from $\mathcal{T}$ as described above. Then,*

$$\mathbb{P}(L_2(\mathcal{T}) = k) = \tfrac{1}{2}\mathbb{P}(L_2(\mathcal{T}') = k) + \tfrac{1}{2}\mathbb{P}(L_2(\mathcal{T}'') = k - 1).$$

Proof. Let $\chi : X \to C$ denote a random binary character on X. The two events $\chi(x) = \chi(y)$ and $\chi(x) \neq \chi(y)$ each have probability $\frac{1}{2}$. Furthermore, conditional on the first event, $l(\chi, \mathcal{T}) = l(\chi|(X - \{x\}), \mathcal{T}')$ with probability 1, while, conditional on the second event, $l(\chi, \mathcal{T}) = l(\chi|(X - \{x, y\}), \mathcal{T}'') + 1$ with probability 1, as $|C| = 2$. The result now follows by the law of total probability. □

The recursion in Lemma 5.6.1 is central to Theorem 5.6.2.

Theorem 5.6.2. *Let $\mathcal{T}$ be a binary phylogenetic X-tree. The probability distribution of $L_2(\mathcal{T})$ depends only on $n = |X|$. Furthermore,*

(i)

$$\mathbb{P}(L_2(\mathcal{T}) = k) = \begin{cases} [(2n - 3k)/k]\binom{n-k-1}{k-1}2^{k-n}, & \text{if } 1 \leq k \leq n/2, \\ 2^{1-n}, & \text{if } k = 0; \\ 0, & \text{otherwise.} \end{cases}$$

(ii) Let $\mu(n) = \mathbb{E}(L_2(\mathcal{T}))$, the mean value of $L_2(\mathcal{T})$, and let $\sigma^2(n) = \mathbb{V}(L_2(\mathcal{T}))$, the variance of $L_2(\mathcal{T})$. Then,

$$\mu(n) = \frac{3n - 2 - (-1/2)^{n-1}}{9} \sim \frac{n}{3}$$

and

$$\sigma^2(n) = \frac{6n + 2 - (6n + 1)(-1/2)^{n-1} - (-1/2)^{2n-2}}{81} \sim \frac{2n}{27}.$$

Proof. We first prove (i). For $n = 1$ or $n = 2$, $L_2(\mathcal{T})$ is, trivially, independent of the shape of $\mathcal{T}$. Referring to Lemma 5.6.1, we note that for $n \geq 3$ both $\mathcal{T}'$ and $\mathcal{T}''$ are binary phylogenetic trees with smaller-sized label sets than $\mathcal{T}$, and so a simple induction argument shows that the probability distribution of $L_2(\mathcal{T})$ is independent of the shape of $\mathcal{T}$. Consequently, we may let $p(n, k)$ denote the probability $\mathbb{P}(L_2(\mathcal{T}) = k)$ for any binary phylogenetic X-tree $\mathcal{T}$ with $n \geq 1$ leaves, and consider the ordinary generating function

$$p(x, y) = \sum_{n\geq 1}\sum_{k\geq 0} p(n, k)x^n y^k.$$

We have

$$p(x, y) = x + \tfrac{1}{2}x^2 + \tfrac{1}{2}x^2 y + p_3(x, y), \tag{5.9}$$

where $p_3(x, y) = \sum_{n\geq 3}\sum_{k\geq 0} p(n, k)x^n y^k$. By Lemma 5.6.1,

$$p(n, k) = \tfrac{1}{2}p(n - 1, k) + \tfrac{1}{2}p(n - 2, k - 1),$$

and so

$$p_3(x, y) = \tfrac{1}{2}x(p(x, y) - x) + \tfrac{1}{2}x^2 y p(x, y). \tag{5.10}$$

Combining eqns (5.9) and (5.10) gives

$$p(x, y) = x + \tfrac{1}{2}x^2 y + \tfrac{1}{2}xp(x, y) + \tfrac{1}{2}x^2 y p(x, y),$$

which in turn implies that

$$p(x, y) = \left(x + \tfrac{1}{2}x^2 y\right)\left(1 - \tfrac{1}{2}x - \tfrac{1}{2}x^2 y\right)^{-1}. \tag{5.11}$$

Now, if we apply the identity $(1 - Q)^{-1} = \sum_{i\geq 0} Q^i$ to $Q = \frac{1}{2}x(1 + xy)$ and expand Q^i in powers of x and y using the binomial expansion of $(1 + xy)^i$, we obtain

$$[x^r y^s](1 - Q)^{-1} = 2^{s-r}\binom{r - s}{s}. \tag{5.12}$$

From eqn (5.11), we have

$$\mathbb{P}(L_2(\mathcal{T}) = k) = [x^n y^k]p(x, y) = [x^{n-1}y^k](1 - Q)^{-1} + \tfrac{1}{2}[x^{n-2}y^{k-1}](1 - Q)^{-1},$$

and part (i) now follows from eqn (5.12).

For (ii), the value $\mu(n)$ is the coefficient of x^n in $(\partial/\partial y)p(x, y)|_{y=1}$ while $\sigma^2(n)$ is the coefficient of x^n in $(\partial^2/\partial y^2)p(x, y)|_{y=1} + \mu(n) - \mu(n)^2$. Routine calculations establish (ii), completing the proof of the theorem. □

If $|C| > 2$ or if the distribution of states to each element of X is non-uniform or if it varies across X, then the distribution of $L_c(\mathcal{T})$ will generally depend on the shape of $\mathcal{T}$. However, it is still possible in these more general cases to recursively compute the distribution of $L_c(\mathcal{T})$, and to establish a central limit theorem when $|X|$ is large (see Steel *et al.* 1996).

We now turn our attention to the second problem. Given a character $\chi : X \to C$, the number of binary phylogenetic X-trees $\mathcal{T}$ for which $l(\chi, \mathcal{T}) = l$ clearly depends only on the sizes of the blocks of the partition $\pi(\chi)$ of X induced by χ. Consequently, if $C = \{\alpha_1, \alpha_2, \ldots, \alpha_c\}$ and, for each $i \in \{1, 2, \ldots, c\}$, a_i denotes the quantity $|\chi^{-1}(\alpha_i)|$, we may let $p_l(a_1, \ldots, a_c)$ denote the proportion of binary phylogenetic X-trees $\mathcal{T}$ for which $l(\chi, \mathcal{T}) = l$. Equivalently, $p_l(a_1, \ldots, a_c)$ is the probability that a binary phylogenetic X-tree $\mathcal{T}$ selected uniformly at random has $l(\chi, \mathcal{T}) = l$. In general, one can compute these probabilities recursively. However, we describe here two cases where exact formulae exist for $p_l(a_1, \ldots, a_c)$, both of which are due to Carter *et al.* (1990).

For the first case, $l = c - 1$, and so $p_l(a_1, a_2, \ldots, a_c)$ is the probability that the underlying character on X is convex on the randomly selected binary phylogenetic X-tree. In the second case, $c = 2$.

Theorem 5.6.3. *Let $a_1, \ldots, a_c$ be positive integers with sum n. Then,*

$$p_{c-1}(a_1, \ldots, a_c) = \frac{1}{b(n - c + 2)} \prod_{i=1}^{c} b(a_i + 1).$$

Proof. To establish this result we will use Theorem 2.8.3. First, note that if one of the a_i values, a_c say, takes the value 1, then $p_{c-1}(a_1, \ldots, a_c) = p_{c-2}(a_1, \ldots, a_{c-1})$. Consequently, it suffices to establish part (i) under the assumption that $a_i > 1$ for each $i \in \{1, \ldots, c\}$. For a character $\chi : X \to \{\alpha_1, \ldots, \alpha_c\}$ let $\mathcal{F}(\chi)$ denote the set of unrooted binary phylogenetic forests $\{\mathcal{T}_1, \ldots, \mathcal{T}_c\}$ on X such that $\mathcal{L}(\mathcal{T}_i) = \chi^{-1}(\alpha_i)$ for each $i \in \{1, \ldots, c\}$. The set of binary phylogenetic X-trees $\mathcal{T}$ for which $l(\chi, \mathcal{T}) = c - 1$ is precisely the set

$$\bigcup_{\mathcal{F} \in \mathcal{F}(\chi)} \mathrm{Ext}(\mathcal{F}).$$

Furthermore, for every pair of distinct elements $\mathcal{F}$ and $\mathcal{F}'$ in $\mathcal{F}(\chi)$, $\mathrm{Ext}(\mathcal{F}) \cap \mathrm{Ext}(\mathcal{F}')$ is empty, and $|\mathrm{Ext}(\mathcal{F})| = |\mathrm{Ext}(\mathcal{F}')| = [b(n)/b(n - c + 2)] \prod_{i=1}^{c} (2a_i - 3)$ by

Theorem 2.8.3. Hence, the number of binary phylogenetic X-trees for which $l(\chi, \mathcal{T}) = c - 1$ is

$$|\mathcal{F}(\chi)| \frac{b(n)}{b(n-c+2)} \prod_{i=1}^{c} (2a_i - 3). \tag{5.13}$$

Now, $|\mathcal{F}(\chi)| = \prod_{i=1}^{c} b(a_i)$ and, for all i, $(2a_i - 3)b(a_i) = b(a_i + 1)$. Substituting these into (5.13) and dividing the resulting quantity by $b(n)$ to get a probability, we obtain the formula for $p_{c-1}(a_1, \ldots, a_c)$ given in the statement of the theorem. □

The second special case is when $c = 2$. Carter *et al.* (1990) gave an exact formula for $p_l(a_1, a_2)$ using an algebraic argument based on generating functions and the multivariate Lagrange inversion formula. The structural proof we present was described by Steel (1993) and refined by Erdös and Székely (1993). Recall from Section 2.8 that $N(n, k)$ denotes the number of rooted binary phylogenetic forests on a set X of size n and consisting of k trees.

Theorem 5.6.4. *For $c = 2$ and $l \geq 1$,*

$$p_l(a_1, a_2) = \frac{(l-1)!(2n-3l)N(a_1, l)N(a_2, l)}{b(n-l+2)}.$$

Equivalently,

$$p_l(a_1, a_2) = 2^l \cdot \frac{(2n-3l)(2a_1 - l - 1)!(2a_2 - l - 1)!(n-l)!}{(a_1 - l)!(a_2 - l)!(l-1)!(2n-2l)!}.$$

Outline of proof. It is convenient to view each element of X as being coloured either blue or red, depending on whether it is an element of $\chi^{-1}(\alpha_1)$ or $\chi^{-1}(\alpha_2)$. For the purposes of the proof, a rooted binary phylogenetic X-tree is a *bichromatic tree* precisely if, under this colouring, one of its two maximal rooted phylogenetic subtrees has an entirely blue leaf set and the other has an entirely red leaf set. For all $l \geq \min\{a_1, a_2\}$, let $\mathcal{F}(a_1, a_2, l)$ denote the collection of phylogenetic forests on X consisting of l bichromatic trees. There is a straightforward bijection between $\mathcal{F}(a_1, a_2, l)$ and the set of triples $(\mathcal{F}_1, \mathcal{F}_2, m)$, where $\mathcal{F}_1$ (resp. $\mathcal{F}_2$) is a rooted binary phylogenetic forest on $\chi^{-1}(\alpha_1)$ [resp. $\chi^{-1}(\alpha_2)$] consisting of l trees, and where m is a matching that associates each tree in $\mathcal{F}_1$ with precisely one tree in $\mathcal{F}_2$. This bijection takes each forest in $\mathcal{F}(a_1, a_2, l)$ and replaces each tree by its two maximal rooted phylogenetic subtrees. It then places the resulting blue subtrees in $\mathcal{F}_1$, the resulting red subtrees in $\mathcal{F}_2$, and the matching m records which pairs of subtrees came from the same tree in $\mathcal{F}(a_1, a_2, l)$. It follows that

$$|\mathcal{F}(a_1, a_2, l)| = l! N(a_1, l) N(a_2, l). \tag{5.14}$$

Now select any forest $\mathcal{F}$ in $\mathcal{F}(a_1, a_2, l)$ and select any element, $\mathcal{T}_1$ say, of $\mathcal{F}$. Let $\mathcal{F}'$ be the phylogenetic forest obtained by 'unrooting' $\mathcal{T}_1$; that is, replacing $\mathcal{T}_1$ by $\mathcal{T}_1^{-\rho}$ in $\mathcal{F}$. Let $\mathcal{T}$ be an element of $\mathrm{Ext}(\mathcal{T}_1^{-\rho}, \mathcal{F}')$ and let $\mathcal{G}_l(a_1, a_2)$ denote the set of triples $(\mathcal{F}, \mathcal{T}, \mathcal{T}_1)$ that can be obtained in this way. By construction, there exists a set $P(\mathcal{T})$ of

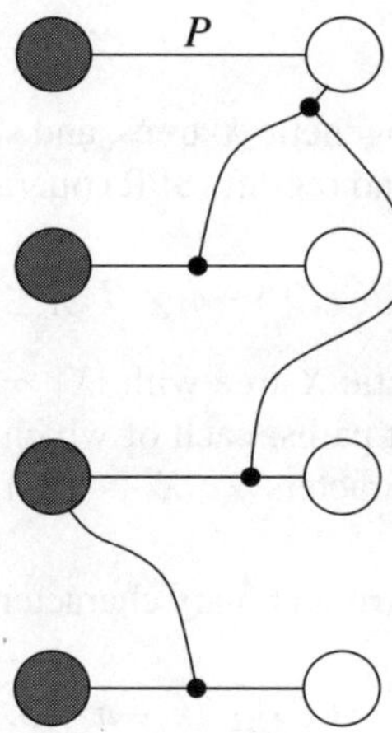

Figure 5.10 Representing $\mathcal{T}$ using four bichromatic trees.

l edge-disjoint paths in $\mathcal{T}$ each of which connects a red and a blue leaf. Furthermore, one of these paths, call it P, is distinguished by virtue of it lying in $\mathcal{T}_1$. This construction is illustrated in Fig. 5.10 for $l = 4$.

It follows by Corollary 5.1.8 that $l(\chi, \mathcal{T}) \geq l$. However, it is also easily seen that $l(\chi, \mathcal{T}) \leq l$ and so $l(\chi, \mathcal{T}) = l$. Consequently, we have a technique for assembling binary phylogenetic X-trees with parsimony score l, which is closely tied to the Ext construction of Section 2.8. Furthermore, provided one selects the set $P(\mathcal{T})$ appropriately, it can be shown that the map $(\mathcal{F}, \mathcal{T}, \mathcal{T}_1) \mapsto (\mathcal{T}, P)$ is a bijection from $\mathcal{G}_l(a_1, a_2)$ to $\{(\mathcal{T}, P)\colon l(\chi, \mathcal{T}) = l, P \in P(\mathcal{T})\}$. Since the image of this map has size $lp_l(a_1, a_2)b(n)$ it follows that

$$p_l(a_1, a_2) = \frac{|\mathcal{G}_l(a_1, a_2)|}{lb(n)}. \tag{5.15}$$

Now, applying Theorem 2.8.3, we have

$$\begin{aligned}
|\mathcal{G}_l(a_1, a_2)| &= \sum_{\mathcal{F} \in \mathcal{F}(a_1, a_2, l)} \sum_{\mathcal{T}_1 \in \mathcal{F}} |\mathrm{Ext}(\mathcal{T}_1^{-\rho}, \mathcal{F}')|, \\
&= \sum_{\mathcal{F} \in \mathcal{F}(a_1, a_2, l)} \sum_{\mathcal{T}_1 \in \mathcal{F}} |E_1| \frac{b(n)}{b(n - l + 2)}, \\
&= (2n - 3l)|\mathcal{F}(a_1, a_2, l)| \frac{b(n)}{b(n - l + 2)},
\end{aligned}$$

where E_1 denotes the set of edges in $\mathcal{T}_1$. The first of the two expressions for $p_l(a_1, a_2)$ now follows by eqns (5.14) and (5.15). The second expression for $p_l(a_1, a_2)$ now follows by invoking Proposition 2.8.1. □

5.7 Exercises

1. Let $\mathcal{T}$ and $\mathcal{T}'$ be binary phylogenetic X-trees, and suppose that $\mathcal{T}'$ is obtained from $\mathcal{T}$ by a single subtree prune and regraft (SPR) operation. Let χ be a character on X. Show that
$$|l(\chi, \mathcal{T}) - l(\chi, \mathcal{T}')| \leq 1.$$
2. Let $\mathcal{T}$ be a binary phylogenetic X-tree with $|X| = 2m$. Show that in $\mathcal{T}$ there is a unique set of m edge-disjoint paths, each of which joins two leaves of $\mathcal{T}$. Deduce that the number of binary characters $\chi : X \rightarrow \{0, 1\}$ with parsimony score m on $\mathcal{T}$ is exactly 2^m.
3. Let $\mathcal{C}$ be a sequence of k non-trivial binary characters on X that induce distinct splits of X. Prove that
$$l(\mathcal{C}) \geq 2k - n + 3,$$
where $n = |X|$.
4. Let χ be a character on X and let $\mathcal{T}$ be a phylogenetic X-tree. Suppose that $\bar{\chi}$ is an extension of χ to $\mathcal{T}$ such that every two edges in $\mathrm{Ch}(\bar{\chi})$ are separated by at least three edges of $\mathcal{T}$. Prove that $\bar{\chi}$ is the unique minimum extension of χ to $\mathcal{T}$.
5. (Bryant, personal communication) Let $\mathcal{C}$ be a collection of binary characters on X, and suppose that χ is a binary character on X that is compatible with every character in $\mathcal{C}$.
   (i) Show that there is a maximum parsimony tree for $\mathcal{C}$ that has the X-split $\pi(\chi)$.
   (ii) Show that if $\chi \in \mathcal{C}$, then every maximum parsimony tree for $\mathcal{C}$ induces the X-split $\pi(\chi)$.
6. The *parsimony operation* $*$ is a commutative, non-associative binary operation on $2^C - \{\emptyset\}$ that is defined, for all $A, B \in 2^C - \{\emptyset\}$, by
$$A * B = \begin{cases} A \cap B, & \text{if } A \cap B \neq \emptyset, \\ A \cup B, & \text{otherwise.} \end{cases}$$
Let χ be a full character on X and let $\mathcal{T}$ be a rooted binary phylogenetic X-tree. Consider the Fitch–Hartigan algorithm applied to χ and $\mathcal{T}$.
   (i) Show that, for all $v \in V(\mathcal{T})$, the set $\psi(v)$ satisfies
$$\psi(v) = \psi(w_1) * \psi(w_2),$$
   where w_1 and w_2 are the immediate descendants of v in $\mathcal{T}$.
   (ii) For all $v \in V(\mathcal{T})$, set $\delta(v) = l(v) - l(w_1) - l(w_2)$. Show that
$$\delta(v) = \begin{cases} 0, & \text{if } A \cap B \neq \emptyset, \\ 1, & \text{otherwise,} \end{cases}$$
   for all $v \in V(\mathcal{T})$.

7. Let g be a generalized character on X and let $\mathcal{T}$ be a phylogenetic X-tree. Show how the Fitch–Hartigan algorithm can be modified to find a minimum extension of g to $\mathcal{T}$.
8. Let $\mathcal{T}$ be the caterpillar tree with label set $\{1, 2, \ldots, n\}$, as shown in Fig. 2.6(a). Suppose that n is even, so that we may write $n = 2m$, and let χ be a binary character on $\{1, 2, \ldots, n\}$ defined by

$$\chi(i) = \begin{cases} \alpha, & \text{if } i = 0, 1 \text{ (mod 4)}, \\ \beta, & \text{otherwise.} \end{cases}$$

Show that the number F_m of minimum extensions of χ to $\mathcal{T}$ satisfies, for all $m \geq 2$, the recurrence relation defined by $F_m = F_{m-1} + F_{m-2}$ and $F_0 = F_1 = 1$. In particular, for all m, F_m is the m-th Fibonacci number.
9. In the metric space setting for parsimony, show that every minimum extension of a character $\chi : X \to C$ to a phylogenetic X-tree $\mathcal{T}$ is a median extension.
10. (Foulds 1984) For any real number $\lambda > \frac{1}{2}$, construct a collection $\mathcal{C}$ of characters for which $\lambda w(\mathcal{C}) > l(\mathcal{C})$, where $w(\mathcal{C})$ is as in Proposition 5.4.1.
11. (Barthélemy 1989) Let Σ be a non-empty set of X-splits. Prove that if v is a degree-three vertex of $G(\Sigma)$ and v is not an element of $\{N_x : x \in X\}$, then v is the median of three vertices in $\{N_x : x \in X\}$.
12. (Barthélemy 1989) Let $G = (V, E)$ be a median graph and let V' be a subset of V. The *median closure* of V', denoted $\mathrm{mcl}(V')$, is the minimal subset of V such that
    (i) V' is a subset of $\mathrm{mcl}(V')$, and
    (ii) if $v \in V$ is the median of three vertices in $\mathrm{mcl}(V')$, then $v \in \mathrm{mcl}(V')$.
    Let Σ be a non-empty set of X-splits and consider $G(\Sigma)$. Prove that the median closure of $\{N_x : x \in X\}$ is equal to the vertex set of $G(\Sigma)$.
13. Suppose that a random character $\chi : X \to \{0, 1\}$ independently assigns each element of X the state 0 with probability p and the state 1 with probability $1 - p$. Let $\mathcal{T}$ be a balanced phylogenetic X-tree of height h, and consider the Fitch–Hartigan algorithm applied to χ and $\mathcal{T}$. For all $A \in 2^{\{0,1\}}$, let $p_A(h) = \mathbb{P}(\psi(\rho) = A)$. For $p = \frac{1}{2}$, show that $p_{\{0\}}(h)$, $p_{\{1\}}(h)$, and $p_{\{0,1\}}(h)$ all converge to $\frac{1}{3}$ as h tends to infinity. What are the corresponding limits if $p > \frac{1}{2}$?
14. Suppose that a random character $\chi : X \to \{0, 1\}$ independently assigns each element of X the state 0 with probability $\frac{1}{2}$ and the state 1 with probability $\frac{1}{2}$. Let $\mathcal{T}$ be the star phylogenetic X-tree. Let $\bar{l}$ denote the expected parsimony score of the random character χ to $\mathcal{T}$. Show that

$$\lim_{n \to \infty} \frac{\bar{l}}{n} = \frac{1}{2},$$

where $n = |X|$. Contrast this with the corresponding limit for binary phylogenetic X-trees in Theorem 5.6.2(ii).
15. Let $\mathcal{T}$ be a binary phylogenetic X-tree and let $k \geq 0$. Let $\mathcal{C}$ be the collection of all binary characters on X that have parsimony score at most k on $\mathcal{T}$. Show that $\mathcal{T}$ is a maximum parsimony tree for $\mathcal{C}$.

6
Subtrees and supertrees

A central task in phylogenetics is to combine phylogenetic trees that classify overlapping sets of species into a single phylogenetic tree classifying the union of the sets of species. If possible, one would like the resulting parent tree to preserve all of the relationships that are shown in each of the initial phylogenetic trees. In this chapter, we formulate and examine this task for both the rooted and unrooted settings. As we shall see, the underlying concepts in this chapter are closely related to notions introduced earlier.

6.1 Restricted subtrees

In this section, we introduce the operation of restriction. This operation formalizes the notion of a semi-labelled tree being a subtree of another semi-labelled tree.

Definition 6.1.1. Let $\mathcal{T}$ be an X-tree and suppose that X' is a non-empty subset of X. The *restriction of* $\mathcal{T}$ *to* X', denoted $\mathcal{T}|X'$, is the X'-tree for which

$$\Sigma(\mathcal{T}|X') = \{A \cap X'|B \cap X' \colon A|B \in \Sigma(\mathcal{T}) \text{ and } A \cap X', B \cap X' \neq \emptyset\}.$$

The X'-tree $\mathcal{T}|X'$ is said to be a *restricted subtree* of $\mathcal{T}$.

Graphically, the restriction of an X-tree $\mathcal{T} = (T; \phi)$ to a non-empty subset X' of X is equivalent to the following. Let $T(X')$ denote the minimal subtree of T induced by the elements of the set $\phi(X')$. Let $T|X'$ denote the tree obtained from $T(X')$ by suppressing all vertices of degree two that are not in $\phi(X')$. Then, $\mathcal{T}|X'$ denotes the X'-tree $(T|X'; \phi|X')$. Figure 6.1 illustrates Definition 6.1.1.

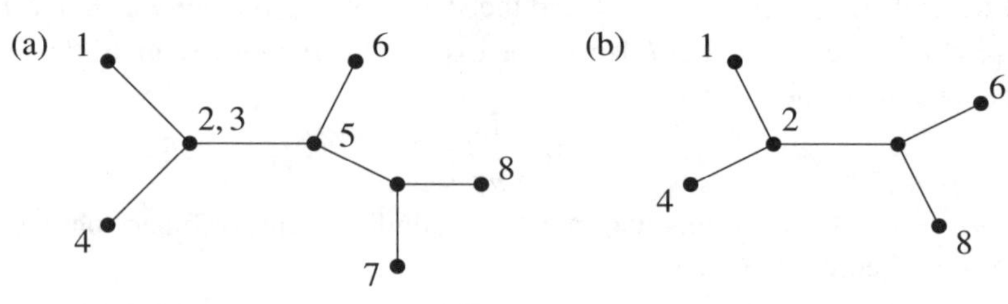

Figure 6.1 (a) An X-tree $\mathcal{T}$. (b) $\mathcal{T}|\{1, 2, 4, 6, 8\}$.

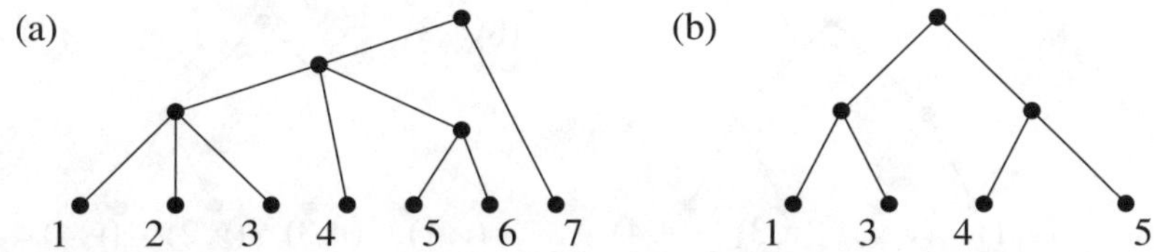

Figure 6.2 (a) A rooted phylogenetic X-tree $\mathcal{T}$. (b) $\mathcal{T}|\{1,3,4,5\}$.

Observe that, if $\mathcal{T}$ is a phylogenetic X-tree and X' is a non-empty subset of X, then $\mathcal{T}|X'$ is a phylogenetic X'-tree. If, in addition, $\mathcal{T}$ is binary, then $\mathcal{T}|X'$ is also binary.

The operation of restriction extends to rooted semi-labelled trees as follows.

Definition 6.1.2. Let $\mathcal{T}$ be a rooted X-tree and suppose that X' is a subset of X. The *restriction of* $\mathcal{T}$ *to* X', denoted $\mathcal{T}|X'$, is the rooted X'-tree for which

$$\mathcal{H}(\mathcal{T}|X') = \{C \cap X' \colon C \in \mathcal{H}(\mathcal{T}) \text{ and } C \cap X' \neq \emptyset\}.$$

The rooted X'-tree $\mathcal{T}|X'$ is said to be a *restricted subtree* of $\mathcal{T}$.

Graphically, the restriction of a rooted X-tree $\mathcal{T} = (T; \phi)$ to a non-empty subset X' of X is equivalent to the following. Let $T(X')$ denote the minimal rooted subtree of T induced by the elements of the set $\phi(X')$. Let $T|X'$ denote the rooted tree that is obtained from $T(X')$ by suppressing all vertices of degree two that are not in $\phi(X')$ other than possibly the root. Then, $\mathcal{T}|X'$ denotes the rooted X'-tree $(T|X'; \phi|X')$. Figure 6.2 illustrates Definition 6.1.2 for a rooted phylogenetic tree.

Biological interpretation of restriction. Let X_2 be a subset of X_1 and suppose that, for each $i \in \{1,2\}$, $\mathcal{T}_i$ is a rooted phylogenetic tree on X_i that describes the evolutionary history of the set X_i of extant species. Then, $\mathcal{T}_2$ is a restriction of $\mathcal{T}_1$ precisely if all of the ancestral relationships shown in $\mathcal{T}_2$ are shown in $\mathcal{T}_1$. This means that if x, y, and z are species in X_2, and the most recent common ancestor of x and y is a proper descendant of the most recent common ancestor of x and z in $\mathcal{T}_2$, then these relationships also hold in $\mathcal{T}_1$.

6.2 Maximum agreement subtrees

Let $\mathcal{P}$ be a collection $\{\mathcal{T}_1, \mathcal{T}_2, \ldots, \mathcal{T}_k\}$ of X-trees and let X' be a maximum-sized subset of X for which

$$\mathcal{T}_1|X' = \mathcal{T}_2|X' = \cdots = \mathcal{T}_k|X'.$$

This restricted X'-tree is called a *maximum agreement subtree* for $\mathcal{P}$. Analogously, we can define a maximum agreement subtree for a collection of rooted X-trees.

In applications, one motivation for finding a maximum agreement subtree for a collection $\mathcal{P}$ of X-trees is that a small number of 'problematic' species may cause estimates of an underlying evolutionary tree to vary greatly between data sets. Finding a maximum agreement subtree may help to identify and thereby exclude such species. Another motivation is that exhibiting a maximum agreement subtree for $\mathcal{P}$ provides

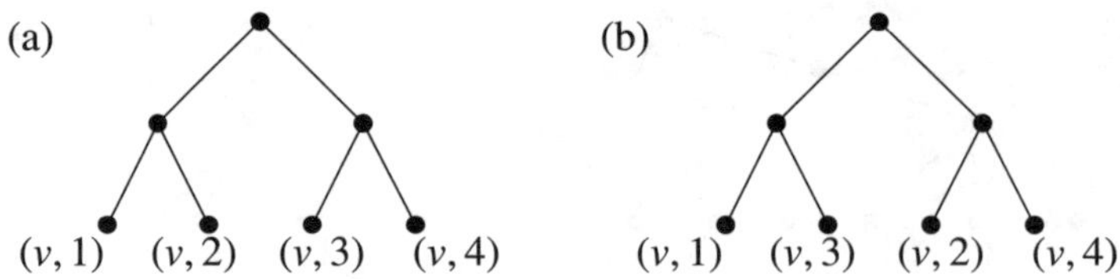

Figure 6.3

an alternative way of combining different and conflicting X-trees to the consensus tree approach described in Section 3.6. In the consensus approach, one generally deletes splits and produces a less refined tree, while in the maximum agreement subtree approach, one deletes species.

Example 6.2.1 shows that even for two rooted binary phylogenetic trees, the number of maximum agreement subtrees can be exponentially large.

Example 6.2.1. Following Kubicka *et al.* (1992), let T be a binary tree and let V' denote the set of leaves of T. We construct two rooted binary phylogenetic X-trees from T, where $X = V' \times \{1, 2, 3, 4\}$, as follows. Let $\mathcal{T}_1$ denote the rooted binary phylogenetic X-tree obtained from T by identifying each $v \in V'$ with the root of the rooted binary phylogenetic tree on $\{(v, 1), (v, 2), (v, 3), (v, 4)\}$ shown in Fig. 6.3(a). Let $\mathcal{T}_2$ denote the rooted binary phylogenetic X-tree obtained from T by identifying each $v \in V'$ with the root of the rooted binary phylogenetic tree on $\{(v, 1), (v, 2), (v, 3), (v, 4)\}$ shown in Fig. 6.3(b). It is easily checked that the maximum-sized subsets X' of X that satisfy $\mathcal{T}_1|X' = \mathcal{T}_2|X'$ are precisely those subsets of X that contain exactly two elements of $\{(v, 1), (v, 2), (v, 3), (v, 4)\}$ for each $v \in V'$. It follows that the number of maximum agreement subtrees for $\mathcal{T}_1$ and $\mathcal{T}_2$ is $\binom{4}{2}^{|Y|} = (6^{1/4})^{|X|}$, which grows exponentially with $|X|$.

Let $\mathcal{P}$ be a collection of X-trees that are either all rooted or all unrooted. If $\mathcal{P}$ consists of two rooted phylogenetic X-trees, then a maximum agreement subtree for $\mathcal{P}$ can be constructed in polynomial time by using a dynamic programming approach. This algorithm can then be adapted to provide a polynomial–time algorithm for constructing a maximum agreement subtree for a pair of unrooted phylogenetic X-trees (Steel and Warnow 1993). However, if $\mathcal{P}$ consists of three rooted phylogenetic X-trees or three unrooted phylogenetic X-trees, then in general, it is an NP-hard problem to compute a maximum agreement subtree for $\mathcal{P}$. Nevertheless, if $\mathcal{P}$ consists of phylogenetic X-trees, d is a fixed positive integer, and at least one of the phylogenetic trees in $\mathcal{P}$ has maximum vertex degree d, then there is an algorithm for constructing a maximum agreement subtree for $\mathcal{P}$ whose running time is polynomial in $|X|$ and $|\mathcal{P}|$ (Amir and Keselman 1997).

6.3 Compatibility of semi-labelled trees

In Chapters 3 and 4, we considered compatibility for a collection of X-trees, where X is fixed, and compatibility for a collection of characters. In this section, we explore a further, related notion of compatiblity that applies to any collection of semi-labelled trees.

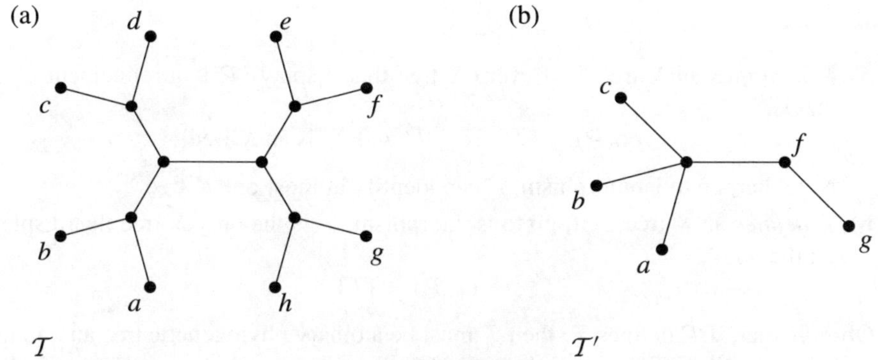

Figure 6.4 $\mathcal{T}$ displays $\mathcal{T}'$.

Definition 6.3.1. Let X' be a subset of X. An X-tree $\mathcal{T}$ *displays an X'-tree* $\mathcal{T}'$ if $\mathcal{T}|X'$ is a refinement of $\mathcal{T}'$; that is, $\mathcal{T}' \leq \mathcal{T}|X'$. Furthermore, a set $\mathcal{P}$ of semi-labelled trees is *compatible* if there exists a semi-labelled tree $\mathcal{T}$ that displays every tree in $\mathcal{P}$, in which case $\mathcal{T}$ *displays* $\mathcal{P}$.

Observe that the concept of compatibility introduced in Definition 6.3.1 extends the definition of compatibility for a set of X-trees introduced in Section 3.3. To illustrate Definition 6.3.1, the semi-labelled tree shown in Fig. 6.4(a) displays the semi-labelled tree shown in Fig. 6.4(b).

Biological interpretation of display. One could impose the simpler and stronger requirement that $\mathcal{T}' = \mathcal{T}|X'$ in place of $\mathcal{T}' \leq \mathcal{T}|X'$ in Definition 6.3.1, in which case one might say that $\mathcal{T}$ 'exactly displays' $\mathcal{T}'$. Indeed, when $\mathcal{P}$ consists of binary phylogenetic trees, there is no distinction between the requirement $\mathcal{T}' = \mathcal{T}|X'$ and $\mathcal{T}' \leq \mathcal{T}|X'$. However, this runs contrary to the usual interpretation in biology where a vertex of a phylogenetic tree of degree at least four is regarded as an uncertainty as to the exact order of speciation, rather than as a certainty of a simultaneous multiple speciation event. Biologists distinguish these two situations by referring to them as a 'soft-polytomy' and a 'hard polytomy', respectively. Adopting the soft polytomy approach gives a clear relationship between the concepts of compatibility for a set of semi-labelled trees and compatibility for a set of characters. This relationship is described at the end of this section.

We next introduce two basic notions that describe the extent to which a set of semi-labelled trees displayed by an X-tree $\mathcal{T}$ recovers $\mathcal{T}$.

Definition 6.3.2. Let $\mathcal{P}$ be a set of semi-labelled trees and let X denote the label set of $\mathcal{P}$. The *compatible span* of $\mathcal{P}$, denoted co($\mathcal{P}$), is the set of X-trees that display $\mathcal{P}$.

Furthermore,

(i) $\mathcal{P}$ *identifies* an X-tree $\mathcal{T}$ if every X-tree that displays $\mathcal{P}$ is a refinement of $\mathcal{T}$; that is,

$$\text{co}(\mathcal{P}) = \{\mathcal{T}' \colon \mathcal{T} \leq \mathcal{T}' \text{ and } \mathcal{T}' \text{ is an } X\text{-tree}\}.$$

Note that, up to isomorphism, $\mathcal{P}$ can identify at most one X-tree.

(ii) $\mathcal{P}$ *defines* an X-tree $\mathcal{T}$ if, up to isomorphism, $\mathcal{T}$ is the only X-tree that displays $\mathcal{P}$; that is,

$$\text{co}(\mathcal{P}) = \{\mathcal{T}\}$$

Observe that, if $\mathcal{P}$ defines $\mathcal{T}$, then $\mathcal{T}$ must be a binary phylogenetic tree and, in this case, $\mathcal{P}$ also identifies $\mathcal{T}$. Conversely, if $\mathcal{P}$ identifies a binary phylogenetic tree $\mathcal{T}$, then $\mathcal{P}$ defines $\mathcal{T}$.

Example 6.3.3. To illustrate Definition 6.3.2, let $\mathcal{P} = \{\mathcal{T}_1, \mathcal{T}_2\}$ and $\mathcal{P}' = \{\mathcal{T}_1, \mathcal{T}_3\}$, where $\mathcal{T}_1$, $\mathcal{T}_2$, and $\mathcal{T}_3$ are shown in Fig. 6.5(a). Then, $\mathcal{P}$ defines the phylogenetic tree shown in Fig. 6.5(b), while $\mathcal{P}'$ identifies—but does not define—the phylogenetic tree shown in Fig. 6.5(c).

Theorem 6.3.5 shows that every semi-labelled tree is identified by its restricted subtrees with label sets of size at most four. The proof of this theorem makes use of Lemma 6.3.4.

Lemma 6.3.4. *Let $\mathcal{T}$ be an X-tree and let $A|B$ be an X-split. Then, $A|B$ is a split of $\mathcal{T}$ if and only if, for all $a, a' \in A$ and $b, b' \in B$,*

$$\{a, a'\}|\{b, b'\} \in \Sigma(\mathcal{T}|\{a, a', b, b'\}). \tag{6.1}$$

Proof. If $A|B \in \Sigma(\mathcal{T})$, then, by definition of restriction,

$$\{a, a'\}|\{b, b'\} = A \cap \{a, a'\}|B \cap \{b, b'\} \in \Sigma(\mathcal{T}|\{a, a', b, b'\})$$

for all $a, a' \in A$ and $b, b' \in B$.

Now suppose that (6.1) holds for all $a, a' \in A$ and $b, b' \in B$, and suppose that $A|B$ is not a split of $\mathcal{T}$. Then either (i) there is a split $A'|B'$ of $\mathcal{T}$ that is incompatible with $A|B$ or (ii) $A|B$ is compatible with every split of $\mathcal{T}$. Assume that (i) holds. Then, there exists elements $a, a', b, b' \in X$ such that $a \in A \cap A', a' \in A \cap B', b \in B \cap A'$, and $b' \in B \cap B'$. But then, $\{a, b\}|\{a', b'\} \in \Sigma(\mathcal{T}|\{a, a', b, b'\})$, which together with (6.1) implies that $\Sigma(\mathcal{T}|\{a, a', b, b'\})$ is pairwise incompatible; a contradiction.

Now assume that (ii) holds, and let T and ϕ denote the underlying tree and labelling map of $\mathcal{T}$, respectively. Then, by Lemma 3.1.7, there is a unique vertex v of T such that, for each component (V', E') of $T\backslash v$, either $\phi^{-1}(V') \subseteq A$ or $\phi^{-1}(V') \subseteq B$. If there are elements $a \in \phi^{-1}(v) \cap A$ and $b \in \phi^{-1}(v) \cap B$, then $\{a\}|\{b\} \notin \Sigma(\mathcal{T}|\{a, b\})$, contradicting (6.1) for $a' = a, b' = b$. If there is an element $a \in \phi^{-1}(v) \cap A$ but $\phi^{-1}(v) \cap B$ is empty, then by the uniqueness property of v, there exists two distinct elements $b, b' \in B$ such that the vertices $\phi(b)$ and $\phi(b')$ are in distinct components of

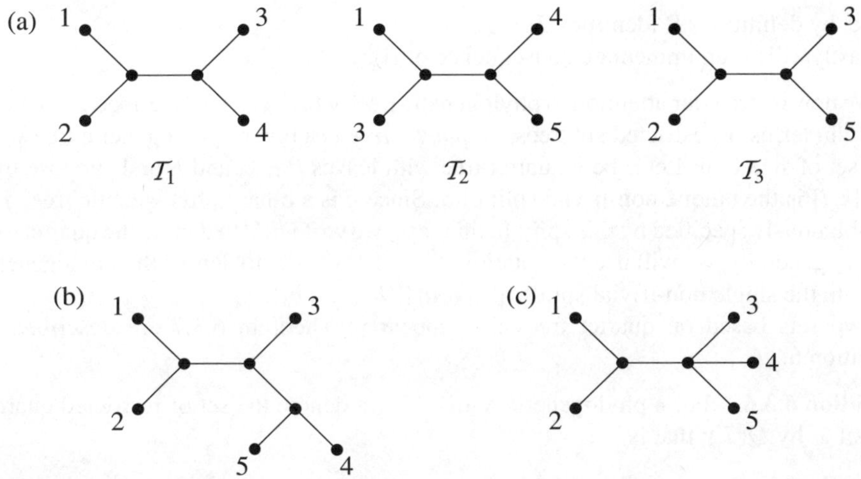

Figure 6.5 (a) Three semi-labelled trees. (b) The semi-labelled tree defined by $\mathcal{T}_1$ and $\mathcal{T}_2$. (c) The semi-labelled tree identified by $\mathcal{T}_1$ and $\mathcal{T}_3$.

$T\backslash v$. But then, $\{a\}|\{b, b'\} \notin \Sigma(\mathcal{T}|\{a, b, b'\})$, again contradicting (6.1). By symmetry, we may assume that both $\phi^{-1}(V') \cap A$ and $\phi^{-1}(V') \cap B$ are empty. Then, by the uniqueness property of v, there exists two distinct elements $a, a' \in A$ and two distinct elements $b, b' \in B$ such that the vertices $\phi(a)$, $\phi(a')$, $\phi(b)$, and $\phi(b')$ are in distinct components of $T\backslash v$. But this implies that $\{a, a'\}|\{b, b'\} \notin \Sigma(\mathcal{T}|\{a, a', b, b'\})$. This last contradiction to (6.1) completes the proof of the lemma. □

Theorem 6.3.5. *Let $\mathcal{T}$ and $\mathcal{T}'$ be two X-trees. Then*

(i) $\mathcal{T} \leq \mathcal{T}'$ if and only if $\mathcal{T}|S \leq \mathcal{T}'|S$ for all subsets S of X of size at most four;

(ii) $\{\mathcal{T}|S\colon S \subseteq X$ and $|S| \leq 4\}$ identifies $\mathcal{T}$;

(iii) $\mathcal{T} \cong \mathcal{T}'$ if and only if $\mathcal{T}|S \cong \mathcal{T}'|S$ for all subsets S of X of size at most four.

Proof. We first prove (i). If $\mathcal{T} \leq \mathcal{T}'$, then $\mathcal{T}|S \leq \mathcal{T}'|S$ for all subsets S of X, in particular for those subsets S of size at most four. Conversely, suppose that $\mathcal{T}|S \leq \mathcal{T}'|S$ for all subsets S of X of size at most four. Let $A|B \in \Sigma(\mathcal{T})$. Then, for all elements $a, a' \in A$ and $b, b' \in B$, we have $\{a, a'\}|\{b, b'\} \in \Sigma(\mathcal{T}|\{a, a', b, b'\})$. By assumption, this implies that

$$\{a, a'\}|\{b, b'\} \in \Sigma(\mathcal{T}'|\{a, a', b, b'\}),$$

for all $a, a' \in A$ and $b, b' \in B$. Thus, by Lemma 6.3.4, $A|B \in \Sigma(\mathcal{T}')$. Consequently, $\Sigma(\mathcal{T}) \subseteq \Sigma(\mathcal{T}')$ and so $\mathcal{T} \leq \mathcal{T}'$, completing the proof of (i).

To prove (ii), let $\mathcal{P} = \{\mathcal{T}|S\colon S \subseteq X \text{ and } |S| \leq 4\}$ and suppose that $\mathcal{T}' \in \text{co}(\mathcal{P})$. Then, for each subset S of X of size at most four, we have $\mathcal{T}|S \leq \mathcal{T}'|S$. By (i), this implies that $\mathcal{T} \leq \mathcal{T}'$. Therefore, as $\mathcal{T} \in \text{co}(\mathcal{P})$,

$$\text{co}(\mathcal{P}) = \{\mathcal{T}'\colon \mathcal{T} \leq \mathcal{T}' \text{ and } \mathcal{T}' \text{ is an } X\text{-tree}\}.$$

Hence, by definition, $\mathcal{P}$ identifies $\mathcal{T}$.

Lastly, (iii) is an immediate consequence of (i). □

We now restrict our attention to phylogenetic trees which can be described even more simply in terms of restricted subtrees. A *quartet tree* is a binary phylogenetic tree with label set of size four. Let q be a quartet tree with leaves i, j, k, and l, and suppose that $\{i,j\}|\{k,l\}$ is the unique non-trivial split of q. Since q is a binary phylogenetic tree, q is unambiguously specified by this split. In this case, we write $ij|kl$ to denote the quartet tree q. More generally, we will use the notation $a_1 \cdots a_r|b_1 \cdots b_s$ to denote the phylogenetic tree with the single non-trivial split $\{a_1, \ldots, a_r\}|\{b_1, \ldots, b_s\}$.

Two sets based on quartet trees that appear in Theorem 6.3.7 are described in Definition 6.3.6.

Definition 6.3.6. For a phylogenetic X-tree $\mathcal{T}$, we denote the set of restricted quartet trees of $\mathcal{T}$ by $\mathcal{Q}(\mathcal{T})$; that is,

$$\mathcal{Q}(\mathcal{T}) = \{\mathcal{T}|S\colon S \subseteq X, |S| = 4, \text{ and } \mathcal{T}|S \text{ is binary}\}.$$

Furthermore, for a set $\mathcal{Q}$ of quartet trees with label set X, let $\Sigma(\mathcal{Q})$ denote the set

$$\Sigma(\mathcal{Q}) = \{A|B \in \Sigma(X)\colon aa'|bb' \in \mathcal{Q} \text{ for all distinct } a, a' \in A \text{ and } b, b' \in B\}$$

of non-trivial X-splits.

Theorem 6.3.7.

(i) *Let $\mathcal{Q}$ be a set of quartet trees with $\mathcal{L}(\mathcal{Q}) = X$. Suppose that, for each subset X' of X of size four, at most one quartet tree in $\mathcal{Q}$ has label set X'. Then, $\Sigma(\mathcal{Q}) = \Sigma(\mathcal{T})$ for some X-tree $\mathcal{T}$. Furthermore, $\Sigma(\mathcal{Q})$ can be constructed from $\mathcal{Q}$ in polynomial time.*

(ii) *Let $\mathcal{T}$ be a phylogenetic X-tree. Then, $\Sigma(\mathcal{Q}(\mathcal{T}))$ is the set of non-trivial X-splits of $\mathcal{T}$.*

Proof. To prove the first part of (i), it suffices to show by the Splits-Equivalence Theorem that $\Sigma(\mathcal{Q})$ is a collection of pairwise compatible X-splits. Let $A_1|B_1$ and $A_2|B_2$ be a pair of elements of $\Sigma(\mathcal{Q})$, and suppose that $A_1|B_1$ and $A_2|B_2$ are not compatible. Then, there exist distinct elements $x, y, w, z \in X$ such that $x \in A_1 \cap A_2$, $y \in A_1 \cap B_2$, $w \in B_1 \cap A_2$, $z \in B_1 \cap B_2$. Since $A_1|B_1 \in \Sigma(\mathcal{Q})$, we have $xy|wz \in \mathcal{Q}$ and since $A_2|B_2 \in \Sigma(\mathcal{Q})$, we have $xw|yz \in \mathcal{Q}$. But then $\mathcal{Q}$ contains two distinct quartet trees on the same label set. This contradiction establishes the first part of (i). The existence of a polynomial–time algorithm (Berry and Gascuel 2000) is left as an exercise.

We next prove (ii). By Lemma 6.3.4, a non-trivial X-split, $A|B$ say, is an X-split of a phylogenetic X-tree $\mathcal{T}$ if and only if, for all distinct $a, a' \in A$ and $b, b' \in B$, $aa'|bb' \in \mathcal{Q}(\mathcal{T})$. Therefore, $\Sigma(\mathcal{Q}(\mathcal{T}))$ consists of every non-trivial X-split of $\mathcal{T}$. Part (ii) of the theorem immediately follows. □

Corollary 6.3.8 is an immediate consequence of Theorem 6.3.7(ii).

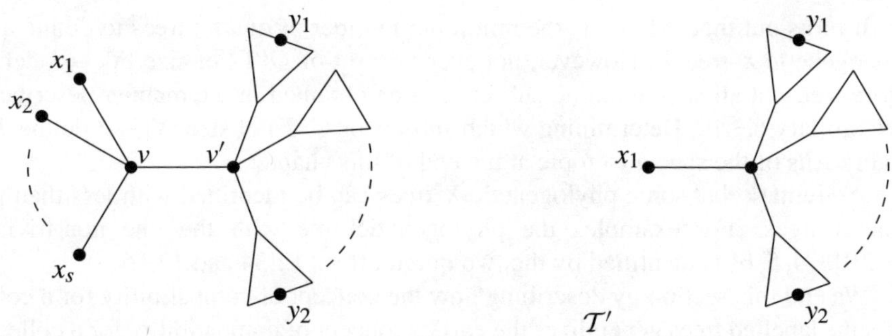

Figure 6.6 The inductive step in the proof of Theorem 6.3.9.

Corollary 6.3.8. *Let $\mathcal{T}$ and $\mathcal{T}'$ be phylogenetic X-trees. Then, $\mathcal{Q}(\mathcal{T}) \subseteq \mathcal{Q}(\mathcal{T}')$ if and only if $\mathcal{T} \leq \mathcal{T}'$. In particular, $\mathcal{Q}(\mathcal{T})$ identifies $\mathcal{T}$.*

Theorem 6.3.9 shows that we generally require only a small subset of $\mathcal{Q}(\mathcal{T})$ to identify $\mathcal{T}$.

Theorem 6.3.9. *Let $\mathcal{T}$ be a phylogenetic X-tree with at least one interior edge. Then, there is a set of $|X| - 3$ quartet trees that identifies $\mathcal{T}$.*

Proof. We prove the theorem by induction on the number of interior edges of $\mathcal{T}$. Suppose that $\mathcal{T}$ has just one interior edge. Then, $\mathcal{T}$ has exactly one non-trivial split $A|B$. In this case, select distinct elements $a_1, a_2 \in A$ and distinct elements $b_1, b_2 \in B$, and let

$$\mathcal{Q} = \{a_1a_2|b_1b\colon b \in B - \{b_1\}\} \cup \{a_1a|b_1b_2\colon a \in A - \{a_1\}\}.$$

One can easily check that $|\mathcal{Q}| = |X| - 3$ and that $\mathcal{Q}$ identifies $\mathcal{T}$.

Now suppose that $\mathcal{T}$ has precisely r interior edges, where $r \geq 2$, and that the result holds for all phylogenetic trees with precisely $r - 1$ interior edges. Let v be an interior vertex of $\mathcal{T}$ that is adjacent to exactly one other interior vertex v' of $\mathcal{T}$, and let y_1 and y_2 be elements of X such that the path from y_1 to y_2 in $\mathcal{T}$ passes through v' but not v. Let $x_1, x_2, \ldots, x_s$ denote the elements of X adjacent to v, where $s \geq 2$, and let $\mathcal{T}' = \mathcal{T}|(X - \{x_2, \ldots, x_s\})$. The phylogenetic trees $\mathcal{T}$ and $\mathcal{T}'$ are shown in Fig. 6.6. Then, $\mathcal{T}'$ is a phylogenetic tree with $r - 1$ interior edges. Therefore, by the induction assumption, there is a set $\mathcal{Q}'$ of $|X| - (s - 1) - 3$ quartet trees that identifies $\mathcal{T}'$. Let $\mathcal{Q} = \mathcal{Q}' \cup \{y_1y_2|x_1x_i\colon i \in \{2, \ldots, s\}\}$. Then, $|\mathcal{Q}| = |X| - 3$ and one can easily check that $\mathcal{Q}$ identifies $\mathcal{T}$. This completes the proof of Theorem 6.3.9. □

For a phylogenetic X-tree $\mathcal{T}$, the inductive proof of Theorem 6.3.9 provides us with a polynomial–time algorithm for constructing a set of $|X| - 3$ quartet trees that identifies $\mathcal{T}$. We leave this as an exercise. Corollary 6.3.10 is an immediate consequence of Theorem 6.3.9.

Corollary 6.3.10. *Let $\mathcal{T}$ be a binary phylogenetic X-tree, where $|X| \geq 4$. Then, there is a set of $|X| - 3$ quartet trees that defines $\mathcal{T}$.*

It turns out that $|X| - 3$ is the minimum number of quartet trees to define a binary phylogenetic X-tree $\mathcal{T}$. However, not every subset of $\mathcal{Q}(\mathcal{T})$ of size $|X| - 3$ defines $\mathcal{T}$. Moreover, not all such defining subsets can be obtained in the manner described prior to Corollary 6.3.10. Determining which subsets of $\mathcal{Q}(\mathcal{T})$ of size $|X| - 3$ define $\mathcal{T}$ is the main focus of the specialist topic at the end of this chapter.

We remark that some phylogenetic X-trees can be identified with less than $|X| - 3$ quartet trees. For example, the phylogenetic tree with the one non-trivial split $\{1, 2\}|\{3, 4, 5, 6\}$ is identified by the two quartet trees $12|34$ and $12|56$.

We end this section by describing how the concept of compatibility for a collection of semi-labelled trees generalizes the earlier concept of compatibility for a collection of characters. For a character χ on X, let $\mathcal{Q}(\chi)$ denote the set

$$\{aa'|bb' \colon \chi(a) = \chi(a') \neq \chi(b) = \chi(b') \text{ for distinct } a, a', b, b' \in X\}.$$

We leave the proof of Proposition 6.3.11 as an exercise.

Proposition 6.3.11. *Let $\mathcal{C}$ be a collection of characters on X. Then, $\mathcal{C}$ is compatible if and only if the collection $\bigcup_{\chi \in \mathcal{C}} \mathcal{Q}(\chi)$ of quartet trees is compatible. Furthermore, a phylogenetic X-tree has the property that each character is convex on $\mathcal{T}$ if and only if $\mathcal{T}$ displays $\bigcup_{\chi \in \mathcal{C}} \mathcal{Q}(\chi)$.*

Conversely, the concept of compatibility for a collection of characters generalizes the concept of compatibility for a collection of semi-labelled trees. For an X-tree $\mathcal{T}$, let $\mathcal{C}(\mathcal{T})$ denote a collection of binary characters on X such that χ is an element of this collection precisely if $\pi(\chi)$ is a non-trivial split of $\mathcal{T}$. The proof of Proposition 6.3.12 is again left as an exercise.

Proposition 6.3.12. *Let $\mathcal{P}$ be a collection of semi-labelled trees with $\mathcal{L}(\mathcal{P}) = X$. Then, $\mathcal{P}$ is compatible if and only if the collection $\bigcup_{\mathcal{T} \in \mathcal{P}} \mathcal{C}(\mathcal{T})$ of two-state characters on X is compatible.*

6.4 A reconstruction algorithm for rooted trees

For a set of unrooted phylogenetic trees, determining compatibility is an NP-complete problem, as we shall show in Section 6.5. However, for a set of rooted phylogenetic trees, Aho *et al.* (1981) give a polynomial–time algorithm for determining compatibility called BUILD, which we describe in this section.

For a subset X' of X, a rooted X-tree $\mathcal{T}$ *displays a rooted X'-tree $\mathcal{T}'$* if $\mathcal{T}|X'$ is a refinement of $\mathcal{T}'$; that is, $\mathcal{T}' \leq \mathcal{T}|X'$. Furthermore, a set $\mathcal{R}$ of rooted semi-labelled trees is *compatible* if there exists a rooted semi-labelled tree $\mathcal{T}$ that displays every tree in $\mathcal{R}$, in which case $\mathcal{T}$ *displays* $\mathcal{R}$.

A *rooted triple* is a rooted binary phylogenetic tree with label set of size three. The rooted triple with leaves a, b, and c is denoted $ab|c$ if the path from a to b does not intersect the path from c to the root. Note that no distinction is made between $ab|c$ and $ba|c$. Corollary 6.3.8 shows that the basic building blocks of phylogenetic trees are

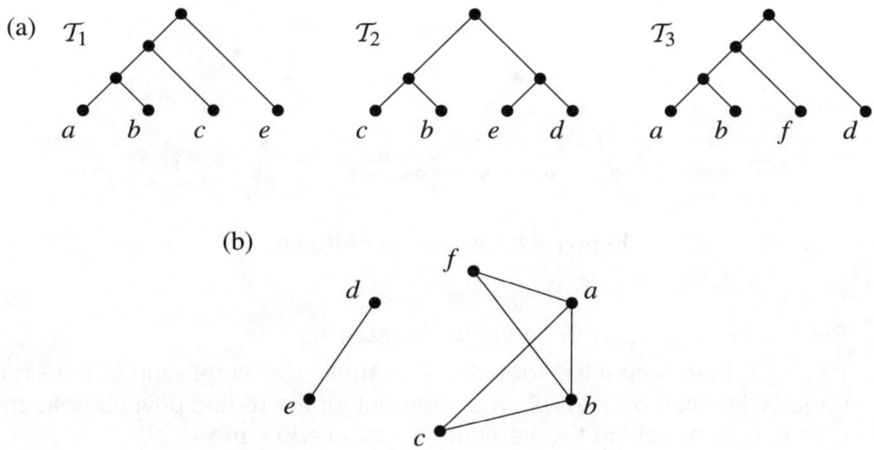

Figure 6.7 (a) $\mathcal{R}$ and (b) $[\mathcal{R}, S]$.

quartet trees. For rooted phylogenetic trees, the basic building blocks are rooted triples. For a rooted phylogenetic X-tree $\mathcal{T}$, we denote the set of restricted rooted triples by $\mathcal{R}(\mathcal{T})$; that is,

$$\mathcal{R}(\mathcal{T}) = \{\mathcal{T}|S: S \subseteq X, |S| = 3, \text{ and } \mathcal{T}|S \text{ is binary}\}.$$

The following result is the rooted analogue of Corollary 6.3.8. We leave the proof to the reader.

Theorem 6.4.1. *Let $\mathcal{T}$ and $\mathcal{T}'$ be rooted phylogenetic X-trees. Then, $\mathcal{R}(\mathcal{T}) \subseteq \mathcal{R}(\mathcal{T}')$ if and only if $\mathcal{T} \leq \mathcal{T}'$. In particular, $\mathcal{R}(\mathcal{T})$ identifies $\mathcal{T}$.*

Before describing BUILD, we define an associated graph that will play a fundamental role in this algorithm. Let $\mathcal{R}$ be a collection of rooted phylogenetic trees, and let S be a subset of the label set of $\mathcal{R}$. Define $[\mathcal{R}, S]$ to be the graph that has vertex set S and an edge joining any two vertices a and b precisely if there exists a $c \in S$ and a $\mathcal{T} \in \mathcal{R}$ such that

$$\mathcal{T}|\{a, b, c\} = ab|c.$$

We call $[\mathcal{R}, S]$ the *clustering graph on S induced by $\mathcal{R}$.*

Example 6.4.2. Suppose that $\mathcal{R} = \{\mathcal{T}_1, \mathcal{T}_2, \mathcal{T}_3\}$ is the collection of rooted phylogenetic trees whose members are shown in Fig. 6.7(a). Then, $[\mathcal{R}, S]$ is the graph shown in Fig. 6.7(b), where $S = \{a, b, \ldots, f\}$.

We now describe BUILD.

Algorithm: BUILD($\mathcal{R}, v, \mathcal{T}$).
Input: A collection $\mathcal{R}$ of rooted phylogenetic trees and a vertex v.
Output: A rooted phylogentic tree $\mathcal{T}$ that displays $\mathcal{R}$ with root vertex v, or the statement $\mathcal{R}$ *is not compatible.*

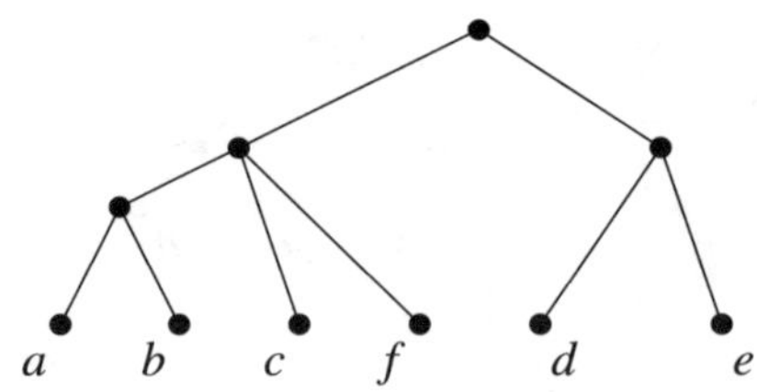

Figure 6.8 An output of BUILD.

1. Set $S = \{x_1, x_2, \dots, x_n\}$ to be the label set of $\mathcal{R}$.
2. If $|S| = 1$, then output the rooted phylogenetic tree consisting of the single vertex v labelled by x_1. If $|S| = 2$, then output the rooted phylogenetic tree obtained by attaching two vertices to v labelled x_1 and x_2.
3. If $|S| \geq 3$, then construct $[\mathcal{R}, S]$.
4. Let $S_1, S_2, \dots, S_k$ denote the vertex sets of the components of $[\mathcal{R}, S]$. If $k = 1$, then stop and output $\mathcal{R}$ *is not compatible*.
5. For each $i \in \{1, 2, \dots, k\}$, call BUILD($\mathcal{R}_i, v_i, \mathcal{T}_i$), where $\mathcal{R}_i$ is the collection of rooted phylogenetic trees obtained from $\mathcal{R}$ by restricting each tree in $\mathcal{R}$ to S_i. If BUILD($\mathcal{R}_i, v_i, \mathcal{T}_i$) outputs a tree, then attach $\mathcal{T}_i$ to v via the edge $\{v_i, v\}$.

To illustrate BUILD, consider the collection $\mathcal{R}$ of rooted phylogenetic trees in Example 6.4.2. Using $\mathcal{R}$ as the input, one can easily check that BUILD outputs the rooted phylogenetic tree shown in Fig. 6.8.

Theorem 6.4.3. *Let $\mathcal{R}$ be a collection of rooted phylogenetic trees. Then* BUILD *applied to $\mathcal{R}$ either*

(i) outputs a rooted phylogenetic tree that displays $\mathcal{R}$ if $\mathcal{R}$ is compatible; or

(ii) outputs the statement '$\mathcal{R}$ is not compatible' *otherwise.*

Proof. First suppose that $\mathcal{R}$ is compatible. We show that under this assumption, BUILD applied to $\mathcal{R}$ outputs a rooted phylogenetic tree. Since $\mathcal{R}$ is compatible, it follows that, for every subset S' of the label set of $\mathcal{R}$, the set

$$\{\mathcal{T}|S' : \mathcal{T} \in \mathcal{R}\}$$

of restricted subtrees is compatible. Therefore, it suffices to show that if $\mathcal{R}'$ is such a set of restricted subtrees with label set S' and $|S'| \geq 2$, then $[\mathcal{R}', S']$ does not consist of a single component.

Assume that $[\mathcal{R}', S']$ consists of a single component. Since $\mathcal{R}'$ is compatible, there exists a rooted phylogenetic S'-tree $\mathcal{T}'$ that displays $\mathcal{R}'$. This, in turn, implies that, as $[\mathcal{R}', S']$ is connected, there exists a pair of elements, a and b say, in S' such that a and b are in distinct maximal clusters of $\mathcal{T}'$, and some tree in $\mathcal{R}'$ induces a rooted triple $ab|c$ where $c \in S'$. But $ab|c$ is not an element of $\mathcal{R}(\mathcal{T}')$; a contradiction. Hence,

$[\mathcal{R}', S']$ has at least two components. It follows that BUILD applied to $\mathcal{R}$ outputs a rooted phylogenetic tree.

Now suppose that BUILD outputs a rooted phylogenetic tree $\mathcal{T}$. In this case, we show that $\mathcal{T}$ displays $\mathcal{R}$. By Theorem 6.4.1, it suffices to show that if $ab|c$ is a restricted rooted triple of a rooted phylogenetic tree in $\mathcal{R}$, then $ab|c$ is a restricted rooted triple of $\mathcal{T}$. Let S be the minimal cluster of $\mathcal{T}$ that contains a, b, and c, and consider $[\mathcal{R}, S]$. Since $ab|c$ is a restricted rooted triple of a rooted phylogenetic tree in $\mathcal{R}$, a and b are in the same component. Furthermore, as S is minimal, c is in a different component of $[\mathcal{R}, S]$ from a and b. Thus, $ab|c$ is a restricted rooted triple of $\mathcal{T}$. This completes the proof of Theorem 6.4.3. □

Following the original approach by Aho *et al.* (1981), the BUILD algorithm can be implemented to run in time $O(nk)$, where $n = |X|$ and $k = |\mathcal{R}|$. However, it is possible to achieve somewhat better running times by a more careful approach, as described by Henzinger *et al.* (1999).

Although the BUILD algorithm applies to rooted phylogenetic trees, it can be modified to apply to certain collections of unrooted phylogenetic trees. A particular case, which we will use later, arises when the label set of each tree in the collection contains a particular element. We leave the proof of the following result as an exercise.

Proposition 6.4.4. *Let $\mathcal{P}$ be a collection of phylogenetic trees and suppose that $x \in \bigcap_{\mathcal{T} \in \mathcal{P}} \mathcal{L}(\mathcal{T})$. For each $\mathcal{T} \in \mathcal{P}$, let $\mathcal{T}_x$ be the rooted phylogenetic tree obtained from $\mathcal{T}$ by deleting the leaf labelled x and its incident edge, and then rooting the resulting tree at the remaining end vertex of this edge. Then, $\mathcal{P}$ is compatible if and only if $\{\mathcal{T}_x \colon \mathcal{T} \in \mathcal{P}\}$ is compatible.*

Lastly, we note that BUILD can also be extended to determine the compatibility of a collection of rooted semi-labelled trees.

6.5 Complexity analysis

The aim of this section is to show that in contrast to the rooted setting, the problem of determining the compatibility of a collection of unrooted phylogenetic trees is an NP-complete problem. To this end, consider the following decision problem.

Decision problem: BI-ROOTED QUARTET COMPATIBILITY.

Instance: A set $\mathcal{Q}$ of quartet trees with the property that the label set of each quartet in $\mathcal{Q}$ contains either x or y, where x and y are distinct elements of $\mathcal{L}(\mathcal{Q})$.

Question: Is $\mathcal{Q}$ compatible?

In view of Proposition 6.4.4, it might be expected that there is an extension of the BUILD algorithm to determine BI-ROOTED QUARTET COMPATIBILITY in polynomial time. However, the next theorem, due to Ng *et al.* (2000), states that this problem is NP-complete.

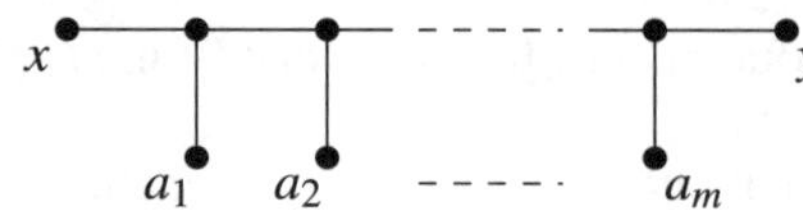

Figure 6.9 An *xy*-caterpillar tree.

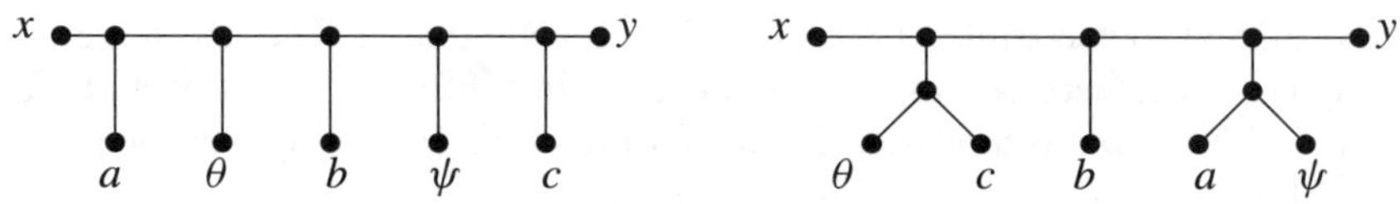

Figure 6.10 The only trees that display $\{x\theta|by, xb|\psi y, xa|c\theta, yc|a\psi\}$.

Theorem 6.5.1. BI-ROOTED QUARTET COMPATIBILITY *is* NP-*complete.*

Before proving Theorem 6.5.1, we present two necessary lemmas on caterpillar trees. The proofs of both lemmas are left as exercises for the reader.

We shall call a caterpillar $\mathcal{T} = (T; \phi)$ an *xy-caterpillar* if $x, y \in \mathcal{L}(\mathcal{T})$, and $\phi(x)$ and $\phi(y)$ are the end vertices of a longest path in T. We write $xa_1|a_2 \cdots a_{m-1}|a_m y$ to denote the *xy*-caterpillar shown in Fig. 6.9. The two caterpillars appearing in the statement of Lemma 6.5.2 are shown in Fig. 6.5. The reason for the two alphabets will be given in the proof of Theorem 6.5.1.

Lemma 6.5.2. *Let* $\mathcal{Q} = \{x\theta|by, xb|\psi y, xa|c\theta, yc|a\psi\}$ *be a collection of quartet trees. Then,*

$$\text{co}(\mathcal{Q}) = \{xa|\theta b\psi|cy, c\theta|xby|a\psi\}.$$

Lemma 6.5.3 is due to Steel (1992).

Lemma 6.5.3. *If a collection* $\mathcal{P}$ *of xy-caterpillars is compatible, then there exists an xy-caterpillar that displays* $\mathcal{P}$.

In the proof of Theorem 6.5.1, we freely use the following two facts. For a pair of elements $x, y \in X$, there is a canonical one-to-one correspondence between the set of *xy*-caterpillar trees on X and the set of linear orderings on $X - \{x, y\}$. Furthermore, if $\mathcal{T}$ is an *xy*-caterpillar with label set X and $A \subseteq X$ with $x, y \in A$, then $\mathcal{T}|A$ is also an *xy*-caterpillar.

Proof of Theorem 6.5.1. BI-ROOTED QUARTET COMPATIBILITY is clearly in NP since, given a semi-labelled tree $\mathcal{T}$, we can verify whether $\mathcal{T}$ displays each quartet in $\mathcal{Q}$ in polynomial time.

Now consider the following NP-complete problem (see Garey and Johnson 1979, p. 279).

Decision problem: BETWEENNESS.
Instance: A finite set A and a collection $\mathcal{I}$ of ordered triples (a, b, c) of distinct elements of A.

Question: Is there a betweenness ordering f of A for $\mathcal{I}$? In other words, is there a one-to-one function $f: A \rightarrow \{1, 2, \ldots, |A|\}$ such that, for each $(a, b, c) \in \mathcal{I}$, either $f(a) < f(b) < f(c)$ or $f(c) < f(b) < f(a)$?

Without loss of generality, we may assume that each element of A appears in at least one triple of $\mathcal{I}$. To complete the proof of Theorem 6.5.1, we show that BETWEENNESS transforms to BI-ROOTED QUARTET COMPATIBILITY.

Suppose that A and $\mathcal{I} = \{(a_i, b_i, c_i): i \in \{1, 2, \ldots, k\}\}$ is a given instance of BETWEENNESS. For all $i \in \{1, 2, \ldots, k\}$, let

$$\mathcal{Q}_i = \{x\theta_i|b_iy, xb_i|\psi_iy, xa_i|c_i\theta_i, yc_i|a_i\psi_i\},$$

where $x, y, \theta_1, \ldots, \theta_k, \psi_1, \ldots, \psi_k$ are distinct elements and

$$\{x, y, \theta_1, \ldots, \theta_k, \psi_1, \ldots, \psi_k\} \cap A$$

is empty. Let $\mathcal{Q}(\mathcal{I}) = \bigcup_{i=1}^{k} \mathcal{Q}_i$. Note that each quartet in $\mathcal{Q}(\mathcal{I})$ has a leaf labelled by either x or y. We now show that there is a betweenness ordering of A for $\mathcal{I}$ if and only if $\mathcal{Q}(\mathcal{I})$ is compatible. The reason for the two alphabets in defining $\mathcal{Q}(\mathcal{I})$ is to emphasize the fact that the elements in $\{\theta_i, \psi_i: i \in \{1, 2, \ldots, k\}\}$ are extraneous labels that are used to ensure the desired betweenness ordering given below.

Suppose that $\mathcal{Q}(\mathcal{I})$ is compatible. Then, there is a semi-labelled tree $\mathcal{T}$ that displays $\mathcal{Q}(\mathcal{I})$. For each $i \in \{1, 2, \ldots, k\}$, let $\mathcal{T}_i = \mathcal{T}|\{x, y, a_i, b_i, c_i\}$. By Lemma 6.5.2, $\mathcal{T}_i$ is either $xa_i|b_i|c_iy$ or $xc_i|b_i|a_iy$. Consider the collection $\mathcal{P} = \{\mathcal{T}_1, \mathcal{T}_2, \ldots, \mathcal{T}_k\}$ of xy-caterpillar trees. Since $\mathcal{T}$ displays $\mathcal{P}$, the set $\mathcal{P}$ is compatible and so, by Lemma 6.5.3, there exists an xy-caterpillar $\mathcal{T}'$ that displays $\mathcal{P}$. One can now easily check that the order of the elements of $\mathcal{L}(\mathcal{T}') - \{x, y\}$ from x to y along $\mathcal{T}'$ provides a betweenness ordering of A for $\mathcal{I}$.

For the converse, suppose that f is a betweenness ordering of A for $\mathcal{I}$. Let $\mathcal{T}'$ be the xy-caterpillar with label set $A \cup \{x, y\}$ such that the order of the elements of A from x to y is as prescribed by f. We next describe the construction of a binary phylogenetic tree from $\mathcal{T}'$ that displays $\mathcal{Q}(\mathcal{I})$.

Let $\mathcal{T}_0 = \mathcal{T}'$ and, for all $i \in \{1, 2 \ldots, k\}$, let $\mathcal{T}_i$ be a binary phylogenetic tree on $\mathcal{L}(\mathcal{T}_{i-1}) \cup \{\theta_i, \psi_i\}$ such that $\mathcal{T}_{i-1}$ is a restricted subtree of $\mathcal{T}_i$, and

(i) if $\mathcal{T}'|\{x, y, a_i, b_i, c_i\} = xa_i|b_i|c_iy$, then $\mathcal{T}_i|\{x, y, a_i, b_i, c_i, \theta_i, \psi_i\}$ is the caterpillar $xa_i|\theta_ib_i\psi_i|c_iy$;

(ii) if $\mathcal{T}'|\{x, y, a_i, b_i, c_i\} = xc_i|b_i|a_iy$, then $\mathcal{T}_i|\{x, y, a_i, b_i, c_i, \theta_i, \psi_i\}$ is the caterpillar $c\theta_i|xby|a\psi_i$.

It is easily checked that, for all i, such a $\mathcal{T}_i$ exists. Furthermore, it follows by Lemma 6.5.2 that $\mathcal{T}_k$ displays $\mathcal{Q}(\mathcal{I})$, and so BETWEENNESS transforms to BI-ROOTED QUARTET COMPATIBILITY. Moreover, this transformation can be done in polynomial time, thus completing the proof of Theorem 6.5.1. □

Now consider the following two decision problems:

Decision problem: TREE COMPATIBILITY.
Instance: A set $\mathcal{P}$ of phylogenetic trees.
Question: Is $\mathcal{P}$ compatible?

and

Decision problem: CHARACTER COMPATIBILITY.
Instance: A set $\mathcal{C}$ of characters
Question: Is $\mathcal{C}$ compatible?

As an immediate consequence of Theorem 6.5.1, TREE COMPATIBILITY is an NP-complete problem (Steel 1992). Furthermore, by combining Proposition 6.3.12 with Theorem 6.5.1, CHARACTER COMPATIBILITY is also NP-complete (Bodlaender *et al.* 1992; Steel 1992). This gives Corollary 6.5.4.

Corollary 6.5.4. TREE COMPATIBILITY *and* CHARACTER COMPATIBILITY *are both* NP-*complete.*

We remark here that based on the BUILD algorithm, TREE COMPATIBILITY has a polynomial–time solution if the cardinality of $\mathcal{P}$ is bounded (Steel 1992).

We end this section by considering the computational complexity of a further problem arising in the reconstruction of phylogenetic trees from restricted subtrees. Using the inductive proof of Theorem 6.3.9, one can reconstruct a phylogenetic X-tree $\mathcal{T}$ using $|X| - 3$ restricted quartet subtrees of $\mathcal{T}$. However, these subtrees are selected on the basis of knowing $\mathcal{T}$ in advance. This suggests the following algorithmic problem. Suppose that $\mathcal{T}$ is unknown and that one wishes to perform a sequence of experiments to determine $\mathcal{T}$. To this end, suppose that one obtains correct answers to a sequence of carefully posed questions of the form

'What is $\mathcal{T}|S$?'

where S is a subset of X of size four chosen according to the information one has so far built up from the answers to the previous questions. The algorithmic problem is to construct a polynomial–time algorithm that determines a sequence of such questions and that simultaneously reconstructs $\mathcal{T}$. Pearl and Tarsi (1986) and Kannan *et al.* (1996) constructed such an algorithm for which the number of questions asked is $O(n^2)$, where $n = |X|$. Furthermore, in the case $\mathcal{T}$ is known to be binary, at most $O(n \log(n))$ questions are required.

6.6 Supertree methods

In Sections 6.4 and 6.5, we were interested in determining the compatibility of a set of phylogenetic trees that are either all rooted or all unrooted. In practice, such sets are often not compatible. Nevertheless, biologists would still like to reconstruct a rooted or an unrooted phylogenetic tree that in some sense best represents the evolutionary information contained in the original set. In this section, we introduce the notion of a 'supertree' method, the collective name for such reconstructions, and discuss some of the inherent limitations of these methods. For a general survey of supertree methods, we refer the reader to Bininda-Emonds *et al.* (2002).

For a collection $\mathcal{P}$ of phylogenetic trees that are either all rooted or all unrooted, a *supertree method* will denote any function that associates to the *input* collection $\mathcal{P}$ an *output* phylogenetic X-tree that is rooted or unrooted, respectively, where $X \subseteq \mathcal{L}(\mathcal{P})$. The output phylogenetic tree is sometimes called a *supertree*. Observe that the consensus and the maximum agreement subtree methods of Sections 3.6 and 6.2 are particular supertree methods. One of the main biological motivations for developing supertree methods is to compare species that have never been collectively included on a single tree.

A popular procedure for constructing supertrees in biology is *matrix representation with parsimony analysis* (MRP). For a collection $\mathcal{P}$ of unrooted phylogenetic trees, let $\mathcal{C}(\mathcal{P})$ denote the concatenation of the sequences $(\chi_\sigma\colon \sigma \in \Sigma(\mathcal{T}))$ of two-state characters on $\mathcal{L}(\mathcal{P})$ over all phylogenetic trees $\mathcal{T}$ in $\mathcal{P}$, where, for all $\sigma = A|B \in \Sigma(\mathcal{T})$, $\chi_\sigma\colon \mathcal{L}(\mathcal{T}) \rightarrow \{0, 1\}$ is defined by

$$\chi_\sigma(x) = \begin{cases} 0, & \text{if } x \in A, \\ 1, & \text{if } x \in B, \end{cases}$$

for all $x \in \mathcal{L}(\mathcal{T})$. The MRP *spectrum* for $\mathcal{P}$ is the collection of maximum parsimony trees for $\mathcal{C}(\mathcal{P})$, and the *strict consensus* MRP *supertree* for $\mathcal{P}$ is the strict consensus tree of the MRP spectrum for $\mathcal{P}$.

If $\mathcal{C}(\mathcal{P})$ is compatible in the sense defined in Section 4.6, then every phylogenetic tree $\mathcal{T}$ on which all the characters of $\mathcal{C}(\mathcal{P})$ are convex is a maximum parsimony tree for $\mathcal{P}$. Thus, in this case, $\mathcal{T}$ is in the MRP spectrum for $\mathcal{P}$. This observation forms the basis of Proposition 6.6.1 (Thorley 2000).

Proposition 6.6.1. *Let $\mathcal{P}$ be a compatible collection of phylogenetic trees. Then, the* MRP *spectrum for $\mathcal{P}$ is precisely* co($\mathcal{P}$).

For a compatible collection $\mathcal{P}$ of phylogenetic trees, the strict consensus tree of co($\mathcal{P}$) has become known in biology as the *strict consensus supertree* for $\mathcal{P}$. The proof of Proposition 6.6.2 (Steel 1992) is based on the BUILD algorithm and Proposition 6.4.4, and it is left as an exercise.

Proposition 6.6.2. *Let $\mathcal{P}$ be a compatible collection of phylogenetic trees such that $\bigcap_{\mathcal{T}\in\mathcal{P}} \mathcal{L}(\mathcal{T}) \neq \emptyset$. Then, the strict consensus supertree for $\mathcal{P}$ can be constructed by a polynomial–time algorithm.*

Propositions 6.6.1 and 6.6.2 have a useful computational consequence. Proposition 6.6.1 implies that when $\mathcal{P}$ is a compatible collection of phylogenetic trees, the strict consensus supertree for $\mathcal{P}$ and the strict consensus MRP supertree for $\mathcal{P}$ are identical. Consequently, whenever $\mathcal{P}$ satisfies the hypothesis of Proposition 6.6.2, one can construct the strict consensus MRP supertree for $\mathcal{P}$ in polynomial time regardless of the (possibly exponential) size of co($\mathcal{P}$).

Any supertree method has an inherent limitation in what it can achieve. In particular, although one can easily list a number of desirable properties for such a method, we demonstrate by a simple example that no method can achieve all of these simultaneously. Some limitations to supertree methods have been described by McMorris (1985), but the ones we describe here, due to Steel *et al.* (2000), are even simpler.

Consider the following three possible properties of a supertree method:

(S1) The supertree method can be applied to an unordered set of input trees.

(S2) If we rename all of the elements of $\mathcal{L}(\mathcal{P})$ and then apply the supertree method to the new set of input trees, the output tree is simply the original output tree but with the elements of $\mathcal{L}(\mathcal{P})$ renamed accordingly.

(S3) If the set of input trees is compatible, then the output tree displays each of the input trees.

Property (S1) specifies that, in case we have equal confidence in all of the input trees, the order in which we input these trees does not matter. Property (S2) specifies that the manner in which we name or label our species does not bias the outcome. Formally, a supertree method satisfying (S1) is said to be *symmetric* and one satisfying (S2) is *equivariant with respect to permutations of* $\mathcal{L}(\mathcal{P})$.

For a supertree method, properties (S1) and (S2) seem quite compelling, while (S3) also appears desirable. However, for a collection of unrooted phylogenetic trees, all three properties cannot be simultaneously achieved.

Proposition 6.6.3. *No supertree method for unrooted phylogenetic trees simultaneously satisfies* (S1), (S2), *and* (S3).

Proof. We prove the proposition by giving an explicit example. Let $\mathcal{Q}_1$ denote the compatible triple $(12|45, 34|16, 56|23)$ of quartet trees, and suppose that a particular supertree method satisfies properties (S1) and (S2). We shall show that (S3) fails for this method when applied to $\mathcal{Q}_1$.

Let $\mathcal{Q}_2$ denote the triple $(16|43, 54|12, 32|65)$ of quartet trees. Observe that $\mathcal{Q}_2$ is obtained from $\mathcal{Q}_1$ by simultaneously interchanging 2 and 6, and interchanging 3 and 5 and, furthermore, up to the ordering of the elements, $\mathcal{Q}_1$ is the same as $\mathcal{Q}_2$. Thus, as (S1) and (S2) hold, the output tree resulting from the application of our particular supertree method to $\mathcal{Q}_1$ must remain unchanged under the simultaneous interchanging of 2 and 6, and the interchanging of 3 and 5. However, if (S3) is to hold, then the output tree must be one of the trees shown in Fig. 6.11, as these are the only phylogenetic trees with label set $\mathcal{L}(\mathcal{Q}_1)$ that display $\mathcal{Q}_1$. Since neither of these trees remain unchanged under the above interchanges, we deduce that our supertree method does not satisfy (S3). □

Proposition 6.6.3 suggests that the 'supertree problem' does not have a satisfactory solution. However, in contrast to the unrooted setting, Semple and Steel (2000) showed that a suitable modification of the BUILD algorithm provides a supertree method for rooted phylogenetic trees that satisfies (S1), (S2), and (S3) as well as the following two desirable properties:

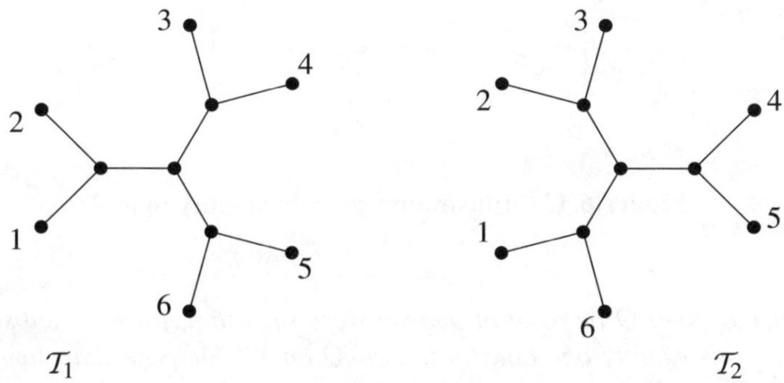

Figure 6.11 Two phylogenetic trees that display 12|45, 34|16, and 56|23.

(S4) Each element that occurs in the label set of at least one input tree occurs in the label set of the output tree.

(S5) The computational complexity of the supertree method is polynomial in $|\mathcal{L}(\mathcal{P})|$.

Property (S4) states that no elements of $\mathcal{L}(\mathcal{P})$ are discarded, even if the input trees are not compatible. Property (S5) is certainly useful if one hopes to compute an output tree in realistic time when the label sets of the input trees are large and the input trees are numerous.

6.7 Closure and correction operations

In this section, we describe two approaches to tree construction that exploit the redundancy that arises by virtue of the fact that, for a phylogenetic tree $\mathcal{T}$, many proper subsets of $\mathcal{Q}(\mathcal{T})$ determine $\mathcal{T}$. These approaches are called 'quartet cleaning' and 'quartet rules'.

Quartet cleaning. Applying some method to data on a set X of species, suppose that we have constructed a set $\mathcal{Q}$ of quartet trees. Furthermore, suppose that, for each subset X' of X of size four, there is exactly one quartet tree in $\mathcal{Q}$ on X'. In an ideal setting, we would like $\mathcal{Q}$ to coincide with $\mathcal{Q}(\mathcal{T})$ for some phylogenetic X-tree $\mathcal{T}$ (which would imply that $\mathcal{T}$ is binary) and that one could easily construct $\mathcal{T}$ from $\mathcal{Q}$. However, this is unlikely to be the case and a more realistic hope is that $\mathcal{Q}$ will contain a sufficiently small number of quartet trees that do not lie in $\mathcal{Q}(\mathcal{T})$, and so redundancy might still allow one to recover $\mathcal{T}$. Implementations of this strategy have become known as the 'quartet cleaning' approach to tree reconstruction, introduced by Jiang *et al.* (1998). The advantage of this approach is that it allows a controlled degree of error in the set of input trees.

Given a phylogenetic X-tree $\mathcal{T}$, let e be an interior edge of $\mathcal{T}$ and let $A(e)|B(e)$ denote the X-split corresponding to edge e. Let

$$\mathcal{Q}_{\mathcal{T}}(e) = \{aa'|bb' \colon \text{for distinct } a, a' \in A(e) \text{ and distinct } b, b' \in B(e)\}.$$

Theorem 6.7.1 is due to Jiang *et al.* (1998).

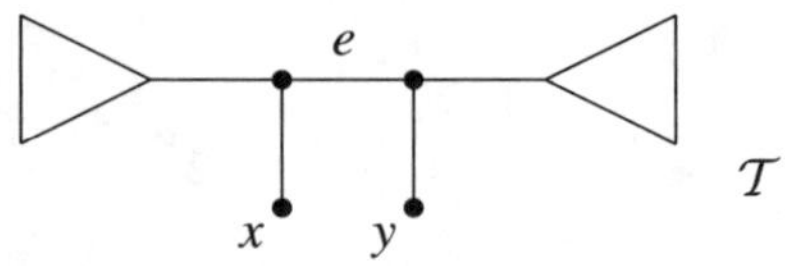

Figure 6.12 Illustrating strict inequality in (6.2).

Theorem 6.7.1. *Let $\mathcal{Q}$ be a set of quartet trees such that, for each subset X' of X of size four, there is exactly one quartet tree in $\mathcal{Q}$ on X'. Suppose that, for some binary phylogenetic X-tree $\mathcal{T}$,*

$$|\mathcal{Q}_{\mathcal{T}}(e) - \mathcal{Q}| < \tfrac{1}{2}(|A(e)| - 1)(|B(e)| - 1), \tag{6.2}$$

for each interior edge e of $\mathcal{T}$. Then, $\mathcal{T}$ is the only binary phylogenetic X-tree satisfying this property. Furthermore, $\mathcal{T}$ can be constructed from $\mathcal{Q}$ by a polynomial–time algorithm.

The inequality in (6.2) must be strict to ensure the validity of Theorem 6.7.1. To see this, let $\mathcal{T}$ be a binary phylogenetic tree with an interior edge e as shown schematically in Fig. 6.12, and suppose that $|A(e)|$ and $|B(e)|$ are both odd. Let $\mathcal{T}'$ denote the tree obtained from $\mathcal{T}$ by interchanging the leaf labels x and y, and let e' denote the interior edge of $\mathcal{T}'$ that separates the leaf labelled x from the leaf labelled y. Clearly, both $\mathcal{Q}_{\mathcal{T}}(e) - \mathcal{Q}_{\mathcal{T}'}(e')$ and $\mathcal{Q}_{\mathcal{T}'}(e') - \mathcal{Q}_{\mathcal{T}}(e)$ have cardinality $(|A(e)| - 1)(|B(e)| - 1)$. Let $\mathcal{Q}$ denote the set obtained from $\mathcal{Q}(\mathcal{T})$ by replacing half of the quartet trees in $\mathcal{Q}_{\mathcal{T}}(e) - \mathcal{Q}_{\mathcal{T}'}(e')$ by half of the quartet trees in $\mathcal{Q}_{\mathcal{T}'}(e') - \mathcal{Q}_{\mathcal{T}}(e)$. Then,

$$|\mathcal{Q}_{\mathcal{T}}(e) - \mathcal{Q}| \leq \tfrac{1}{2}(|A(e)| - 1)(|B(e)| - 1),$$

for all interior edges e of $\mathcal{T}$, and

$$|\mathcal{Q}_{\mathcal{T}'}(e') - \mathcal{Q}| \leq \tfrac{1}{2}(|A(e')| - 1)(|B(e')| - 1),$$

for all interior edges e' of $\mathcal{T}'$. It follows that the strict bound in Theorem 6.7.1 cannot be weakened to an inequality.

Quartet rules. Quartet rules are based on the observation that any phylogenetic tree that displays a certain collection of quartet trees must necessarily display other quartet trees that are not in the collection. Consequently, given an arbitrary collection $\mathcal{Q}$ of quartet trees, it is possible to enlarge $\mathcal{Q}$ to provide more information about phylogenetic trees that display $\mathcal{Q}$. In particular, this enlarged set of quartet trees may provide information on the compatibility of $\mathcal{Q}$.

For a set $\mathcal{Q}$ of quartet trees, we write $\mathcal{Q} \vdash ab|cd$ precisely if every phylogenetic tree that displays $\mathcal{Q}$ also displays ab|cd. We call the statement $\mathcal{Q} \vdash ab|cd$ a *quartet rule*.

Colonius and Schulze (1981) originally introduced quartet rules in the context of psychology; they were also used for tree reconstruction in stemmatology, a branch of linguistics that deals with the filiation of manuscripts (Dekker 1986). Subsequently, quartet rules have been developed, studied, and applied by a number of authors.

Three of the simplest quartet rules are listed below, each of which can be easily checked.

$$\{ab|cd, ab|ce\} \vdash ab|de \qquad \text{(dc1)}$$

$$\{ab|cd, ac|de\} \vdash ab|ce \qquad \text{(dc2)}$$

$$\{ab|cd, ab|ef, ce|df\} \vdash ab|df \qquad \text{(tc)}$$

Observe that rule (dc2) can also be used to derive the quartet trees $ab|de$ and $bc|de$.

If $\mathcal{Q} \vdash ab|cd$ and no proper subset $\mathcal{Q}'$ of $\mathcal{Q}$ has the property that $\mathcal{Q}' \vdash ab|cd$, then the *order* of the rule $\mathcal{Q} \vdash ab|cd$ is equal to the cardinality of $\mathcal{Q}$. Thus, the order of (dc1) and (dc2) are both two, while the order of (tc) is three.

The *dyadic closure* of an arbitrary collection $\mathcal{Q}$ of quartet trees, denoted $\mathrm{cl}_2(\mathcal{Q})$, is the minimal set of quartet trees that contains $\mathcal{Q}$ and has the property that if $\mathrm{cl}_2(\mathcal{Q}) \vdash ab|cd$ using either rule (dc1) or (dc2), then $ab|cd \in \mathrm{cl}_2(\mathcal{Q})$. The *semi-dyadic closure* of $\mathcal{Q}$, denoted $\mathrm{scl}_2(\mathcal{Q})$, is defined in the analogous way as $\mathrm{cl}_2(\mathcal{Q})$ except that we only allow (dc2) to be used. The dyadic closure of $\mathcal{Q}$ can be constructed in polynomial time from $\mathcal{Q}$ by repeatedly applying the quartet rules (dc1) and (dc2) until the sequence of sets so obtained stabilizes. Similarly, the semi-dyadic closure can be efficiently obtained by repeatedly applying (dc2).

The motivation for using quartet rules on a set $\mathcal{Q}$ of quartet trees is that they may either reveal that $\mathcal{Q}$ is incompatible by exhibiting two or more quartet trees on the same label set (e.g. $xy|wz$ and $xz|wy$), or generate enough quartet trees to easily construct a phylogenetic tree that displays $\mathcal{Q}$. This is explored further in the specialist topic of this chapter.

In addition to the 'triadic' quartet rule (tc), Dekker (1986) described some other, more complicated rules, and conjectured the existence of rules of arbitrarily high order and the non-existence of a finite set of rules for determining compatibility. Bryant and Steel (1995) settled both conjectures affirmatively.

For a compatible set $\mathcal{Q}$ of quartet trees, let

$$\mathrm{cl}(\mathcal{Q}) = \bigcap_{\mathcal{T} \in \mathrm{co}(\mathcal{Q})} \mathcal{Q}(\mathcal{T}).$$

Thus, $\mathcal{Q} \vdash ab|cd$ is a quartet rule if and only if $ab|cd \in \mathrm{cl}(\mathcal{Q})$. The set $\mathrm{cl}(\mathcal{Q})$ is called the *closure* of $\mathcal{Q}$. A compatible set of quartet trees $\mathcal{Q}$ is *closed* if $\mathrm{cl}(\mathcal{Q}) = \mathcal{Q}$.

Using the theory surrounding the BUILD algorithm, Bryant and Steel (1995) established the following result, which we state without proof.

Theorem 6.7.2. *For each $k > 2$, there exists a compatible set $\mathcal{Q}$ of k quartet trees that is not closed, even though every proper subset is closed. Thus, for all $k > 2$, there exists a quartet rule of order k.*

By Theorem 6.5.1, determining the compatibility of a collection $\mathcal{Q}$ of quartet trees is an NP-complete problem. Consequently, one cannot expect a polynomial–time algorithm for this determination. However, using (dc1), the next result shows that we can certainly do much better than checking all possible binary phylogenetic trees to see if one of them displays $\mathcal{Q}$.

Proposition 6.7.3. *Let $\mathcal{Q}$ be a set of k quartet trees. Then, the compatibility of $\mathcal{Q}$ can be determined by an algorithm that runs in $O(nk2^k)$ time, where $n = |\mathcal{L}(\mathcal{Q})|$.*

Proof. Let x be an element not in $\mathcal{L}(\mathcal{Q})$, and, for each $q = ab|cd$ in $\mathcal{Q}$, let $S_1(q) = \{ax|cd, bx|cd\}$ and $S_2(q) = \{ab|cx, ab|dx\}$. It follows from (dc1) that, for each $q \in \mathcal{Q}$, every phylogenetic tree that displays either $S_1(q)$ or $S_2(q)$ also displays q. Using BUILD one can determine the compatibility of

$$\bigcup_{q \in \mathcal{Q}} S_{\pi(q)}(q)$$

for each of the 2^k functions $\pi\colon \mathcal{Q} \rightarrow \{1, 2\}$. Furthermore, it is easily checked that $\mathcal{Q}$ is compatible if and only if $\bigcup_{q \in \mathcal{Q}} S_{\pi(q)}(q)$ is compatible for some choice of π. The proposition now follows. □

We end this section by describing a closure operation that is defined on collections of splits of subsets of X, due to Meacham (1983). A split of a subset X' of X is called a *partial split* of X. A closure operation on partial splits of X can be obtained by repeatedly applying the following rule: we replace the pair of partial splits $A_1|B_1$ and $A_2|B_2$ by the pair of enlarged partial splits $(A_1 \cup A_2)|B_1$ and $A_2|(B_1 \cup B_2)$, whenever $A_1 \cap A_2$, $A_1 \cap B_2$, and $B_1 \cap B_2$ are all non-empty. The main property of Meacham's closure operation is described in Exercise 17. There is a close connection between this operation and the semi-dyadic closure operation. For details of this connection, see Semple and Steel (2001).

6.8 Defining a tree by subtrees

In this section, we consider the problem of determining if, for a collection $\mathcal{P}$ of phylogenetic trees, $|\text{co}(\mathcal{P})| = 1$. This is a natural problem in applications if one hopes to identify some 'true' underlying tree based on a strategy of combining trees using compatibility.

Definition 6.8.1. A collection $\mathcal{P}$ of semi-labelled trees is *definitive* if $|\text{co}(\mathcal{P})| = 1$.

Observe that in Definition 6.8.1, if $\text{co}(\mathcal{P}) = \{\mathcal{T}\}$ for some semi-labelled tree $\mathcal{T}$, then $\mathcal{T}$ is a binary phylogenetic tree and $\mathcal{L}(\mathcal{T}) = \mathcal{L}(\mathcal{P})$.

Example 6.8.2. Consider the phylogenetic tree $\mathcal{T}$ shown in Fig. 6.13. The set

$$\mathcal{Q}_1 = \{12|36, 13|46, 24|57, 25|67\}$$

defines $\mathcal{T}$. To see this, first observe that $\mathcal{T}$ displays $\mathcal{Q}_1$. Therefore, it suffices to show that $\mathcal{Q}_1$ is definitive. Both $\{12|36, 13|46\}$ and $\{24|57, 25|67\}$ are definitive. Furthermore, it is

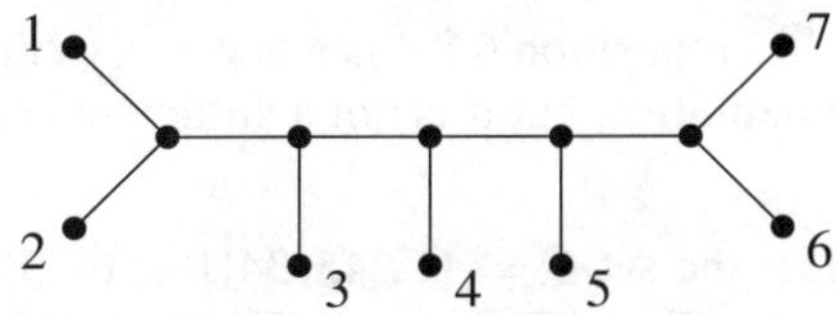

Figure 6.13 The phylogenetic tree defined by $\mathcal{Q}_1$.

easily seen that $\mathcal{Q}_1$ is definitive if and only if the set consisting of the binary phylogenetic trees defined by $\{12|36, 13|46\}$ and $\{24|57, 25|67\}$ is definitive. A straightforward check shows that the latter holds, so $\mathcal{Q}_1$ is indeed definitive and therefore defines $\mathcal{T}$.

A necessary condition for a collection of phylogenetic trees to be definitive is stated after Definition 6.8.3.

Definition 6.8.3. Let $\mathcal{T} = (T; \phi)$ be a phylogenetic tree and let $xy|wz \in \mathcal{Q}(\mathcal{T})$. An interior edge e of T is *distinguished* by $xy|wz$ if e is the only interior edge of T that has the property that $\{\phi(x), \phi(y)\}$ and $\{\phi(w), \phi(z)\}$ are subsets of the vertex sets of different components of $T \backslash e$. More generally, let $\mathcal{T}'$ be a phylogenetic tree that is displayed by $\mathcal{T}$. An interior edge e' of $\mathcal{T}'$ *distinguishes* an interior edge e of $\mathcal{T}$ if e and e' are both distinguished by a particular quartet tree.

To illustrate Definition 6.8.3, the quartet tree $12|35$ distinguishes the edge e_1 in the phylogenetic tree $\mathcal{T}$ shown in Fig. 6.13, but $12|45$ does not distinguish any edge in $\mathcal{T}$.

Proposition 6.8.4. *Let $\mathcal{P}$ be a collection of phylogenetic trees and suppose that $\mathcal{P}$ defines $\mathcal{T}$. Then, every interior edge of $\mathcal{T}$ is distinguished by an interior edge of at least one tree in $\mathcal{P}$. Moreover, if e is an interior edge of a tree in $\mathcal{P}$, then e distinguishes at most one interior edge of $\mathcal{T}$.*

Proof. Suppose that some interior edge e of $\mathcal{T}$ is not distinguished by at least one interior edge of some tree in $\mathcal{P}$. Then, it is easily checked that $\mathcal{T}/e$ also displays $\mathcal{P}$, thereby contradicting the assumption that $\mathcal{P}$ defines $\mathcal{T}$. We leave the proof of the second part of the lemma as an exercise for the reader. □

Proposition 6.8.4 provides useful information about definitive collections of phylogenetic trees. Suppose that $\mathcal{P} = \{\mathcal{T}_1, \mathcal{T}_2, \ldots, \mathcal{T}_k\}$ is a collection of phylogenetic trees that defines $\mathcal{T}$. It follows by Proposition 6.8.4 that

$$\sum_{i=1}^{k} |\mathring{E}(\mathcal{T}_i)| \geq |\mathring{E}(\mathcal{T})|.$$

Furthermore, provided $|\mathcal{L}(\mathcal{T})| \geq 4$, it follows by Corollary 6.3.10 that there is a set of $|\mathring{E}(\mathcal{T})|$ quartet trees that defines $\mathcal{T}$. Hence, the size of a minimum set of quartet trees that defines $\mathcal{T}$ is equal to $|\mathring{E}(\mathcal{T})|$. Thus, the set $\mathcal{Q}_1$ in Example 6.8.2 is a minimum-sized set of quartet trees that defines the binary phylogenetic tree shown in Fig. 6.13. We shall return to the latter consequence of Proposition 6.8.4 at the end of this section.

The condition given in Proposition 6.8.4 is a necessary condition for a collection of phylogenetic trees to be definitive, but it is not a sufficient condition as Example 6.8.5 shows.

Example 6.8.5. Consider the set $\mathcal{Q} = \{12|45, 34|16, 56|23\}$ of quartet trees, and the two binary phylogenetic trees $\mathcal{T}_1$ and $\mathcal{T}_2$ shown in Fig. 6.11. Now, $\mathcal{T}_1$ displays $\mathcal{Q}$, $\mathcal{L}(\mathcal{T}) = \mathcal{L}(\mathcal{Q})$, and each interior edge of $\mathcal{T}_1$ is distinguished by a quartet tree in $\mathcal{Q}$. However, $\mathcal{Q}$ does not define $\mathcal{T}_1$, since $\mathcal{T}_2$ also displays $\mathcal{Q}$.

We now describe a sufficient—but not necessary—condition for a collection of phylogenetic trees to define a binary phylogenetic X-tree $\mathcal{T}$. Letting $n = |X|$, a subset $\mathcal{Q}$ of $\mathcal{Q}(\mathcal{T})$ is called a *leaf elimination basis* for $\mathcal{T}$ if the following three conditions hold:

(i) $\mathcal{Q}$ has size $n - 3$;
(ii) $\mathcal{Q}$ distinguishes each interior edge of $\mathcal{T}$; and
(iii) $\mathcal{Q}$ can be ordered $q_1, q_2, \ldots, q_{n-3}$ so that, for each $i \in \{2, 3, \ldots, n-3\}$,

$$\left| \mathcal{L}(q_i) - \bigcup_{j<i} \mathcal{L}(q_j) \right| = 1. \tag{6.3}$$

Applying the inductive construction used in the proof of Theorem 6.3.9, one can establish the following result.

Proposition 6.8.6. *Let $\mathcal{T}$ be a binary phylogenetic tree and let $\mathcal{Q}$ be a subset of $\mathcal{Q}(\mathcal{T})$. If $\mathcal{Q}$ contains a leaf elimination basis for $\mathcal{T}$, then $\mathcal{Q}$ defines $\mathcal{T}$.*

For a binary phylogenetic X-tree $\mathcal{T}$, not every subset of $\mathcal{Q}(\mathcal{T})$ of size $|X| - 3$ that defines $\mathcal{T}$ need be a leaf elimination basis for $\mathcal{T}$; for in Example 6.8.2, there is no ordering of the elements of $\mathcal{Q}_1$ that satisfies (6.3).

Using results from Section 4.7, it is possible to give a graph-theoretic characterization for when a collection of phylogenetic trees is definitive, as we now describe.

For a collection $\mathcal{P}$ of phylogenetic trees, let $X = \mathcal{L}(\mathcal{P})$ and let $\mathcal{C}(\mathcal{P})$ denote the set

$$\mathcal{C}(\mathcal{P}) = \bigcup_{\mathcal{T} \in \mathcal{P}} \{\chi_\sigma : \sigma \in \Sigma(\mathcal{T})\}$$

of two-state characters on X, where, for each $\sigma = A|B \in \Sigma(\mathcal{T})$, the character $\chi_\sigma : X \to \{A, B\}$ is defined by

$$\chi_\sigma(x) = \begin{cases} A, & \text{if } x \in A, \\ B, & \text{if } x \in B, \end{cases}$$

for all $x \in \mathcal{L}(\mathcal{T})$. Clearly, an X-tree $\mathcal{T}$ displays $\mathcal{P}$ if and only if each character in $\mathcal{C}(\mathcal{P})$ is convex on $\mathcal{T}$. Combining this observation with Theorem 4.7.1, we get the following characterization.

Theorem 6.8.7. *Let $\mathcal{P}$ be a collection of phylogenetic trees and let $X = \mathcal{L}(\mathcal{P})$. Then, $\mathcal{P}$ defines $\mathcal{T}$ if and only if the following two conditions are satisfied:*

(i) *$\mathcal{T}$ displays $\mathcal{P}$ and each interior edge of $\mathcal{T}$ is distinguished by an interior edge of at least one tree in $\mathcal{P}$; and*

(ii) *there is a unique minimal restricted chordal completion of* $\text{int}(\mathcal{C}(\mathcal{P}))$.

The difficulty of verifying whether condition (ii) of Theorem 6.8.7 holds suggests that this characterization may not lead to a polynomial–time algorithm for when a collection of phylogenetic trees $\mathcal{P}$ is definitive. In contrast, Theorem 6.8.8 (Steel 1992) leads to such an algorithm based on BUILD. However, a particular condition is imposed on $\mathcal{P}$. We omit the inductive proof of this theorem.

Theorem 6.8.8. *Let $\mathcal{P}$ be a collection of phylogenetic trees and suppose that $\bigcap_{\mathcal{T}\in\mathcal{P}} \mathcal{L}(\mathcal{T}) \neq \emptyset$. Then, $\mathcal{P}$ defines $\mathcal{T}$ if and only if $\mathcal{T}$ displays $\mathcal{P}$ and each interior edge of $\mathcal{T}$ is distinguished by an interior edge of at least one tree in $\mathcal{P}$.*

Using Theorem 6.8.8, one can easily obtain a minimum-sized set of restricted quartet trees of a binary phylogenetic X-tree $\mathcal{T}$ that defines $\mathcal{T}$ as follows. Select an element x of X and then, for each interior edge e of $\mathcal{T}$, select a restricted quartet tree q of $\mathcal{T}$ with $x \in \mathcal{L}(q)$ that distinguishes e. By Theorem 6.8.8, this selection of quartet trees defines $\mathcal{T}$.

Theorem 6.8.8 provides a convenient characterization when a set consisting of two phylogenetic trees is definitive.

Corollary 6.8.9. *Let $\mathcal{P} = \{\mathcal{T}_1, \mathcal{T}_2\}$ be a set of two phylogenetic trees and let $\mathcal{L}_{12} = \mathcal{L}(\mathcal{T}_1) \cap \mathcal{L}(\mathcal{T}_2)$. Then*

(i) *$\mathcal{P}$ is compatible if and only if $\{\mathcal{T}_1|\mathcal{L}_{12}, \mathcal{T}_2|\mathcal{L}_{12}\}$ is compatible.*

(ii) *Let $X = \mathcal{L}(\mathcal{P})$, and suppose that $\mathcal{T}$ is a binary phylogenetic X-tree that displays $\mathcal{P}$. Then, $\mathcal{P}$ defines $\mathcal{T}$ if and only if each interior edge of $\mathcal{T}$ is distinguished by an interior edge of at least one tree in $\mathcal{P}$.*

Proof. It is easily seen that (i) holds. The 'only if' part of (ii) follows from Proposition 6.8.4. We now establish the 'if' part of (ii).

Assume each interior edge of $\mathcal{T}$ is distinguished by an interior edge of at least one tree in $\mathcal{P}$. Counting the number of interior edges of $\mathcal{T}$, $\mathcal{T}_1$, and $\mathcal{T}_2$ and noting, by our assumption, $|\mathring{E}(\mathcal{T})| \leq |\mathring{E}(\mathcal{T}_1)| + |\mathring{E}(\mathcal{T}_2)|$, we deduce that

$$\begin{aligned} |\mathcal{L}(\mathcal{T}_1) \cup \mathcal{L}(\mathcal{T}_2)| &= |\mathcal{L}(\mathcal{T})| = |\mathring{E}(\mathcal{T})| + 3 \\ &\leq |\mathring{E}(\mathcal{T}_1)| + |\mathring{E}(\mathcal{T}_2)| + 3 \\ &\leq |\mathcal{L}(\mathcal{T}_1)| + |\mathcal{L}(\mathcal{T}_2)| - 3 \\ &= |\mathcal{L}(\mathcal{T}_1) \cup \mathcal{L}(\mathcal{T}_2)| + |\mathcal{L}_{12}| - 3. \end{aligned}$$

Therefore, $|\mathcal{L}_{12}| \geq 3$ and, in particular, $\mathcal{L}_{12}$ is non-empty. Choose an element x in $\mathcal{L}_{12}$, and consider the set $\mathcal{Q}_x$ of all restricted quartet trees q of either $\mathcal{T}_1$ or $\mathcal{T}_2$ such that $x \in \mathcal{L}(q)$. For every interior edge e of $\mathcal{T}$, there is a restricted quartet tree of either $\mathcal{T}_1$ or $\mathcal{T}_2$ that distinguishes e, which means that there must also be a quartet tree in $\mathcal{Q}_x$ that distinguishes e. It now follows, by Theorem 6.8.8, that $\mathcal{P}$ defines $\mathcal{T}$. This completes the proof of the corollary. □

As Example 6.8.5 illustrates, Corollary 6.8.9(b) cannot be extended to a collection of more than two phylogenetic trees.

An immediate consequence of Corollary 6.8.9(b) is that one can determine if a collection $\{\mathcal{T}_1, \mathcal{T}_2\}$ of two phylogenetic trees is definitive in $O(|\mathcal{L}_1| + |\mathcal{L}_2|)$ time (Böcker *et al.* 2000). Furthermore, if $\mathcal{T}_1$ and $\mathcal{T}_2$ are binary phylogenetic trees, then $\{\mathcal{T}_1, \mathcal{T}_2\}$ is compatible if and only if $\mathcal{T}_1|\mathcal{L}_{12} \cong \mathcal{T}_2|\mathcal{L}_{12}$, since two binary phylogenetic trees on the same label set are compatible if and only if they are isomorphic.

To end this section, reconsider Example 6.8.2. We mentioned earlier in this section that $\mathcal{Q}_1$ is a minimum-sized subset of $\mathcal{Q}(\mathcal{T})$ that defines $\mathcal{T}$. However, not every minimal subset of $\mathcal{Q}(\mathcal{T})$ that defines $\mathcal{T}$ need be of minimum size. For example, we leave it to the reader to check that

$$\mathcal{Q}_2 = \{12|35, 24|57, 13|47, 34|56, 15|67\}$$

is such a set. Note that the details of this check are not as straightforward as those done earlier for $\mathcal{Q}_1$. In Section 6.9, we present a characterization for when a minimum-sized subset of $\mathcal{Q}(\mathcal{T})$, for some binary phylogenetic tree $\mathcal{T}$, defines $\mathcal{T}$. However, as yet there is no such characterization for when a minimal subset of $\mathcal{Q}(\mathcal{T})$ defines $\mathcal{T}$. Indeed, the computational complexity of the following decision problem is still open:

Decision problem: DEFINING A TREE BY QUARTETS.
Instance: A binary phylogenetic tree $\mathcal{T}$ and a subset $\mathcal{Q}$ of $\mathcal{Q}(\mathcal{T})$.
Question: Does $\mathcal{Q}$ define $\mathcal{T}$?

6.9 Specialist topic: phylogenetic patchworks

For a collection $\mathcal{P}$ of phylogenetic trees, the *excess of* $\mathcal{P}$, denoted $\text{exc}(\mathcal{P})$, is the value

$$\text{exc}(\mathcal{P}) = |\mathcal{L}(\mathcal{P})| - 3 - \sum_{\mathcal{T} \in \mathcal{P}} |\mathring{E}(\mathcal{T})|. \tag{6.4}$$

Note that if $\mathcal{P}$ is definitive, then, by Proposition 6.8.4,

$$\text{exc}(\mathcal{P}) \leq 0.$$

The collection $\mathcal{P}$ is said to be *excess-free* if $\text{exc}(\mathcal{P}) = 0$.

In this section, we characterize when an excess-free collection of quartet trees is definitive. This characterization is based on an elegant 'patchwork' structure on the set of excess-free subsets of such a collection. As a consequence of this characterization, we present a polynomial–time algorithm for determining whether an excess-free collection of binary phylogenetic trees is definitive. We also describe a further algorithm for a more general tree compatibility problem. The material we present in this section is based on Böcker *et al.* (1999, 2000). In most cases, we omit the proofs of the results and refer the reader to these two papers.

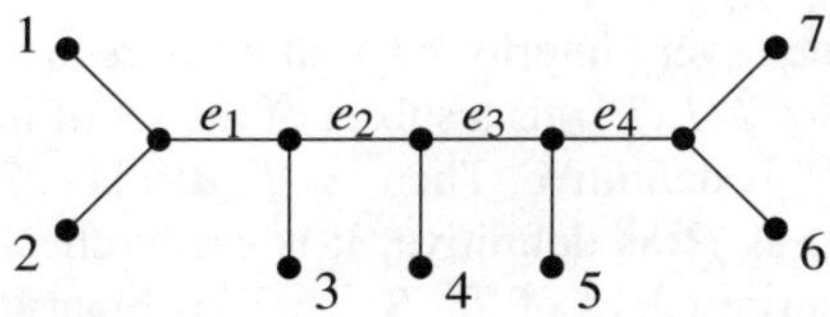

Figure 6.14 The X-tree defined by $\mathcal{Q}_1$.

A fundamental property of excess-free definitive collections of phylogenetic trees is summarized in the following proposition, whose straightforward proof we leave to the reader.

Proposition 6.9.1. *Let $\mathcal{P}$ be a collection of phylogenetic trees that defines $\mathcal{T}$. Then the following are equivalent:*

(i) $\mathcal{P}$ is excess-free;

(ii) $\sum_{\mathcal{T}' \in \mathcal{P}} |\mathring{E}(\mathcal{T}')| = |\mathring{E}(\mathcal{T})|$; and

(iii) for every interior edge e of each tree in $\mathcal{P}$, there is exactly one interior edge of $\mathcal{T}$ distinguished by e and, moreover, every interior edge of $\mathcal{T}$ is distinguished by exactly one interior edge of some tree in $\mathcal{P}$.

For a collection $\mathcal{P}$ of phylogenetic trees, we denote the collection of excess-free subsets of $\mathcal{P}$ by $\text{Exc}(\mathcal{P})$. Note that $\emptyset \notin \text{Exc}(\mathcal{P})$, since $\text{exc}(\emptyset) = -3$.

Example 6.9.2. From Example 6.8.2, recall the minimum-sized set

$$\mathcal{Q}_1 = \{12|36, 13|46, 24|57, 25|67\}$$

of quartet trees that defines the binary phylogenetic X-tree shown in Fig. 6.14. Denoting each element of $\mathcal{Q}_1$ by the unique interior edge of $\mathcal{T}$ that it distinguishes, the collection of excess-free subsets of $\mathcal{Q}_1$ is

$$\text{Exc}(\mathcal{Q}_1) = \{\{e_1\}, \{e_2\}, \{e_3\}, \{e_4\}, \{e_1, e_2\}, \{e_3, e_4\}, \{e_1, e_2, e_3, e_4\}\}.$$

In Example 6.9.2, $\mathcal{Q}_1$ is excess-free and definitive, and every element of $\text{Exc}(\mathcal{Q}_1)$ is definitive. Furthermore, one can easily check that every subset of $\mathcal{Q}_1$ that is not excess-free is not definitive. This is no coincidence as Theorem 6.9.3 illustrates.

Theorem 6.9.3. *Let $\mathcal{P}$ be an excess-free collection of phylogenetic trees that defines $\mathcal{T}$, let $\mathcal{P}'$ be a subset of $\mathcal{P}$, and let $\mathcal{T}' = \mathcal{T}|\mathcal{L}(\mathcal{P}')$. Then, $\mathcal{P}'$ is excess-free if and only if $\mathcal{P}'$ is definitive, in which case $\mathcal{P}'$ defines $\mathcal{T}'$.*

Proof. Let $\mathring{E}(\mathcal{P}')$ denote the set of interior edges of the trees in $\mathcal{P}'$. Since $\mathcal{P}$ is excess-free, there is a bijective map ψ from the set of interior edges of the phylogenetic trees in $\mathcal{P}$ into the set of interior edges of $\mathcal{T}$ as described by the equivalence of (i) and (iii) in Proposition 6.9.1. Since every interior edge of each tree in $\mathcal{P}'$ distinguishes a unique interior edge of $\mathcal{T}$, the set $\psi(\mathring{E}(\mathcal{P}'))$ is a subset of the set of interior edges of $\mathcal{T}'$.

Now suppose that $\mathcal{P}'$ is definitive. Then, as $\mathcal{T}'$ displays $\mathcal{P}'$, the collection $\mathcal{P}'$ must define $\mathcal{T}'$. Furthermore, as $\mathcal{P}'$ is definitive, it is easily checked that the set $\psi(\mathring{E}(\mathcal{P}'))$ is equal to the set of interior edges of $\mathcal{T}'$. Since $\mathcal{T}'$ is binary, it follows that $|\mathring{E}(\mathcal{P}')| = |\mathcal{L}(\mathcal{P}')| - 3$, thus establishing $\text{exc}(\mathcal{P}') = 0$.

Conversely, suppose that $\mathcal{P}'$ is excess-free. Since $\text{exc}(\mathcal{P}') = 0$ we have,

$$|\psi(\mathring{E}(\mathcal{P}'))| = |\mathring{E}(\mathcal{P}')| = |\mathcal{L}(\mathcal{P}')| - 3.$$

Therefore, as $\mathcal{T}'$ is binary, the set $\psi(\mathring{E}(\mathcal{P}'))$ is the set of interior edges of $\mathcal{T}'$. This in turn implies that $\mathcal{T}$ can be obtained from $\mathcal{T}'$ by adjoining restricted subtrees of $\mathcal{T}$ to pendant edges of $\mathcal{T}'$.

We complete the 'if' part of the proof of Theorem 6.9.3 by showing that $\mathcal{P}'$ defines $\mathcal{T}'$. To obtain a contradiction, assume that this is not the case. Then, there exists a binary phylogenetic tree $\mathcal{T}''$ on $\mathcal{L}(\mathcal{P}')$ that displays $\mathcal{P}'$ and is distinct from $\mathcal{T}'$. Now, consider the procedure by which $\mathcal{T}$ can be obtained from $\mathcal{T}'$ by adjoining restricted subtrees of $\mathcal{T}$ onto the pendant edges of $\mathcal{T}'$. Let $\mathcal{T}_1$ be the binary phylogenetic tree on $\mathcal{L}(\mathcal{T})$ that is obtained by applying this procedure to $\mathcal{T}''$ instead of $\mathcal{T}'$. Observe that as $\mathcal{T}'$ and $\mathcal{T}''$ are distinct, $\mathcal{T}$ and $\mathcal{T}_1$ are distinct. It is now easily checked using the bijection ψ that $\mathcal{T}_1$ displays $\mathcal{P}$, contradicting the fact that $\mathcal{P}$ is definitive. Thus, $\mathcal{P}'$ defines $\mathcal{T}'$. This completes the proof of Theorem 6.9.3. □

Some properties of excess-free collections of quartet trees are given in Lemma 6.9.4, whose routine proof is omitted.

Lemma 6.9.4. *Let $\mathcal{T}$ be a binary phylogenetic tree and let $\mathcal{Q}$ be an excess-free collection of quartet trees of $\mathcal{Q}(\mathcal{T})$ such that every interior edge of $\mathcal{T}$ is distinguished by an element of $\mathcal{Q}$. Suppose that $\mathcal{Q}_1$ and $\mathcal{Q}_2$ are non-empty subsets of $\mathcal{Q}$. Then,*

(i) $\text{exc}(\mathcal{Q}_1) \geq 0$;
(ii) $\text{exc}(\mathcal{Q}_1 \cup \mathcal{Q}_2) + \text{exc}(\mathcal{Q}_1 \cap \mathcal{Q}_2) \leq \text{exc}(\mathcal{Q}_1) + \text{exc}(\mathcal{Q}_2)$.

Lemma 6.9.4 imposes an interesting structure on certain excess-free subsets of $\mathcal{Q}(\mathcal{T})$. To explore this further, Böcker and Dress (2001) introduces the following definition. Given a collection $\mathcal{W}$ of subsets of a finite set M, the pair $(M, \mathcal{W})$ is called a *patchwork* if it satisfies the following property:

> If A and B are elements of $\mathcal{W}$ and $A \cap B$ is non-empty, then $A \cap B$ and $A \cup B$ are both elements of $\mathcal{W}$.

Proposition 6.9.5. *Let $\mathcal{T}$ be a binary phylogenetic tree and let $\mathcal{Q}$ be an excess-free subset of $\mathcal{Q}(\mathcal{T})$ such that every interior edge of $\mathcal{T}$ is distinguished by an element of $\mathcal{Q}$. Then, $(\mathcal{Q}, \text{Exc}(\mathcal{Q}))$ is a patchwork.*

Proof. Suppose $\mathcal{Q}_1, \mathcal{Q}_2 \in \text{Exc}(\mathcal{Q})$ and that $\mathcal{Q}_1 \cap \mathcal{Q}_2 \neq \emptyset$. By Lemma 6.9.4(ii),

$$\text{exc}(\mathcal{Q}_1 \cup \mathcal{Q}_2) + \text{exc}(\mathcal{Q}_1 \cap \mathcal{Q}_2) \leq 0. \tag{6.5}$$

Furthermore, Lemma 6.9.4(i) implies $\text{exc}(\mathcal{Q}_1 \cap \mathcal{Q}_2)$ and $\text{exc}(\mathcal{Q}_1 \cup \mathcal{Q}_2)$ are both non-negative. Combining this observation with (6.5) ensures that $\text{exc}(\mathcal{Q}_1 \cap \mathcal{Q}_2) = \text{exc}(\mathcal{Q}_2 \cup \mathcal{Q}_2) = 0$, and so $\mathcal{Q}_1 \cap \mathcal{Q}_2$ and $\mathcal{Q}_1 \cup \mathcal{Q}_2$ are both elements of $\text{Exc}(\mathcal{Q})$, as required. □

A patchwork $(M, \mathcal{W})$ is said to be *ample* if, for every two elements A and B of $\mathcal{W}$ such that $A \subset B$ and there is no element $D \in \mathcal{W}$ with

$$A \subset D \subset B,$$

we have $B - A \in \mathcal{W}$. Böcker and Dress (2001) provide a variety of characterizations for when a patchwork is ample. One in particular leads to useful algorithmic consequences, and it is described in Lemma 6.9.6.

A hierarchy $\mathcal{H}$ on X is *maximal* if there is no element H of $2^X - \{\emptyset\}$ such that $\mathcal{H} \cup H$ is a hierarchy. Equivalently, $\mathcal{H}$ is maximal if it consists of X and the clusters of a binary phylogenetic X-tree.

Lemma 6.9.6. *Let $(M, \mathcal{W})$ be a patchwork. Suppose that M is an element of $\mathcal{W}$ and, for all $m \in M$, $\{m\}$ is an element of $\mathcal{W}$. Then, $(M, \mathcal{W})$ is ample if and only if $\mathcal{W}$ contains a maximal hierarchy on M.*

Theorem 6.9.7 is the main result of Böcker *et al.* (1999) and is the basis for much of the remainder of this section. An outline of the proof of this theorem is given in their paper; for full details the reader is referred to Böcker (1999).

Theorem 6.9.7. *Let $\mathcal{T}$ be a binary phylogenetic X-tree and let $\mathcal{Q}$ be a subset of $\mathcal{Q}(\mathcal{T})$ of size $|X| - 3$. Then, $\mathcal{Q}$ defines $\mathcal{T}$ if and only if the following two conditions hold:*

(i) every interior edge of $\mathcal{T}$ is distinguished by $\mathcal{Q}$; and
(ii) $(\mathcal{Q}, \text{Exc}(\mathcal{Q}))$ is an ample patchwork or, equivalently, $\text{Exc}(\mathcal{Q})$ contains a maximal hierarchy on $\mathcal{Q}$.

To illustrate Theorem 6.9.7, reconsider Example 6.8.5. The reason that the excess-free collection $\mathcal{Q}$ of quartet trees given in this example is not definitive is because $\text{Exc}(\mathcal{Q})$ is not ample. On the other hand, the set $\mathcal{Q}_1$ in Example 6.9.2 is definitive because $(\mathcal{Q}_1, \text{Exc}(\mathcal{Q}_1))$ is an ample patchwork. A maximal hierarchy on the set $\mathcal{Q}_1$ using only elements of $\text{Exc}(\mathcal{Q}_1)$ is shown in Fig. 6.15. In this last example, $\text{Exc}(\mathcal{Q}_1)$ itself is a maximal hierarchy. However, this need not be so for a general excess-free definitive set $\mathcal{Q}$ of quartet trees as Example 6.9.8 shows.

Example 6.9.8. Let $\mathcal{Q} = \{12|35, 34|25, 56|13\}$ be a set of quartet trees. Then, one can easily check that $\mathcal{Q}$ is excess-free and definitive. Therefore, by Theorem 6.9.7, $\text{Exc}(\mathcal{Q})$ contains a maximal hierarchy on $\mathcal{Q}$. Now, every subset of $\mathcal{Q}$ is excess-free other than the empty set and $\{34|25, 56|13\}$. We leave it to the reader to check that $\text{Exc}(\mathcal{Q})$ contains precisely two maximal hierarchies on $\mathcal{Q}$.

Theorem 6.9.7 can be extended from the context of collections of quartet trees to collections of binary phylogenetic trees. To do this, we invoke a second property of patchworks (Böcker and Dress 2001).

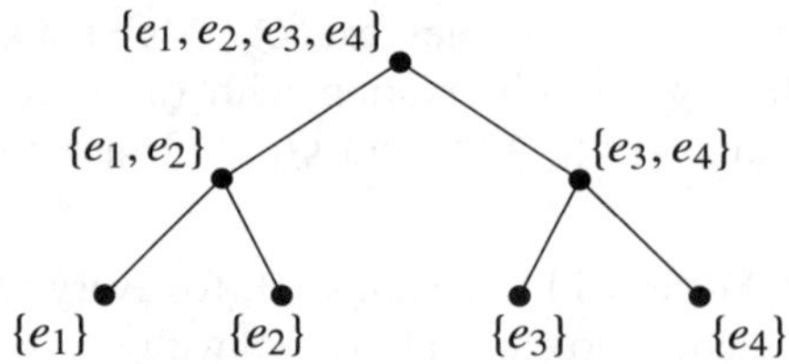

Figure 6.15 A maximal hierarchy on $\mathcal{Q}_1$ using only elements of $\mathrm{Exc}(\mathcal{Q}_1)$.

Lemma 6.9.9. *Let $(M, \mathcal{W})$ be an ample patchwork. Let $W \in \mathcal{W}$, and suppose that $\{W_1, W_2, \ldots, W_r\}$ is a partition of W such that $r \geq 2$, and $W_i \in \mathcal{W}$ for each $i \in \{1, 2, \ldots, r\}$. Let $I = \{1, 2, \ldots, r\}$, and let*

$$\mathcal{J} = \left\{ J \subseteq I : \bigcup\nolimits_{j \in J} W_j \in \mathcal{W} \right\}.$$

Then, $(I, \mathcal{J})$ is an ample patchwork.

Theorem 6.9.10. *Let $\mathcal{P}$ be an excess-free definitive collection of at least two binary phylogenetic trees. Then*

(i) $(\mathcal{P}, \mathrm{Exc}(\mathcal{P}))$ *is an ample patchwork;*
(ii) $\mathrm{Exc}(\mathcal{P})$ *contains a maximal hierarchy on $\mathcal{P}$;*
(iii) $\mathcal{P}$ *contains an excess-free definitive subset of size two.*

Proof. By the definition of excess-free, it suffices to establish the theorem in the case where all of the phylogenetic trees in $\mathcal{P}$ have label sets of size at least four. Let $\mathcal{P} = \{\mathcal{T}_1, \mathcal{T}_2, \ldots, \mathcal{T}_r\}$, where $r \geq 2$. By Corollary 6.3.10, for each $i \in \{1, 2, \ldots, r\}$, we can choose an excess-free collection $\mathcal{Q}_i$ of quartet trees of $\mathcal{Q}(\mathcal{T}_i)$ that defines $\mathcal{T}_i$. Let $\mathcal{Q} = \bigcup_{i=1}^{r} \mathcal{Q}_i$.

Since $\mathcal{P}$ is definitive, it is clear that $\mathcal{Q}$ is definitive and, indeed, $\mathcal{P}$ and $\mathcal{Q}$ define the same binary phylogenetic tree. Therefore, by Proposition 6.9.1,

$$\sum_{i=1}^{r} |\mathring{E}(\mathcal{T}_i)| = |\mathcal{Q}|. \tag{6.6}$$

Now using (6.4) and (6.6), and the fact that $\mathcal{P}$ is excess-free, it is straightforward to deduce that $\mathcal{Q}$ is also excess-free. Therefore, by Theorem 6.9.7, $(\mathcal{Q}, \mathrm{Exc}(\mathcal{Q}))$ is an ample patchwork. We now apply Lemma 6.9.9 taking $W = \mathcal{Q}$, $\mathcal{W} = \mathrm{Exc}(\mathcal{Q})$, and letting $W_i = \mathcal{Q}_i$ for each $i \in \{1, 2, \ldots, r\}$. By Proposition 6.9.1, the sets $W_1, W_2, \ldots, W_r$ are disjoint, and therefore, by Lemma 6.9.9, $(I, \mathcal{J})$ is an ample patchwork. Now, by Theorem 6.9.3, for all $J \in \mathcal{J}$, a subset $\{\mathcal{T}_j : j \in J\}$ of $\mathcal{P}$ is excess-free if and only if $\bigcup_{j \in J} \mathcal{Q}_j$ is excess-free. Thus, $(\mathcal{P}, \mathrm{Exc}(\mathcal{P}))$ forms an ample patchwork, thereby establishing part (i) of the theorem.

Part (ii) now follows from Lemma 6.9.6. Part (iii) follows from part (ii) by noting that a maximal hierarchy on $\mathcal{P}$ in Exc($\mathcal{P}$) necessarily includes a set of size two and, by Theorem 6.9.3, this excess-free subset is definitive. □

The restriction that $\mathcal{P}$ be excess-free in the statement of Theorem 6.9.10 is necessary. To see this, recall the definitive set

$$\mathcal{Q}_2 = \{12|35, 24|57, 13|47, 34|56, 15|67\}$$

of quartet trees at the end of Section 6.8. Now, $\mathcal{Q}_2$ is not excess-free and a routine check shows that no subset of $\mathcal{Q}_2$ of size two is definitive. Similarly, the requirement in the statement of Theorem 6.9.10 that each of the trees in $\mathcal{P}$ is binary can also be shown to be necessary.

Algorithmic implications. We turn now to the algorithmic implications of Theorem 6.9.10. We begin with a preliminary lemma, whose straightforward proof we leave to the reader.

Lemma 6.9.11. *Let $\mathcal{P}$ be a collection of phylogenetic trees. Let $\mathcal{P}'$ be a subset of $\mathcal{P}$, and suppose that $\mathcal{P}'$ defines $\mathcal{T}'$. Let $\mathcal{P}_1 = (\mathcal{P} - \mathcal{P}') \cup \{\mathcal{T}'\}$. Then, $\mathcal{P}$ is definitive if and only if $\mathcal{P}_1$ is definitive. Furthermore, if $\mathcal{P}$ is definitive and excess-free, then $\mathcal{P}_1$ is definitive and excess-free.*

Suppose that $\mathcal{P}$ is an excess-free collection of binary phylogenetic trees. We now describe a polynomial–time algorithm for determining whether $\mathcal{P}$ is definitive and, if so, constructing the binary phylogenetic tree defined by $\mathcal{P}$. The algorithm proceeds recursively as follows. First, find two distinct members, $\mathcal{T}_1$ and $\mathcal{T}_2$ say, of $\mathcal{P}$ such that $\{\mathcal{T}_1, \mathcal{T}_2\}$ is excess-free and definitive. If there is no such pair, then, by Theorem 6.9.10, $\mathcal{P}$ is not definitive. Let $\mathcal{T}_{12}$ denote the binary phylogenetic tree that is defined by $\{\mathcal{T}_1, \mathcal{T}_2\}$ and let $\mathcal{P}_{12} = (\mathcal{P} - \{\mathcal{T}_1, \mathcal{T}_2\}) \cup \{\mathcal{T}_{12}\}$. Then, by Lemma 6.9.11, $\mathcal{P}$ is definitive precisely if $\mathcal{P}_{12}$ is definitive, in which case $\mathcal{P}_{12}$ is also excess-free. Repeating this procedure on $\mathcal{P}_{12}$, we eventually decide on one of the following options:

(i) $\mathcal{P}$ is not definitive because at some stage we cannot find two distinct members of our current collection of binary phylogenetic trees with the desired properties; or

(ii) $\mathcal{P}$ is definitive as we finally reduce $\mathcal{P}$ to a set consisting of one binary phylogenetic tree, namely the tree defined by $\mathcal{P}$.

Example 6.9.12. Recall the set $\mathcal{Q}_1 = \{12|36, 13|46, 24|57, 25|67\}$ and the collection Exc($\mathcal{Q}_1$) of excess-free subsets given in Example 6.9.2. An application of the above algorithm to reconstruct the binary phylogenetic tree defined by $\mathcal{Q}_1$ is shown in Fig. 6.16.

A further application of Theorem 6.9.10 that is more useful in practice is to provide a polynomial–time algorithm that applies to an arbitrary collection $\mathcal{P}$ of binary phylogenetic trees. We would like such an algorithm to have the property that, when $\mathcal{P}$ contains an (unknown) excess-free definitive subset $\mathcal{P}'$ with $\mathcal{L}(\mathcal{P}') = \mathcal{L}(\mathcal{P})$, the algorithm will

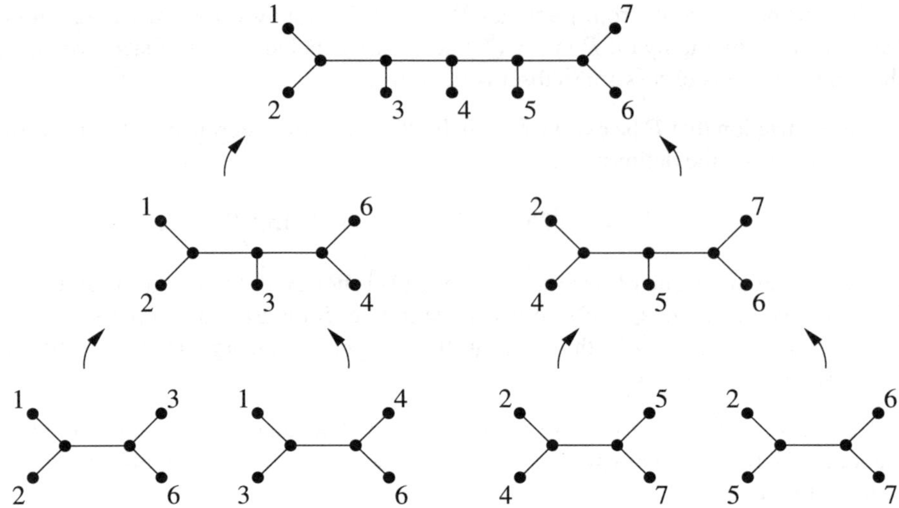

Figure 6.16 An application of the algorithm described prior to Example 6.9.12.

determine if $\mathcal{P}$ is incompatible or definitive, and in the latter case reconstruct the binary phylogenetic tree defined by $\mathcal{P}$. We next outline how this is possible.

For a collection $\mathcal{P}$ of phylogenetic trees, let $\mathcal{Q}(\mathcal{P})$ denote the set

$$\mathcal{Q}(\mathcal{P}) = \bigcup_{\mathcal{T} \in \mathcal{P}} \mathcal{Q}(\mathcal{T})$$

of quartet trees. Note that $\mathcal{P}$ defines $\mathcal{T}$ if and only if $\mathcal{Q}(\mathcal{P})$ defines $\mathcal{T}$. Lemma 6.9.13 and Theorem 6.9.14 are due to Böcker *et al.* (2000). Recall that scl_2 denotes the semi-dyadic closure operator.

Lemma 6.9.13. *Let $\mathcal{P} = \{\mathcal{T}_1, \mathcal{T}_2\}$ be a collection of two binary phylogenetic trees, and suppose that $\mathcal{T}$ is a binary phylogenetic tree that displays both $\mathcal{T}_1$ and $\mathcal{T}_2$. Then, $\mathcal{P}$ defines $\mathcal{T}$ if and only if*

$$\mathrm{scl}_2(\mathcal{Q}(\mathcal{P})) = \mathcal{Q}(\mathcal{T}).$$

The proof of the last lemma is straightforward and we leave it as an exercise. It is an important tool for establishing Theorem 6.9.14.

Theorem 6.9.14. *Let $\mathcal{P}$ be a compatible collection of binary phylogenetic trees. Suppose that a subset $\mathcal{P}'$ of $\mathcal{P}$ has the properties that $\mathcal{L}(\mathcal{P}') = \mathcal{L}(\mathcal{P})$, and $\mathcal{P}'$ is excess-free and defines a binary phylogenetic tree $\mathcal{T}$. Then,*

$$\mathrm{scl}_2(\mathcal{Q}(\mathcal{P})) = \mathcal{Q}(\mathcal{T}).$$

Proof. As $\mathcal{P}$ is compatible and $\mathcal{P}'$ defines $\mathcal{T}$, it is clear that $\mathrm{scl}_2(\mathcal{Q}(\mathcal{P})) \subseteq \mathcal{Q}(\mathcal{T})$. To establish the converse, it follows by Theorem 6.9.10 that we can construct a sequence

$\mathcal{P}_1, \mathcal{P}_2, \ldots, \mathcal{P}_m$ of collections of binary phylogenetic trees such that $\mathcal{P}_1 = \mathcal{P}'$, $m = |\mathcal{P}'|$, $\mathcal{P}_m = \{\mathcal{T}\}$, and, for each $i \in \{1, 2, \ldots, m-1\}$, $\mathcal{P}_{i+1}$ is obtained from $\mathcal{P}_i$ by replacing a definitive subset $\mathcal{P}'_i$ of $\mathcal{P}_i$ of size two by the unique binary phylogenetic tree $\mathcal{T}'_i$ defined by $\mathcal{P}'_i$.

For each $i \in \{1, 2, \ldots, m-1\}$, it follows by Lemma 6.9.14 that $\mathrm{scl}_2(\mathcal{Q}(\mathcal{P}'_i)) = \mathcal{Q}(\mathcal{T}'_i)$, and hence $\mathcal{Q}(\mathcal{P}_{i+1}) \subseteq \mathrm{scl}_2(\mathcal{Q}(\mathcal{P}_i))$. Thus,

$$\mathcal{Q}(\mathcal{T}) = \mathcal{Q}(\mathcal{P}_m) \subseteq \mathrm{scl}_2(\mathcal{Q}(\mathcal{P}_1)) = \mathrm{scl}_2(\mathcal{Q}(\mathcal{P}')) \subseteq \mathrm{scl}_2(\mathcal{Q}(\mathcal{P})),$$

completing the proof of the theorem. □

The converse of Theorem 6.9.14 does not hold for arbitrary collections of phylogenetic trees (see Exercise 21). Whether it holds for collections of quartet trees is unknown.

Analogous to the previous algorithm, Theorem 6.9.14 leads directly to a polynomial–time algorithm that takes as its input a collection $\mathcal{P}$ of binary phylogenetic trees and, whenever $\mathcal{P}$ contains an excess-free definitive subset $\mathcal{P}'$ with $\mathcal{L}(\mathcal{P}') = \mathcal{L}(\mathcal{P})$, it either returns the binary phylogenetic tree defined by $\mathcal{P}$ or determines that $\mathcal{P}$ is incompatible. It would be useful to determine in advance whether an arbitrary collection $\mathcal{P}$ of binary phylogenetic trees contains such a subset $\mathcal{P}'$. However, Böcker *et al.* (2000) showed that this decision problem is NP-complete.

6.10 Exercises

1. Let $\mathcal{T}'$ be binary phylogenetic X'-tree and let X be a set containing X'. Show that the number of binary phylogenetic X-trees $\mathcal{T}$ for which $\mathcal{T}|X' = \mathcal{T}'$ is exactly
$$\frac{b(|X|)}{b(|X'|)}.$$
2. Show that a collection $\mathcal{R}$ of rooted phylogenetic trees is compatible if and only if, for every subset S of $\mathcal{L}(\mathcal{R})$ of size at least two, the graph $[\mathcal{R}, S]$ is disconnected.
3. Let $\mathcal{P} = \{\mathcal{T}_1, \ldots, \mathcal{T}_k\}$ be a collection of semi-labelled trees with $\mathcal{L}(\mathcal{P}) = X$. For all i, let $\mathcal{P}_i$ denote a collection of restricted subtrees of $\mathcal{T}_i$ that identifies $\mathcal{T}_i$. Show that
   (i) an X-tree $\mathcal{T}$ displays $\mathcal{P}$ if and only if $\mathcal{T}$ displays $\bigcup_{i=1}^k \mathcal{P}_i$;
   (ii) $\mathcal{P}$ is compatible if and only if $\bigcup_{i=1}^k \mathcal{P}_i$ is compatible;
   (iii) $\mathrm{co}(\mathcal{P}) = \mathrm{co}(\bigcup_{i=1}^k \mathcal{P}_i)$.
4. Using the BUILD algorithm and Exercise 7 in Chapter 4, describe a polynomial–time algorithm for determining the compatibility of a collection of generalized characters on X with state set $\{0, 1\}$ relative to the directed graph $0 \rightarrow 1$.
5. (Berry and Gascuel 2000) Let $\mathcal{Q}$ be a set of quartet trees such that, for each subset X' of $\mathcal{L}(\mathcal{Q})$ of size four, there is at most one quartet tree in $\mathcal{Q}$ that has label set X'. Describe a polynomial–time algorithm for reconstructing $\Sigma(\mathcal{Q})$ from $\mathcal{Q}$.

6. Let $\mathcal{Q}$ be a set of quartet trees such that, for each subset X' of $\mathcal{L}(\mathcal{Q})$ of size four, there are at most two quartet trees in $\mathcal{Q}$ that have label set X'. Show that $\Sigma(\mathcal{Q})$ is weakly compatible.
7. Let $\mathcal{T}$ be a phylogenetic X-tree with at least one interior edge. Using the inductive proof of Theorem 6.3.9, describe a polynomial–time algorithm for constructing a set of $|X| - 3$ quartet trees that identifies $\mathcal{T}$.
8. Prove Lemmas 6.5.2 and 6.5.3.
9. Prove Propositions 6.3.11 and 6.3.12.
10. Let $\mathcal{T}$ and $\mathcal{T}'$ be rooted semi-labelled trees, and suppose that $\mathcal{T}$ displays $\mathcal{T}'$. Show that if $C \in \mathcal{H}(\mathcal{T})$ and $C' \in \mathcal{H}(\mathcal{T}')$, then

$$C \cap C' \in \{\emptyset, C', C \cap \mathcal{L}(\mathcal{T}')\}.$$

11. (Constantinescu and Sankoff 1995; Ng and Wormald 1996) By extending BUILD, construct an algorithm that, when applied to a compatible collection $\mathcal{R}$ of rooted phylogenetic trees, outputs each rooted phylogenetic tree that displays $\mathcal{R}$ and requires polynomial time for each such tree.
12. (Semple 2002) Let $\mathcal{R}$ be a compatible collection of rooted phylogenetic trees, and let $\text{co}(\mathcal{R})$ denote the collection of rooted phylogenetic trees that display $\mathcal{R}$ and have label set $\mathcal{L}(\mathcal{R})$. Let $\mathcal{T}$ be a rooted phylogenetic tree in $\text{co}(\mathcal{R})$. Prove that $\mathcal{T}$ is isomorphic to the rooted phylogenetic tree that BUILD produces when applied to $\mathcal{R}$ if and only if, for each element $\mathcal{T}'$ in $\text{co}(\mathcal{R})$, the mapping $\psi : \mathcal{H}(\mathcal{T}) \to \mathcal{H}(\mathcal{T}')$ defined by $\psi(C) = C'$, where C' is the (unique) minimal cluster of $\mathcal{T}'$ containing C, satisfies the following property:

$$\text{if } C_1, C_2 \in \mathcal{H}(\mathcal{T}) \quad \text{and} \quad C_1 \subset C_2, \text{ then } \psi(C_1) \subset \psi(C_2).$$

13. (Colonius and Shultze 1981) An X-tree $\mathcal{T}$ induces a 'quaternary' relation $\|$ on $X \times X$ as follows. For all elements $x_1, x_2, x_3, x_4 \in X$, which may not all be distinct, write $x_1x_2\|x_3x_4$ if $\{x_1, x_2\}|\{x_3, x_4\}$ is a split of $\mathcal{T}|\{x_1, x_2, x_3, x_4\}$.
    (i) Show that this relation satisfies the following properties:
        (I) at most one of $x_1x_2\|x_3x_4$, $x_1x_3\|x_2x_4$, and $x_1x_4\|x_2x_3$ holds;
        (II) $x_1x_2\|x_3x_4$ implies $x_2x_1\|x_3x_4$ and $x_3x_4\|x_1x_2$;
        (III) $x_1x_2\|x_3x_4$ implies either $x_1x_2\|x_3x_5$ or $x_1x_5\|x_3x_4$.
    (ii) Show that a relation $\|$ on $X \times X$ is realized by an X-tree $\mathcal{T}$ if and only if $\|$ satisfies properties (I), (II), and (III), in which case $\mathcal{T}$ is unique up to isomorphism.
14. Let $\chi : X' \to C$ be a character on X and let $\mathcal{T}$ be a phylogenetic X-tree. Show that $l(\chi, \mathcal{T}) = l(\chi, \mathcal{T}|X')$, where we recall that l denotes the parsimony score function.
15. Prove Propositions 6.6.1 and 6.6.2.
16. For each $n \geq 5$, construct a subset $\mathcal{P}$ of $B(n)$ of size n such that, for every supertree method that satisfies (S1), (S2), and (S4), the output tree obtained by applying the method to $\mathcal{P}$ is the star tree on $\{1, 2, \ldots, n\}$.

17. (Meacham 1983) An X-tree $\mathcal{T}$ *displays a partial split* $A|B$ of X if $A|B$ is a split of $\mathcal{T}|(A \cup B)$. Suppose that an X-tree $\mathcal{T}$ displays the partial splits $A_1|B_1$ and $A_2|B_2$ of X, where $A_1 \cap A_2$, $A_1 \cap B_2$, and $B_1 \cap B_2$ are all non-empty. Show that $\mathcal{T}$ also displays the partial splits $(A_1 \cup A_2)|B_1$ and $A_2|(B_1 \cup B_2)$ of X.
18. (Bryant and Steel 1995) Let $\mathcal{Q}$ be a collection of quartet trees such that

$$\bigcap_{q \in \mathcal{Q}} \mathcal{L}(q) \neq \emptyset.$$

   (i) If $\mathcal{Q}$ is compatible, describe a polynomial–time algorithm for constructing $\mathrm{cl}(\mathcal{Q})$.
   (ii) If $|\mathcal{Q}| \geq 3$, and every proper subset of $\mathcal{Q}$ is compatible and closed, show that $\mathcal{Q}$ is compatible.
19. Let $\mathcal{Q}$ be a set of quartet trees such that, for each subset X' of $\mathcal{L}(\mathcal{Q})$ of size four, there is exactly one quartet tree in $\mathcal{Q}$ that has label set X'. Prove that the following statements are equivalent:
   (i) $\mathcal{Q}$ is compatible;
   (ii) $\mathrm{scl}_2(\mathcal{Q}) = \mathcal{Q}$;
   (iii) every subset of $\mathcal{Q}$ of size three is compatible;
   (iv) $\mathcal{Q}$ is definitive.
20. Show that there exists a binary phylogenetic tree $\mathcal{T}$ with $|\mathcal{L}(\mathcal{T})| = 9$ that is not defined by any pair $\{\mathcal{T}_1, \mathcal{T}_2\}$ of phylogenetic trees with $|\mathcal{L}(\mathcal{T}_1)| = |\mathcal{L}(\mathcal{T}_2)| = 6$.
21. Let $\mathcal{P} = \{123|47, 45|16, 67|25, 345|12\}$ be a collection of phylogenetic trees and let $\mathcal{T}$ be the binary phylogenetic tree shown in Fig. 6.17.
   (i) Show that $\mathcal{P}$ defines $\mathcal{T}$.
   (ii) Although $\mathcal{Q}(\mathcal{P})$ defines $\mathcal{T}$, show that no excess-free subset of $\mathcal{Q}(\mathcal{P})$ defines $\mathcal{T}$.

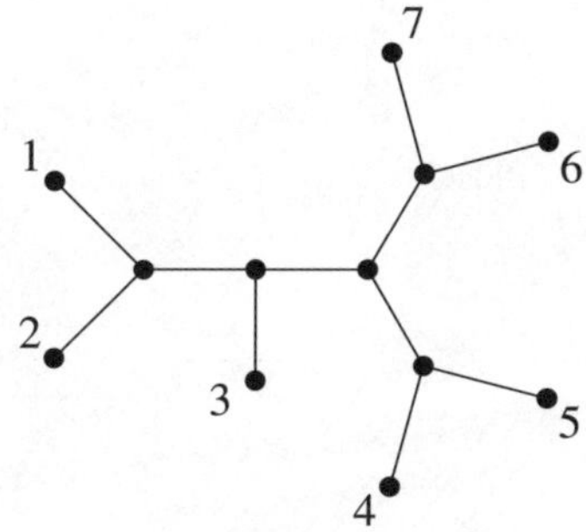

Figure 6.17

22. For a tree $T = (V, E)$, let $\mathcal{W}$ denote the collection of subsets of V that induce connected subgraphs of T. Prove that $(V, \mathcal{W})$ is a patchwork.
23. Let $\mathcal{P}$ be a collection of phylogenetic trees and let $\mathcal{P}'$ be a subset of $\mathcal{P}$ that defines a binary phylogenetic tree $\mathcal{T}'$. Let $\mathcal{P}_1 = (\mathcal{P} - \mathcal{P}') \cup \{\mathcal{T}'\}$. Show that

$$\text{exc}(\mathcal{P}) = \text{exc}(\mathcal{P}') + \text{exc}(\mathcal{P}_1).$$

7

Tree-based metrics

Up to this point, we have regarded the edges of an X-tree $\mathcal{T}$ as just a pair of vertices. However, it also makes sense in certain applications to regard each edge of $\mathcal{T}$ as an interval of a certain length. Adopting this view induces a 'distance' between all pairs of elements x and y of X. This distance is the sum of the weights on the edges in the path joining x and y. In evolutionary biology, such a distance might measure the time two species have been separated from their most recent common ancestor.

In this chapter, we characterize when an arbitrary distance function on X can be represented by an X-tree with positive real-valued weighted edges, and show how one can recover the underlying X-tree and edge weighting when such a representation exists. These results form the basis of tree reconstruction methods that take data-derived measures of distances between species as their input. Results in this chapter will also play an important role in Chapter 8.

7.1 Dissimilarity maps and tree metrics

Let $T = (V, E)$ be a tree and suppose that $w : E \to \mathbb{R}$ is a map that assigns real-valued weights to the edges of T. This edge-weighting of T induces the following map from $V \times V$ into $\mathbb{R}$. For all $u, v \in V$, let $P(T; u, v)$ denote the (unique) path in T from u to v. We define the map $d_{(T;w)} : V \times V \to \mathbb{R}$ by setting, for all $u, v \in V$,

$$d_{(T;w)}(u, v) = \begin{cases} \sum\limits_{e \in P(T;u,v)} w(e), & \text{if } u \neq v, \\ 0, & \text{otherwise.} \end{cases}$$

If $\mathcal{T} = (T; \phi)$ is an X-tree, we define the map $d_{(\mathcal{T};w)} : X \times X \to \mathbb{R}$ by setting $d_{(\mathcal{T};w)}(x, y) = d_{(T;w)}(\phi(x), \phi(y))$ for all $x, y \in X$.

Example 7.1.1. Let w denote the edge-weighting of the semi-labelled tree $\mathcal{T}$ shown in Fig. 7.1(a). Then, $d_{(\mathcal{T};w)}(a, c) = 8$, $d_{(\mathcal{T};w)}(f, d) = 7$, and $d_{(\mathcal{T};w)}(b, b) = 0$.

Definition 7.1.2. An arbitrary function $\delta : X \times X \to \mathbb{R}$ is called a *dissimilarity map* if, for all $x, y \in X$, $\delta(x, x) = 0$ and $\delta(x, y) = \delta(y, x)$. In addition, δ is a *tree metric* (*on* X) if there exists an X-tree $\mathcal{T} = (T; \phi)$ and a positive real-valued weighting $w : E(T) \to \mathbb{R}^{>0}$ such that, for all $x, y \in X$,

$$\delta(x, y) = d_{(\mathcal{T};w)}(\phi(x), \phi(y)).$$

(a)

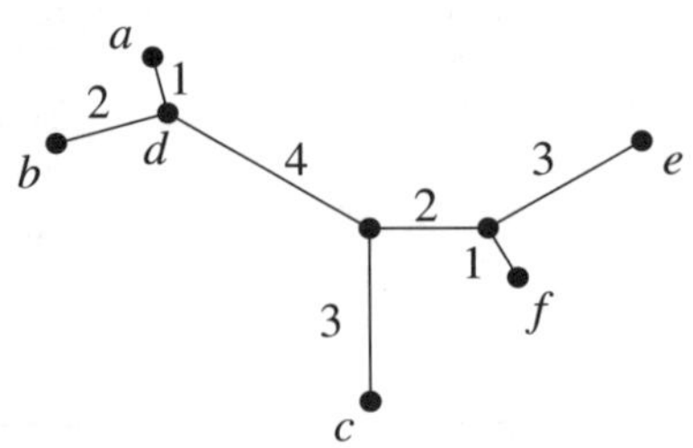

(b)

	a	b	c	d	e	f
a	0	3	8	1	10	8
b		0	9	2	11	9
c			0	7	8	6
d				0	9	7
e					0	4
f						0

Figure 7.1 (a) An edge-weighted X-tree. (b) A dissimilarity map on X.

In this case, we also call δ a $\mathcal{T}$-*metric* (*on* X) if we wish to make $\mathcal{T}$ explicit. Furthermore, we say that $(\mathcal{T};w)$ is a *tree metric representation of* δ and that $\mathcal{T}$ *supports* a tree metric representation of δ.

Example 7.1.3. Let $X = \{a,b,c,d,e,f\}$ and let δ denote the dissimilarity map on X that is determined by the portion of the table shown in Fig. 7.1(b), where the (i,j)-th entry is equal to $\delta(i,j)$. The map δ is a tree metric as the edge-weighted X-tree shown in Fig. 7.1(a) is a tree metric representation of δ.

Lemma 7.1.4 shows that there is no serious loss of generality in enforcing that the edge weights in a tree metric representation are non-zero.

Lemma 7.1.4. *Let δ be a dissimilarity map on X. If $\delta = d_{(\mathcal{T};w)}$ for some X-tree $\mathcal{T}$ and non-negative edge weighting w, then δ is a tree metric. Conversely, if δ is a tree metric with tree metric representation $(\mathcal{T};w)$, then, for every phylogenetic X-tree $\mathcal{T}'$ with $\mathcal{T} \leq \mathcal{T}'$, there exists a non-negative edge-weighting w' such that $\delta = d_{(\mathcal{T};w')}$.*

Proof. Suppose that $\delta = d_{(\mathcal{T};w)}$ for some X-tree $\mathcal{T}$ and non-negative edge weighting w. Let $\mathcal{T}'$ be the X-tree obtained from $\mathcal{T}$ by contracting each edge e for which $w(e) = 0$. Then, $\delta = d_{(\mathcal{T}';w')}$ for the strictly positive edge-weighting w' that is defined, for all $e' \in E(\mathcal{T}')$, by setting $w'(e') = w(e)$, where e is the edge of $\mathcal{T}$ that induces the same X-split in $\mathcal{T}$ as e' induces in $\mathcal{T}'$.

To prove the converse, now suppose that $\delta = d_{(\mathcal{T};w)}$ where w is strictly positive. Now assume that $\mathcal{T}'$ is an X-tree such that $\mathcal{T} \leq \mathcal{T}'$, and let e' be an edge of $\mathcal{T}'$. For an edge weighting w' of $\mathcal{T}'$, set $w'(e') = 0$ if e' does not induce an X-split of $\mathcal{T}$, otherwise set $w'(e') = w(e)$ where e is the edge of $\mathcal{T}$ that induces the same X-split in $\mathcal{T}$ as e' induces in $\mathcal{T}'$. Then, $\delta = d_{(\mathcal{T}';w')}$ as required. □

The main theorem of the first two sections of this chapter, Theorem 7.2.6, is a classical result that characterizes when an arbitrary dissimilarity map is a tree metric. The statement of this theorem uses Definition 7.1.5.

Definition 7.1.5. A dissimilarity map δ on X satisfies the *four-point condition* if, for every four (not necessarily distinct) elements $w,x,y,z \in X$,

$$\delta(w,x) + \delta(y,z) \leq \max\{\delta(w,y) + \delta(x,z), \delta(w,z) + \delta(x,y)\}.$$

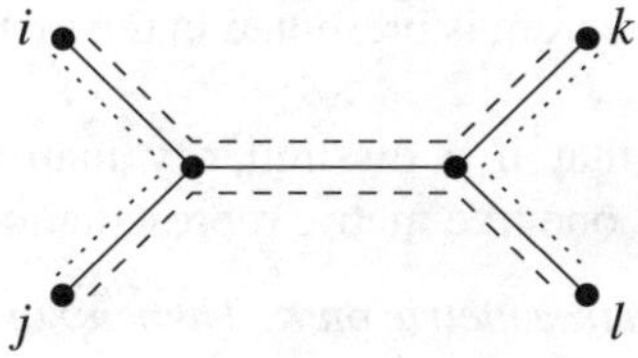

Figure 7.2 The four-point condition.

Since the elements $w, x, y, z \in X$ in Definition 7.1.5 need not be distinct, every dissimilarity map on X satisfying the four-point condition is a metric on X. Moreover, it is easily seen that a dissimilarity map δ satisfies the four-point condition exactly if, for all $w, x, y, z \in X$, two of the sums $\delta(w,x) + \delta(y,z)$, $\delta(w,y) + \delta(x,z)$, and $\delta(w,z) + \delta(x,y)$ are equal and not less than the third.

Example 7.1.6. Let $\mathcal{T}$ be the quartet tree shown in Fig. 7.2 and suppose that w is a strictly positive weighting on the edges of $\mathcal{T}$. By considering the dotted and dashed paths, respectively, in the figure it is clear that

$$d_{(\mathcal{T};w)}(i,j) + d_{(\mathcal{T};w)}(k,l) < d_{(\mathcal{T};w)}(i,k) + d_{(\mathcal{T};w)}(j,l).$$

Furthermore,

$$d_{(\mathcal{T};w)}(i,k) + d_{(\mathcal{T};w)}(j,l) = d_{(\mathcal{T};w)}(i,l) + d_{(\mathcal{T};w)}(j,k).$$

Thus, the tree metric on $\{i,j,k,l\}$ induced by $\mathcal{T}$ and w satisfies the four-point condition. Indeed, this property holds for all tree metrics as Lemma 7.1.7 shows.

Lemma 7.1.7. *If δ is a tree metric on X, then δ satisfies the four-point condition.*

Proof. Suppose that δ is a tree metric on X. Then, by Lemma 7.1.4, there exists a binary phylogenetic X-tree $\mathcal{T} = (T;\phi)$ and a weight function $w : E(T) \rightarrow \mathbb{R}^{\geq 0}$ such that, for all $x, y \in X$,

$$\delta(x,y) = d_{(\mathcal{T};w)}(\phi(x), \phi(y)).$$

Let i, j, k, and l be any four elements of X. If i,j,k,l are distinct, we may assume that $T|\{i,j,k,l\} = ij|kl$, in which case it is easy to check that

$$d(i,j) + d(k,l) \leq d(i,k) + d(j,l) = d(i,l) + d(j,k),$$

so that d satisfies the four-point condition. A simpler argument applies if i,j,k,l are not distinct. This completes the proof of the lemma. □

Remark. The converse of Lemma 7.1.7 is also true and this characterization of tree metrics (Theorem 7.2.6) may be regarded as the 'fundamental theorem' for this chapter. Although one can establish this converse directly, it is somewhat tedious and we have chosen to present it as a consequence of a result derived in Section 7.2. A considerably more general form of this theorem is presented in the first specialist topic of this chapter (see Theorem 7.6.4).

Theorem 7.1.8 shows that if a dissimilarity map is a tree metric, then, up to isomorphism, it has exactly one tree metric representation.

Theorem 7.1.8. *Let δ be a tree metric on X. Then, up to isomorphism, there is a unique tree metric representation of δ.*

Proof. Suppose that $(\mathcal{T}; w)$ is a tree metric representation of δ. Let X' be a subset of X of size at most 4. Let Σ denote the following set of X'-splits. We put $\{x\}|(X' - \{x\}) \in \Sigma$ precisely if

$$\delta(x, y) + \delta(x, z) - \delta(y, z) > 0$$

for all $y, z \in X' - \{x\}$, and we put $\{x, y\}|\{w, z\} \in \Sigma$ precisely if

$$\delta(x, w) + \delta(y, z) - \delta(x, y) - \delta(w, z) > 0.$$

Then, it is easily checked that $\Sigma = \Sigma(\mathcal{T}|X')$ and so δ determines $\mathcal{T}|X'$ for all subsets of X' of size at most 4. Therefore, by Theorem 6.3.5(iii), δ determines $\mathcal{T}$ up to isomorphism. We now show that δ also determines w.

Let T and ϕ denote the underlying tree and labelling map of $\mathcal{T}$. Let $e = \{u, v\}$ be an edge of T. There are two cases to consider depending upon whether e is an interior or a pendant edge of T. First, suppose that e is an interior edge of T. Choose two elements i and j of X so that the path in $T - v$ from $\phi(i)$ to $\phi(j)$ passes through u. Note that if u has degree two, then $\phi^{-1}(u)$ is non-empty, and so even in this case such a choice can be made. Similarly, choose two elements k and l of X so that in $T - u$ the path from $\phi(k)$ to $\phi(l)$ passes through v. Since $(\mathcal{T}; w)$ is a tree metric representation of δ, the weight $w(e)$ is equal to

$$\tfrac{1}{2}(\delta(i, k) + \delta(j, l) - \delta(i, j) - \delta(k, l)).$$

Thus, the weighting of all of the interior edges of T are determined by δ.

Now, suppose that e is a pendant edge of T. Without loss of generality, we may assume that u is a leaf of T. Choose an element i of X such that $\phi(i) = u$, and choose two elements j and k of X so that the path in $T - u$ from $\phi(j)$ to $\phi(k)$ passes through v. Since $(\mathcal{T}; w)$ is a tree metric representation of δ, it follows that $w(e)$ is equal to

$$\tfrac{1}{2}(\delta(i, j) + \delta(i, k) - \delta(j, k)).$$

Thus, the weighting of all of the pendant edges of T are determined by δ. This completes the proof of Theorem 7.1.8. □

In biology, a dissimilarity map is generally a measure of the evolutionary distance between species. To illustrate, consider a sequence $\mathcal{C} = (\chi_1, \chi_2, \ldots, \chi_k)$ of full characters on X. A crude measure of distance between a pair of species x and y in X is the proportion $\delta_{\mathcal{C}}(x, y)$ of characters of $\mathcal{C}$ that assign different character states to x and y. That is,

$$\delta_{\mathcal{C}}(x, y) = \tfrac{1}{k}|\{i : \chi_i(x) \neq \chi_i(y) \text{ and } i \in \{1, 2, \ldots, k\}\}|.$$

A natural question is how the compatibility of $\mathcal{C}$ relates to whether or not $\delta_{\mathcal{C}}$ is a tree metric. One can construct a sequence $\mathcal{C}$ of binary characters on X so that $\delta_{\mathcal{C}}$ is a tree metric on X, yet $\mathcal{C}$ is not compatible. Moreover, it is easy to construct a sequence $\mathcal{C}$ of full characters on X that is compatible, yet $\delta_{\mathcal{C}}$ is not a tree metric on X. However, in the special case of binary characters, we have the following proposition whose proof is left as an exercise.

Proposition 7.1.9. *Let $\mathcal{C}$ be a compatible sequence of binary characters on X. Then, $\delta_{\mathcal{C}}$ is a tree metric. Moreover, if, under $\leq$, $\mathcal{T}$ is the minimally refined X-tree on which all of the characters in $\mathcal{C}$ are convex, then $\mathcal{T}$ supports a tree metric representation of $\delta_{\mathcal{C}}$.*

Other dissimilarity maps that measure some aspect of evolutionary distance between species will be considered in the next chapter.

7.2 Ultrametrics

Definition 7.2.1. A dissimilarity map $\delta : X \times X \to \mathbb{R}$ is called an *ultrametric* (on X) if, for every three distinct elements $x, y, z \in X$,

$$\delta(x,y) \leq \max\{\delta(x,z), \delta(y,z)\}.$$

Equivalently, a dissimilarity map δ is an ultrametric if two of the values $\delta(x,y)$, $\delta(x,z)$, and $\delta(y,z)$ are equal and not less than the third.

Ultrametrics and tree metrics are closely related. First, it is easily verified that every non-negative dissimilarity map that satisfies the ultrametric condition also satisfies the four-point condition and thus is a tree metric. On the other hand, a tree metric on X induces a natural ultrametric for each element of X as we describe next.

Given a dissimilarity map $\delta : X \times X \to \mathbb{R}$ and an element $r \in X$, define a map $\delta_r : (X - \{r\}) \times (X - \{r\}) \to \mathbb{R}$ by setting, for all $x, y \in X - \{r\}$,

$$\delta_r(x,y) = \begin{cases} \frac{1}{2}(\delta(x,y) - \delta(r,x) - \delta(r,y)), & \text{if } x \neq y, \\ 0, & \text{if } x = y. \end{cases}$$

The map $(x, y) \mapsto \delta_r(x, y)$ is sometimes called the *Gromov product* of x and y; while a slight modification of this transformation (adding a constant to make the δ_r values non-negative) is called the *Farris transform* of δ. Lemma 7.2.2 shows that the four-point condition on δ is equivalent to the ultrametric condition on δ_r for all $r \in X$. Its straightforward proof is left to the reader.

Lemma 7.2.2. *A non-negative dissimilarity map δ on X satisfies the four-point condition if and only if δ_r is an ultrametric for all $r \in X$.*

Ultrametrics arise naturally in the context of rooted phylogenetic trees by assigning weights to the vertices of the tree, rather than to the edges. In particular, suppose that

$\mathcal{T} = (T;\phi)$ is a rooted phylogenetic X-tree with root ρ. Let $h : \mathring{V}(T) \rightarrow \mathbb{R}$ be a map such that, for all $u, v \in \mathring{V}(T)$, $h(u) \geq h(v)$ if $u \leq_T v$. Let $d_{(\mathcal{T};h)}$ be the dissimilarity map on X defined, for all $x, y \in X$, by setting

$$d_{(\mathcal{T};h)}(x,y) = \begin{cases} h(\mathrm{lca}_{\mathcal{T}}(\phi(x),\phi(y))), & \text{if } x \neq y, \\ 0, & \text{if } x = y. \end{cases}$$

One can easily check that $d_{(\mathcal{T};h)}$ is an ultrametric on X. Given a dissimilarity map δ on X, we say that $(\mathcal{T};h)$ is a *vertex representation of* δ if

$$\delta(x,y) = d_{(\mathcal{T};h)}(x,y)$$

for all $x, y \in X$.

Example 7.2.3. The table in Fig. 7.3(a) describes a dissimilarity map δ on $\{a,b,c,d,e\}$, where the (i,j)-th entry of the table is equal to $\delta(i,j)$. Figure 7.3(b) illustrates a vertex representation of δ.

An equivalent way that ultrametrics arise in the setting of rooted phylogenetic trees is as follows. Let $\mathcal{T} = (T;\phi)$ be a rooted phylogenetic X-tree with root ρ. An edge-weighting $w : E(T) \rightarrow \mathbb{R}$ of $\mathcal{T}$ is *equidistant* if it satisfies the following two properties:

(i) for all $x, y \in X$,

$$d_{(T;w)}(\rho,\phi(x)) = d_{(T;w)}(\rho,\phi(y));$$

(ii) if $u \leq_T v \leq_T \phi(x)$, for some $u, v \in \mathring{V}(T)$ and $x \in X$, then

$$d_{(T;w)}(\phi(x),v) \leq d_{(T;w)}(\phi(x),u).$$

(a)

	a	b	c	d	e
a	0	-4	2	8	8
b		0	2	8	8
c			0	8	8
d				0	6
e					0

(b) (c)

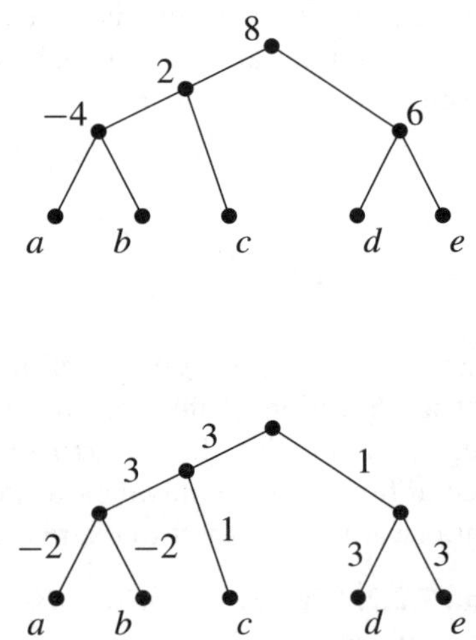

Figure 7.3 (a) A dissimilarity map δ, (b) a vertex representation of δ, and (c) an equidistant representation of δ.

We make two immediate observations concerning equidistant edge-weightings that we will freely use in this section. First, it follows by (i) that if $u \leq_T \phi(x)$ and $u \leq_T \phi(y)$ for some $u \in V(T)$ and $x, y \in X$, we have $d_{(\mathcal{T};w)}(u, \phi(x)) = d_{(\mathcal{T};w)}(u, \phi(y))$. Second, because of (ii), the weighting of an interior edge must be non-negative. Combining these two observations, we immediately obtain Lemma 7.2.4.

Lemma 7.2.4. *If w is an equidistant edge-weighting on a phylogenetic X-tree $\mathcal{T}$, then $d_{(\mathcal{T};w)}$ is an ultrametric on X.*

Let δ be a dissimilarity map on X. If $\mathcal{T}$ is a rooted phylogenetic X-tree and w is an equidistant weighting of the edges of $\mathcal{T}$ such that $\delta(x, y) = d_{(\mathcal{T};w)}(x, y)$ for all $x, y \in X$, then $(\mathcal{T}; w)$ is an *equidistant representation of* δ. As an example, Fig. 7.3(c) shows an equidistant representation of the dissimilarity map given by the table in Fig. 7.3(a).

There is a fundamental correspondence between vertex representations and equidistant representations of a dissimilarity map. Let δ be a dissimilarity map on X and suppose that $(\mathcal{T}; h)$ is a vertex representation of δ. Let T denote the underlying tree of $\mathcal{T}$ and let $w_h : E(T) \to \mathbb{R}$ be the map defined by setting, for all edges $\{u, v\}$ with $u \leq_T v$,

$$w_h(u, v) = \begin{cases} \frac{1}{2}(h(u) - h(v)), & \text{if } \{u, v\} \text{ is an interior edge of } T, \\ \frac{1}{2}h(u), & \text{if } \{u, v\} \text{ is a pendant edge of } T. \end{cases}$$

Then, the pair $(\mathcal{T}; w_h)$ is an equidistant representation of δ. Conversely, suppose that $(\mathcal{T}'; w)$ is an equidistant representation of a dissimilarity map δ' on X. Let T' denote the underlying tree of $\mathcal{T}'$ and let $h_w : \mathring{V}(T') \to \mathbb{R}$ be the map defined by setting, for all $u \in \mathring{V}(T')$,

$$h_w(u) = 2d_{(\mathcal{T}';w)}(u, \phi(x)),$$

where $u \leq_{T'} \phi(x)$ for some element x of X. Then, the pair $(\mathcal{T}'; h_w)$ is a vertex representation of δ'. Furthermore, the two processes just described are inverses of each other.

Remark. In defining a vertex representation and an equidistant representation of a dissimilarity map δ on X, we could have omitted the requirement that the associated rooted X-tree is phylogenetic. However, this would render the above correspondence between the two representations less transparent.

The significance of a vertex and an equidistant representation of a dissimilarity map is highlighted in Theorem 7.2.5.

Theorem 7.2.5. *The following statements are equivalent for a dissimilarity map δ on X:*

(i) δ is an ultrametric;
(ii) there exists a vertex representation of δ;
(iii) there exists an equidistant representation of δ.

Proof. The equivalence of (ii) and (iii) was shown above. The fact that (iii) implies (i) is given by Lemma 7.2.4. We complete the proof of Theorem 7.2.5 by showing that (i) implies (iii).

The proof that (i) implies (iii) is by induction on $|X|$. Evidently, the implication holds for $|X| \in \{1, 2\}$. Suppose that δ is an ultrametric on $X = \{x, y, z\}$ such that $\delta(x, y) \le \delta(x, z) = \delta(y, z)$. Let $\mathcal{T} = (T; \phi)$ be the rooted binary phylogenetic X-tree with root ρ such that $\mathrm{lca}(x, z) = \mathrm{lca}(y, z) = \rho$. Let u denote the interior vertex of T that is not equal to ρ. Let $w : E(T) \to \mathbb{R}$ be the map defined by $w(u, \phi(x)) = w(u, \phi(y)) = \frac{1}{2}\delta(x, y)$, $w(\rho, \phi(z)) = \frac{1}{2}\delta(x, z)$, and $w(\rho, u) = \frac{1}{2}(\delta(x, z) - \delta(x, y))$. It is easily seen that $(\mathcal{T}; w)$ is an equidistant representation of δ. Therefore, the implication holds for $|X| = 3$.

Now suppose that $|X| = n$, where $n \ge 4$, and assume that the implication holds for all smaller values of n. Let a and b be distinct elements of X such that $\delta(a, b)$ is minimized. Then, as δ is an ultrametric, $\delta(a, b) \le \delta(a, x) = \delta(b, x)$ for all $x \in X - \{a, b\}$. Let l_{ab} be an element not in X, and define a dissimilarity map δ' on $X' = (X - \{a, b\}) \cup \{l_{ab}\}$ by setting

$$\delta'(x, y) = \begin{cases} \delta(x, y), & \text{if } x, y \in X, \\ \delta(a, y), & \text{if } x = l_{ab}, \\ \delta(x, a), & \text{if } y = l_{ab}. \end{cases}$$

Since δ is an ultrametric, δ' is clearly an ultrametric on X'. Therefore, by the induction assumption, there exists an equidistant representation $(\mathcal{T}'; w')$ of δ', where $\mathcal{T}' = (T'; \phi')$. Let u and v denote the vertices of T' such that u is the immediate ancestor of $v = \phi'(l_{ab})$ under the partial order $\le_{T'}$. Let $(\mathcal{T}; w)$ be the edge-weighted rooted phylogenetic X-tree obtained from $(\mathcal{T}'; w')$ by adjoining two leaves, one labelled a and the other labelled b, to v, by weighting each of the adjoined edges $\frac{1}{2}\delta(a, b)$, and by re-weighting the edge $\{u, v\}$ with the value

$$\tfrac{1}{2} \min\{\delta'(l_{ab}, x) - \delta(a, b) : x \in X' - \{l_{ab}\}\}.$$

By our choice of a and b, $w(u, v) \ge 0$. Using the fact that $\delta(a, x) = \delta(b, x)$ for all $x \in X - \{a, b\}$ and that $(\mathcal{T}'; w')$ is an equidistant representation of δ', it is easily checked that $(\mathcal{T}; w)$ is an equidistant representation of δ. This completes the proof of Theorem 7.2.5. □

Theorem 7.2.5 provides a convenient way to establish the following fundamental characterization of tree metrics. It is due independently to Zaretskii (1965), Simões-Pereira (1969), and Buneman (1971, 1974*b*).

Theorem 7.2.6 (Tree-Metric Theorem). *Let δ be a dissimilarity map on X. Then, δ is a tree metric on X if and only if δ satisfies the four-point condition.*

Proof. If δ is a tree metric, then, by Lemma 7.1.7, δ satisfies the four-point condition. To establish the converse, suppose that δ satisfies the four-point condition. Let r be an element of X. By Lemma 7.2.2, δ_r is an ultrametric on $X' = X - \{r\}$ and so, by Theorem 7.2.5, there exists an equidistant representation $(\mathcal{T}'; w')$ of δ_r. Let $\mathcal{T}' = (T'; \phi')$ and, for any element $x \in X'$, let $p = -d_{(\mathcal{T}'; w')}(\rho, \phi'(x))$. Note that by (i) in the definition of an equidistant representation, p is independent of the choice of x. Let $\mathcal{T}$ be the phylogenetic X-tree obtained from $\mathcal{T}'$ by adding a new leaf adjacent to the root of $\mathcal{T}'$ via an edge, e_r say, and labelling the new leaf r. Let w be the edge-weighting on $\mathcal{T}$ defined as follows:

$$w(e) = \begin{cases} 2w'(e), & \text{if } e \in \mathring{E}(\mathcal{T}'), \\ 2w'(e) + \delta(x,r), & \text{if } e = \{\phi'(x), u\} \text{ for some } u \in V(\mathcal{T}') \text{ and } x \in X', \\ 2p, & \text{if } e = e_r. \end{cases}$$

Since δ satisfies the four-point condition, δ satisfies the triangle inequality. Using the definition of δ_r, it is now readily checked that $w(e) \geq 0$ for every edge e of $\mathcal{T}$. In particular, $w(e_r) = 2p \geq 0$ since if we select two elements $x, y \in X'$ for which ρ lies on the path connecting $\phi(x)$ and $\phi(y)$, we have $\delta_r(x,y) = d_{(\mathcal{T}';w')}(x,y) = -2p$ and, by the triangle inequality, $\delta_r(x,y) \leq 0$.

We next show that $\delta(x,y) = d_{(\mathcal{T};w)}(x,y)$ for all $x, y \in X$. Evidently, this holds if $x = y$. Therefore, we may assume that $x \neq y$. Suppose that $r \notin \{x,y\}$. Then, by definition of δ_r, we have $\delta(x,y) = \delta(r,x) + \delta(r,y) + 2\delta_r(x,y)$, and so

$$\delta(x,y) = \delta(r,x) + \delta(r,y) + d_{(\mathcal{T}';w')}(x,y) = d_{(\mathcal{T};w)}(x,y).$$

Now suppose that $r \in \{x,y\}$. Without loss of generality we may assume that $y = r$. Then,

$$d_{(\mathcal{T};w)}(x,r) = 2p + 2d_{(\mathcal{T}';w')}(\rho, \phi(x)) + \delta(x,r) = \delta(x,r).$$

Hence, $\delta = d_{(\mathcal{T};w)}$, where w is non-negative. It now follows by Lemma 7.1.4 that δ is a tree metric. This completes the proof of Theorem 7.2.6. □

Corollary 7.2.7 is a reformulation of the Tree-Metric Theorem and is an immediate consequence of it.

Corollary 7.2.7. *Let δ be a dissimilarity map on X. Then, δ is a tree metric on X if and only if, for every subset X' of X of size at most four, δ restricted to $X' \times X'$ is a tree metric.*

The next result shows that under a mild restriction an ultrametric has a unique equidistant representation and a unique vertex representation on a rooted phylogenetic tree. The proof of this result is analogous to Theorem 7.1.8 and is omitted.

Theorem 7.2.8. *Let δ be an ultrametric on X. Then, up to isomorphism,*

(i) there is a unique equidistant representation $(\mathcal{T}; w)$ of δ provided $w(e)$ is non-zero for each interior edge e of $\mathcal{T}$; and

(ii) there is a unique vertex representation $(\mathcal{T}; h)$ of δ provided $h(u) \neq h(v)$ for each interior edge $\{u, v\}$ of $\mathcal{T}$.

We end this section by describing, for an arbitrary dissimilarity map δ on X, an associated canonical ultrametric, called the *sub-dominant ultrametric* of δ, which is denoted δ_U. Let $G_\delta(X)$ denote the edge-weighted complete graph on X, where each edge $\{x, y\}$ is assigned weight $\delta(x, y)$. For each path P in this graph, let $w(P)$ denote the maximum weight of an edge in P. Define $\delta_U : X \times X \to \mathbb{R}$ by setting, for all $x, y \in X$,

$$\delta_U(x, y) = \min_P\{w(P)\colon\ P \text{ is a path in } G_\delta(X) \text{ from } x \text{ to } y\}.$$

Theorem 7.2.9 lists some basic properties of δ_U.

Theorem 7.2.9. *Let δ be a dissimilarity map on X. Then,*

(i) δ_U is an ultrametric on X.

(ii) $\delta_U(x, y) \leq \delta(x, y)$ for all $x, y \in X$.

(iii) δ_U is the (pointwise) maximal ultrametric satisfying property (ii)*; that is, if δ' is an ultrametric on X that satisfies $\delta'(x, y) \leq \delta(x, y)$ for all $x, y \in X$, then $\delta'(x, y) \leq \delta_U(x, y)$ for all $x, y \in X$.*

(iv) δ is an ultrametric on X precisely if $\delta = \delta_U$.

Proof. We first prove (i). Let $x, y, z \in X$, and let P_1 and P_2 be paths in $G_\delta(X)$ from x to z and from z to y, respectively, such that $w(P_1) = \delta_U(x, z)$ and $w(P_2) = \delta_U(y, z)$. Let P_{12} denote the (unique) path from x to y consisting of P_1 followed by P_2. Then,

$$\delta_U(x, y) \leq w(P_{12}) \leq \max\{w(P_1), w(P_2)\} = \max\{\delta_U(x, z), \delta_U(y, z)\}.$$

It follows that δ_U is an ultrametric on X.

We next prove (ii). For all $x, y \in X$, the two-vertex path P in $G_\delta(X)$ consisting of x and y satisfies $w(P) = \delta(x, y)$. Therefore, by definition, $\delta_U(x, y) \leq w(P) = \delta(x, y)$ for all $x, y \in X$, completing the proof of (ii).

To prove (iii), suppose that δ' is an ultrametric on X that satisfies $\delta'(x, y) \leq \delta(x, y)$ for all $x, y \in X$. Let P be a path in $G_\delta(X)$ from t to z with $w(P) = \delta_U(t, z)$; we will write P as $t = v_0, v_1, \ldots, v_k = z$. Using induction on the length of P and the fact that δ' is an ultrametric on X, it straightforward to show that

$$\delta'(t, z) \leq \max_{0 \leq i \leq k-1}\{\delta'(v_i, v_{i+1})\}. \tag{7.1}$$

Therefore, as $\delta_U(t, z) = \max_{0 \leq i \leq k}\{\delta(v_i, v_{i+1})\}$ and as $\delta'(x, y) \leq \delta(x, y)$ for all $x, y \in X$, the right-hand side of (7.1) is at most $\delta_U(t, z)$. Part (iii) immediately follows.

Lastly, we prove (iv). If $\delta = \delta_U$ then, by (i), δ is an ultrametric on X. Conversely, if δ is an ultrametric on X, then taking $\delta' = \delta$ in (iii), we have $\delta(x, y) \leq \delta_U(x, y)$ for all $x, y \in X$. However, the reverse inequality also holds by (ii). Hence $\delta = \delta_U$, completing the proof of (iv). □

For a dissimilarity map δ on X, there is a more constructive way to describe δ_U. Let T be a minimum-weight spanning tree of $G_\delta(X)$. For all $x, y \in X$, let $\delta_T(x, y)$ denote the maximum weight of an edge in the (unique) path in T from x to y. Proposition 7.2.10 shows that, regardless of the particular choice of a minimum-weight spanning tree, $\delta_T(x, y) = \delta_U(x, y)$ for all $x, y \in X$.

Proposition 7.2.10. *Let δ be a dissimilarity map on X and let T be a minimum-weight spanning tree of $G_\delta(X)$. Then, for all $x, y \in X$, $\delta_T(x, y) = \delta_U(x, y)$.*

Proof. Let x and y be elements of X. By definition, $\delta_U(x, y) \leq \delta_T(x, y)$. Assume that $\delta_U(x, y) < \delta_T(x, y)$. We obtain a contradiction by showing that under this assumption T is not a minimum-weight spanning tree of $G_\delta(X)$. Let e_0 denote an edge in the path in T from x to y of maximum weight. Since $\delta_U(x, y) < \delta_T(x, y)$, there exists a path P in $G_\delta(X)$ from x to y such that $w(e) < w(e_0)$ for all edges e in P. Now, deleting e_0 from T partitions X into two blocks with x being in one block and y being in the other block. Since P connects x and y in $G_\delta(X)$, there must be an edge e_1 (distinct from e_0) in P whose end vertices lie in different blocks of this partition. Let T' be the spanning tree obtained from T by deleting e_0 and adding e_1. Since $w(e_0) > w(e_1)$, the total weight of T' is less than T; a contradiction. We conclude that $\delta_T(x, y) = \delta_U(x, y)$ for all $x, y \in X$. □

Proposition 7.2.10 provides us with an efficient method for calculating δ_U, since one can construct a minimum-weight spanning tree of any graph in polynomial time (Kruskal 1956).

A further way to describe the sub-dominant ultrametric of a dissimilarity map δ on X is to make explicit the set of clusters of an equidistant representation of δ_U (this is related to the 'single linkage method' in clustering theory). For each fixed $t \in \mathbb{R}$, let G_t denote the graph that has vertex set X and that has an edge between two vertices x and y precisely if $\delta(x, y) \leq t$. Let $\Delta = \{\delta(x, y)\colon\ x, y \in X\}$. It follows from Proposition 7.2.10 that, for all $x, y \in X$,

$$\delta_U(x, y) = \min_{t \in \Delta}\{t\colon x \text{ and } y \text{ are in the same component of } G_t\}.$$

Furthermore, it is easily seen that the collection $\mathcal{H}$ of subsets of X that comprise the vertices of the connected components in the collection of graphs $\{G_t : t \in \Delta\}$ is a hierarchy on X. We leave it as an exercise to show that the rooted phylogenetic X-tree $\mathcal{T}$ whose set of clusters is $\mathcal{H} - \{X\}$ provides an equidistant representation of δ_U.

7.3 Reconstruction methods

In this section, we describe some methods for constructing an edge-weighted X-tree from an arbitrary dissimilarity map on X.

A desirable feature of such a construction is that it satisfies the following *consistency* property: when applied to a tree metric $d_{(\mathcal{T};w)}$ on X, the construction returns the X-tree $\mathcal{T}$ and its weighting w. For some methods, this property is guaranteed only when $\mathcal{T}$ is a binary phylogenetic tree. However, this consistency property always holds for the first method that we discuss (Buneman 1971). In this method, the given dissimilarity map on X associates a real number to every split of X as follows. Let $\delta : X \times X \rightarrow \mathbb{R}^{\geq 0}$ be a dissimilarity map. For all X-splits $\sigma = A|B$, set

$$B_\delta(\sigma) = \tfrac{1}{2}\min\{\delta(a,b) + \delta(a',b') - \delta(a,a') - \delta(b,b')\colon a, a' \in A \text{ and } b, b' \in B\}.$$

The quantity $B_\delta(\sigma)$ is called the *Buneman index* of split σ. In order to exploit this index, we first establish two preliminary lemmas.

Lemma 7.3.1. *Let $\delta : X \times X \to \mathbb{R}^{\geq 0}$ be a dissimilarity map, and let σ_1 and σ_2 be X-splits. If $B_\delta(\sigma_1) > 0$ and $B_\delta(\sigma_2) > 0$, then σ_1 and σ_2 are compatible.*

Proof. Let $\sigma_1 = A_1|B_1$ and $\sigma_2 = A_2|B_2$. Suppose that σ_1 and σ_2 are not compatible. Then, there exist elements $a \in A_1 \cap A_2$, $a' \in A_1 \cap B_2$, $b \in B_1 \cap A_2$, and $b' \in B_1 \cap B_2$. Since $B_\delta(\sigma_1) > 0$,

$$\delta(a,b) + \delta(a',b') - \delta(a,a') - \delta(b,b') > 0. \tag{7.2}$$

Furthermore, as $B_\delta(\sigma_2) > 0$,

$$\delta(a,a') + \delta(b,b') - \delta(a,b) - \delta(a',b') > 0. \tag{7.3}$$

But (7.3) contradicts (7.2). Hence, σ_1 and σ_2 are compatible. □

The proof of our second preliminary result is straightforward and left as an exercise.

Lemma 7.3.2. *Let $\delta = d_{(\mathcal{T};w)}$ be a tree metric on X and let σ be an X-split. Then, $B_\delta(\sigma) > 0$ if and only if σ is a split of X induced by $\mathcal{T}$, in which case $B_\delta(\sigma) = w(e)$ for the edge e of $\mathcal{T}$ corresponding to σ.*

For a dissimilarity map $\delta : X \times X \to \mathbb{R}^{\geq 0}$, let

$$\Sigma_B(\delta) = \{\sigma\colon \sigma \text{ is an } X\text{-split and } B_\delta(\sigma) > 0\}.$$

By Lemma 7.3.1, $\Sigma_B(\delta)$ is pairwise compatible, and therefore, by the Splits-Equivalence Theorem, there exists a unique X-tree $\mathcal{T}_\delta$ whose associated set of X-splits is equal to $\Sigma_B(\delta)$. Let $w : E(\mathcal{T}_\delta) \to \mathbb{R}^{>0}$ be the map defined, for all $e \in E(\mathcal{T}_\delta)$, by setting $w(e) = B_\delta(\sigma_e)$, where σ_e is the X-split of $\mathcal{T}_\delta$ induced by e. Let δ_B denote the tree metric $d_{(\mathcal{T}_\delta;w)}$. Theorem 7.3.3 describes two fundamental properties of this construction.

Theorem 7.3.3. *Let $\delta : X \times X \to \mathbb{R}^{\geq 0}$ be a dissimilarity map. Then*

(i) $\delta_B(x,y) \leq \delta(x,y)$, *for all $x, y \in X$; and*
(ii) $\delta_B(x,y) = \delta(x,y)$, *for all $x, y \in X$, if and only if δ satisfies the four-point condition.*

Proof. To prove (i), let x and y be elements of X. Let $e_1, e_2, \ldots, e_k$ denote the ordered edges in the path from x to y in $\mathcal{T}_\delta$. Then,

$$\delta_B(x,y) = B(\sigma_{e_1}) + B(\sigma_{e_2}) + \cdots + B(\sigma_{e_k}). \tag{7.4}$$

For all $i \in \{1, 2, \ldots, k\}$, write $\sigma_{e_i} = A_i|B_i$ where $x \in A_i$. Since $\mathcal{T}_\delta$ is an X-tree, there exist distinct elements $z_1, z_2, \ldots, z_{k-1}$ of $X - \{x, y\}$ such that, for each $i \in \{1, 2, \ldots, k-1\}$, z_i is an element of both $\bigcap_{j \leq i} B_j$ and $\bigcap_{j > i} A_j$. It follows that

$$B(\sigma_{e_1}) \leq \tfrac{1}{2}(\delta(x, y) + \delta(x, z_1) - \delta(x, x) - \delta(y, z_1)),$$
$$B(\sigma_{e_k}) \leq \tfrac{1}{2}(\delta(x, y) + \delta(z_{k-1}, y) - \delta(x, z_{k-1}) - \delta(y, y)),$$

and, for all $j \in \{2, 3, \ldots, k-1\}$,

$$B(\sigma_{e_j}) \leq \tfrac{1}{2}(\delta(x, z_j) + \delta(z_{j-1}, y) - \delta(x, z_{j-1}) - \delta(z_j, y)).$$

Summing these inequalities and applying (7.4) gives $\delta_B(x, y) \leq \delta(x, y)$ as required.

We next prove (ii). If $\delta_B(x, y) = \delta(x, y)$ for all $x, y \in X$, then, as δ_B is a tree metric, it follows by Lemma 7.1.7 that δ_B, and hence δ, satisfies the four-point condition.

Conversely, suppose that δ satisfies the four-point condition. Then, by the Tree-Metric Theorem, δ is a tree metric. Let $(\mathcal{T}; w)$ be a tree metric representation of δ. For each edge e of $\mathcal{T}$, Lemma 7.3.2 implies that

$$B_\delta(\sigma_e) = w(e). \tag{7.5}$$

Thus, since $w(e) > 0$, σ_e is an element of $\Sigma(\mathcal{T}_\delta)$. Furthermore, as this holds for all edges of $\mathcal{T}$, we have $\Sigma(\mathcal{T}) \subseteq \Sigma(\mathcal{T}_\delta)$ and so $\mathcal{T}_\delta$ refines $\mathcal{T}$. Combining this observation with (7.5) it follows that, for all $x, y \in X$,

$$\delta_B(x, y) \geq \delta(x, y).$$

But by (i), $\delta_B(x, y) \leq \delta(x, y)$ for all $x, y \in X$. Hence, $\delta_B(x, y) = \delta(x, y)$ for all $x, y \in X$. □

Given a dissimilarity map δ on X, there is an algorithm that runs in polynomial time in $|X|$ to construct δ_B, and thereby a tree-metric representation of δ_B. This algorithm is based on the idea of building up splits of subsets of X with positive Buneman index by the sequential addition of elements of X (in any order) to the domain of δ_B. We leave the details as an exercise. In applications, δ_B tends to yield highly unrefined trees. However, it does have several desirable properties; for example, the map $\delta \to \delta_B$ is continuous. We will explore this further in the second specialist topic of this chapter.

The Buneman construction has a modification that is based on a 'refined' Buneman index. The set of X-splits for which this index is positive is the set of splits of an X-tree that refines $\mathcal{T}_\delta$. For details, see Moulton and Steel (1999).

Two other techniques, which we only briefly describe, are tailored to the reconstruction of binary phylogenetic trees.

Neighbour Joining. *Neighbour joining* has become a popular tree reconstruction method in biology, particularly since it is computationally fast. Neighbour joining is a recursive method for reconstructing an edge-weighted binary phylogenetic tree from an arbitrary dissimilarity map. It is based on the inductive step of replacing a cherry in a binary phylogenetic tree by a single leaf.

For a tree metric $\delta = d_{(\mathcal{T};w)}$ on X, where $\mathcal{T}$ is a phylogenetic X-tree, one can readily identify a pair of elements $x, y \in X$ for which the pair $\{\phi(x), \phi(y)\}$ forms a cherry of $\mathcal{T}$. For example, we can obtain such a pair x, y by first choosing any element $r \in X$, and then selecting $x, y \in X$ to maximize the quantity $\delta_r(x, y) = \frac{1}{2}(\delta(r, x) + \delta(r, y) - \delta(x, y))$. The inductive step of the neighbour-joining method is a slight refinement of this observation to remove dependence on a specific choice of r. In particular, given an arbitrary dissimilarity map δ on X, neighbour joining selects a pair x and y from X so as to maximize the quantity

$$D(x, y) = \delta(x, y) + \sum_{r \neq x,y} \delta_r(x, y).$$

The proof of Proposition 7.3.4 is left as an exercise.

Proposition 7.3.4. *Let δ be a $\mathcal{T}$-metric on X, where $\mathcal{T}$ is a phylogenetic tree. If $x, y \in X$ are chosen to maximize $D(x, y)$, then $\{\phi(x), \phi(y)\}$ forms a cherry of $\mathcal{T}$.*

Once x, y have been chosen, the method proceeds as follows. A dissimilarity map δ' is defined on $X' = (X - \{x, y\}) \cup \{v_{xy}\}$, where v_{xy} is not an element of X. Since $|X'| < |X|$, we may assume, by induction, that we have already constructed an edge-weighted phylogenetic tree $\mathcal{T}'$ on X' by applying neighbour joining to δ'. We now add two new leaves adjacent to v_{xy}, one labelled x and the other labelled y, to obtain a phylogenetic X-tree $\mathcal{T}$. Finally, we assign weights by an appropriate formula to the two new added edges to obtain an edge-weighting of $\mathcal{T}$.

Using Proposition 7.3.4, one can establish a consistency property for the neighbour-joining method by showing that if $\delta = d_{(\mathcal{T};w)}$ is a tree metric, where $\mathcal{T}$ is a binary phylogenetic tree and w is strictly positive, then the method recovers the pair $(\mathcal{T}; w)$. For details see Gascuel (1997).

Optimization-based approaches (minimum evolution). Given a dissimilarity map δ on X, one may seek an edge-weighted phylogenetic X-tree, so that the induced tree metric on X provides a 'best fit' to δ under some optimality criterion.

This approach for reconstructing phylogenetic trees is analogous to maximum parsimony, but it uses dissimilarities on X rather than characters on X. As in the parsimony setting, we face a two-step optimization process: first, for each phylogenetic X-tree, we select optimal edge-weights; and second, we select an optimal edge-weighted tree. Consequently, it is helpful to discuss these two optimization processes separately.

Given a dissimilarity map δ on X and a phylogenetic X-tree $\mathcal{T} = (T; \phi)$, let $w_{(\mathcal{T};\delta)} : E(T) \to \mathbb{R}$ be chosen so that $w = w_{(\mathcal{T};\delta)}$ minimizes the quantity

$$\Delta(\mathcal{T}, \delta) = \sum_{x,y \in X} |\delta(x, y) - d_{(\mathcal{T};w)}(x, y)|^2. \tag{7.6}$$

The real-valued edge-weighting of $\mathcal{T}$ that minimizes the expression in (7.6) is unique and can be described explicitly as follows. Let $K_{\mathcal{T}}$ denote the 0–1 matrix that has its rows indexed by the two-element subsets of X, its columns indexed by the edges of $\mathcal{T}$ and all its entries defined by

$$(K_{\mathcal{T}})_{\{x,y\},e} = \begin{cases} 1, & \text{if } e \in P(T;\phi(x),\phi(y)), \\ 0, & \text{otherwise.} \end{cases}$$

Now, if we write $d_{(\mathcal{T};w)}$ as a vector indexed over the two-element subsets of X and w as a vector indexed over the edges of $\mathcal{T}$, then

$$d_{(\mathcal{T};w)} = K_{\mathcal{T}}w. \tag{7.7}$$

Consequently, by Theorem 7.1.8, the mapping $w \mapsto K_{\mathcal{T}}w$ is one-to-one, and so $K_{\mathcal{T}}$ has full rank. Thus, by standard techniques from linear algebra, the unique vector $w_{(\mathcal{T};\delta)}$ of edge weights for $\mathcal{T}$ that minimizes $\Delta(\mathcal{T},\delta)$ is

$$w_{(\mathcal{T};\delta)} = (K_{\mathcal{T}}^{\mathrm{t}}K_{\mathcal{T}})^{-1}K_{\mathcal{T}}^{\mathrm{t}}\delta, \tag{7.8}$$

where, for a square matrix A, the matrix transpose and inverse of A is denoted A^{t} and A^{-1} respectively, and δ is regarded as a vector indexed over the two-element subsets of X. Notice that if we take $\delta = d_{(\mathcal{T};w)}$, then, by (7.7) and (7.8), we have

$$w_{(\mathcal{T};\delta)} = (K_{\mathcal{T}}^{\mathrm{t}}K_{\mathcal{T}})^{-1}K_{\mathcal{T}}^{\mathrm{t}}K_{\mathcal{T}}w = w. \tag{7.9}$$

Equation (7.8) can be developed further to provide a more transparent closed-form expression of $w_{(\mathcal{T};\delta)}$ (Vach 1989; Rzhetsky and Nei 1993), or to give a fast recursive approach to computing $w_{(\mathcal{T};\delta)}$. The latter recursive approach is essentially equivalent to the recursive determination of edge weights in the neighbour-joining method. Gascuel (1997) and Denis and Gascuel (2002) provide further details.

We now turn to the second step in the optimization process: the selection of an optimal edge-weighted phylogenetic tree. One approach is to select a phylogenetic X-tree $\mathcal{T}$ so as to minimize $\Delta(\mathcal{T},w_{(\mathcal{T};\delta)})$. By Theorem 7.1.8, this approach will recover a binary phylogenetic tree $\mathcal{T}$ when applied to a dissimilarity map that is a $\mathcal{T}$-metric. However, the problem of finding a phylogenetic X-tree $\mathcal{T}$ to minimize $\Delta(\mathcal{T},w_{(\mathcal{T};\delta)})$ for an arbitrary dissimilarity map δ on X is known to be NP-hard (Day 1987).

Another approach, closer in spirit to the maximum parsimony method, is to find a phylogenetic X-tree $\mathcal{T}$ that minimizes the quantity

$$L(\mathcal{T},\delta) = \sum_{e \in E(\mathcal{T})} w_{(\mathcal{T};\delta)}(e), \tag{7.10}$$

where $w_{(\mathcal{T};\delta)}$ is the edge-weighted function described by (7.8). This approach is known as the *minimum evolution* method for dissimilarities. By (7.8), we may write $L(\mathcal{T},\delta) = \sum_{\{x,y\}} k_{\{x,y\}}\delta(x,y)$, where k is a vector $(K_{\mathcal{T}}^{\mathrm{t}}K_{\mathcal{T}})^{-1}K_{\mathcal{T}}^{\mathrm{t}}$; in particular, $L(\mathcal{T},\delta)$ is a linear function of δ. When $\delta = d_{(\mathcal{T};w)}$, (7.9) gives

$$L(\mathcal{T},\delta) = \sum_{e \in E(\mathcal{T})} w(e). \tag{7.11}$$

It is not immediately apparent that the minimum evolution method satisfies the consistency property for binary phylogenetic trees. However, Rzhetsky and Nei (1993) established this, and Denis and Gascuel (2002) subsequently found a much simpler proof of the following result.

Theorem 7.3.5. *Let $\mathcal{T}$ be a binary phylogenetic X-tree, let w be a positive weighting of the edges of $\mathcal{T}$, and let $\delta = d_{(\mathcal{T};w)}$. Suppose that $\mathcal{T}'$ is any phylogenetic X-tree that differs from $\mathcal{T}$. Then,*

$$L(\mathcal{T}', \delta) > L(\mathcal{T}, \delta).$$

Consequently, the minimum evolution method recovers $\mathcal{T}$ when applied to any $\mathcal{T}$-metric.

Sometimes in applications, the function L in (7.10) is replaced by either

$$L^{\geq 0}(\mathcal{T}, \delta) = \sum_e \max\{w_{(\mathcal{T};\delta)}(e), 0\} \quad \text{or} \quad L^{+}(\mathcal{T}, \delta) = \sum_e |w_{(\mathcal{T};\delta)}(e)|.$$

Using (7.11), it is easily checked that Theorem 7.3.5 holds when L is substituted for either of these two functions. The problem of finding a minimum evolution tree from an arbitrary dissimilarity map appears to be difficult, and no exact polynomial–time algorithm is known.

We end this section by noting that several modified versions of both neighbour joining and minimum evolution have been developed and applied. For example, it is helpful for statistical applications to generalize the expression in (7.6) from a simple sum of squares to a general quadratic form. However, in this case, Theorem 7.3.5 may fail to hold for the corresponding minimum evolution approach. For further details the reader is referred to Denis and Gascuel (2002).

7.4 Split decomposition theory

In Section 7.3, we encountered the Buneman index and saw how it formed the basis of a natural approach to tree reconstruction from a dissimilarity map. In this section, we describe a simple modification to this index called the 'isolation' index. This modification is the basis for a particularly elegant decomposition of any dissimilarity map introduced by Bandelt and Dress (1992).

To make this modification somewhat more transparent, we begin by expressing the Buneman index as a double minimization. Let $\delta : X \times X \to \mathbb{R}^{\geq 0}$ be a dissimilarity map. For an X-split $\sigma = A|B$, the Buneman index of σ on δ can be expressed as follows:

$$B_\delta(\sigma) = \frac{1}{2} \min_{\substack{a,a' \in A \\ b,b' \in B}} \left\{ \min \left\{ \begin{array}{c} \delta(a,b) + \delta(a',b') \\ \delta(a,b') + \delta(a',b) \end{array} \right\} - \delta(a,a') - \delta(b,b') \right\}.$$

Now, let δ be an arbitrary dissimilarity map on X. Then, for an X-split $\sigma = A|B$, the *isolation index* of σ on δ is

$$\max\{0, i_\delta(\sigma)\},$$

where

$$i_\delta(\sigma) = \frac{1}{2} \min_{\substack{a,a' \in A \\ b,b' \in B}} \left\{ \max \left\{ \begin{array}{c} \delta(a,b) + \delta(a',b') \\ \delta(a,b') + \delta(a',b) \end{array} \right\} - \delta(a,a') - \delta(b,b') \right\}.$$

Observe that the expression for $i_\delta(\sigma)$ differs from the Buneman index only in the replacement of the inner minimization by a maximization.

Like the Buneman index, we are particularly interested in the isolation index when it is strictly positive. One indication of this interest is given in Lemma 7.4.1 which parallels the property of the Buneman index highlighted in Lemma 7.3.2. The straightforward proof of this lemma is left to the reader.

Lemma 7.4.1. *Let $\delta = d_{(\mathcal{T};w)}$ be a tree metric on X and let σ be an X-split. Then, $i_\delta(\sigma) > 0$ if and only if σ is a split of X induced by $\mathcal{T}$, in which case $i_\delta(\sigma) = w(e)$ for the edge e of $\mathcal{T}$ corresponding to σ.*

For a dissimilarity map δ on X, let

$$\Sigma_i(\delta) = \{\sigma \colon i_\delta(\sigma) > 0 \text{ and } \sigma \text{ is an } X\text{-split}\}.$$

Theorem 7.4.2 is due to Bandelt and Dress (1992).

Theorem 7.4.2. *Every dissimilarity map δ on X can be expressed as*

$$\delta = \delta_0 + \sum_{A|B \in \Sigma_i(\delta)} i_\delta(A|B) \cdot \delta_{A|B}, \tag{7.12}$$

where $\delta_{A|B}$ is the pseudometric defined, for all $x, y \in X$, by setting

$$\delta_{A|B}(x,y) = \begin{cases} 0, & \text{if } x,y \in A \text{ or } x,y \in B, \\ 1, & \text{otherwise,} \end{cases}$$

and where δ_0 is a pseudometric that satisfies $\Sigma_i(\delta_0) = \emptyset$.

The expression on the right-hand side of (7.12) is called the *split decomposition* of δ. The pseudometric δ_0 in the statement of Theorem 7.4.2 is called the *split-prime residue* (of δ). When δ_0 is identically zero, the dissimilarity map δ is said to be *totally decomposable*, in which case δ can be decomposed into the sum of scalar multiples of elements of $\Sigma_i(\delta)$.

A fundamental example of a totally decomposable dissimilarity map δ on X—and indeed one of the motivations for the theory of split decompositions—is when δ is a tree metric on X. In particular, if $\delta = d_{(\mathcal{T};w)}$ for some X-tree $\mathcal{T}$ and positive edge-weighting w, then it follows by Lemma 7.4.1 that

$$\delta = \sum_{\sigma \in \Sigma(\mathcal{T})} w(e_\sigma) \cdot \delta_\sigma = \sum_{\sigma \in \Sigma_i(\delta)} i_\delta(\sigma) \cdot \delta_\sigma,$$

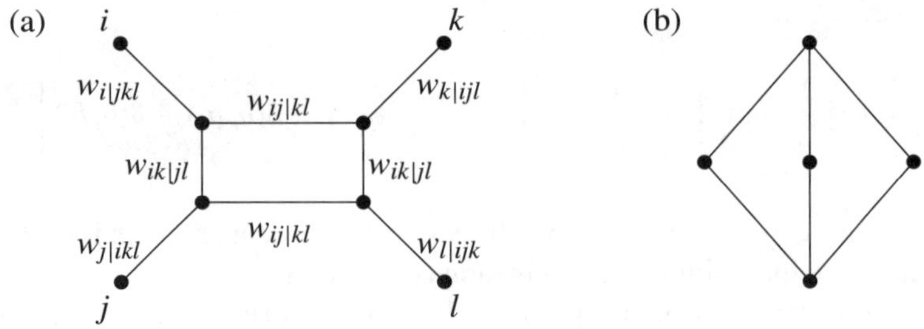

Figure 7.4 (a) A graph representing δ. (b) $K_{2,3}$.

where e_σ is the edge of $\mathcal{T}$ corresponding to σ. However, not all totally decomposable dissimilarity maps are tree metrics as Example 7.4.3 shows.

Example 7.4.3. Let δ be a dissimilarity map on $X = \{i, j, k, l\}$ and suppose that

$$\max\{\delta(i,j) + \delta(k,l), \delta(i,k) + \delta(j,l)\} < \delta(i,l) + \delta(j,k).$$

Because of this inequality, δ does not satisfy the four-point condition and so is not a tree metric on X. Now,

$$\{ij|kl, ik|jl\} \subseteq \Sigma_i(\delta)$$

and $il|jk \notin \Sigma_\sigma(\delta)$. Moreover, regardless of which trivial splits of X are contained in $\Sigma_\sigma(\delta)$, a routine check shows that

$$\delta = \sum_{\sigma \in \Sigma_i(\delta)} i_\delta(\sigma) \cdot \delta_\sigma .$$

Thus, δ is totally decomposable. In fact, we can represent δ by the type of (weighted) graph shown in Fig. 7.4(a), where $\delta(x, y)$ is the sum of the weights on the shortest path from x to y for all $x, y \in X$, and $w_\sigma = i_\delta(\sigma)$ for all $\sigma \in \Sigma_i(\delta)$. Note that the weights of the edges corresponding to trivial splits may be zero depending on whether this split is in $\Sigma_\sigma(\delta)$. Observe that a 'rectangle' has replaced the interior edge of a quartet tree and the non-trivial splits in $\Sigma_i(\delta)$ correspond to the two splits obtained by cutting a pair of parallel edges in the rectangle.

Using the ideas of Example 7.4.3, it is not difficult to show that every dissimilarity map on X, where $|X| \leq 4$, is totally decomposable. However, this does not extend to $|X| = 5$ as Example 7.4.4 shows. Indeed, this example also shows that it is possible for $\Sigma_i(\delta)$ to be empty for some dissimilarity map δ.

Example 7.4.4. Let δ be the dissimilarity map on X described by the graphical distance of the bipartite graph $K_{2,3} = (X, E)$ shown in Fig. 7.4(b). A routine check shows that $\Sigma_i(\delta)$ is empty, and so $\delta = \delta_0$.

Bandelt and Dress (1992) showed that a dissimilarity map δ on X is totally decomposable precisely if every five elements of X satisfy an inequality condition based on δ (see Exercise 13).

Remark. The representation provided by Theorem 7.4.2 underpins the theoretical basis for representing dissimilarities by edge-weighted graphs with classes of parallel edges, as illustrated in Fig. 7.4(a). This method has been implemented in the computer package *SplitsTree* and has proved useful in the analysis of dissimilarity data in molecular biology and psychology (see Dress *et al.* 1996). Bandelt and Dress (1992) provide further details of the rich theory surrounding split decompositions.

We complete our brief introduction into the theory of split decompositions by outlining some of the properties of $\Sigma_i(\delta)$ for a dissimilarity map δ on X. In contrast to the Buneman index, the collection $\Sigma_i(\delta)$ of X-splits does not in general form a compatible collection of splits, and thereby does not correspond to the set of X-splits of an X-tree. Nevertheless, a collection Σ of X-splits of the form $\Sigma_i(\delta)$ for some dissimilarity δ on X must satisfy the following property:

> For any three splits $A_1|B_1$, $A_2|B_2$, and $A_3|B_3$ in Σ, there are no four elements $a, a_1, a_2, a_3 \in X$ such that $\{a, a_1, a_2, a_3\} \cap A_i = \{a, a_i\}$ for all $i \in \{1, 2, 3\}$.

Or, equivalently,

> $\mathcal{Q}(\Sigma)$ contains at most two of the three quartet trees $ij|kl$, $ik|jl$, and $il|jk$ for any four elements $i, j, k, l \in X$.

A collection Σ of X-splits that satisfies the above property is said to be *weakly compatible*—a property that is closely related to weak hierarchies as we now explain.

Let Σ be a collection of X-splits. Select some reference element $r \in X$ and associate each X-split $A|B$ in Σ with the block (A or B) that does not contain r. Then, it is straightforward to show that Σ is weakly compatible if and only if the resulting set of subsets of $X - \{r\}$ is a weak hierarchy on $X - \{r\}$. It was noted in Section 3.7 that the size of a weak hierarchy on $n - 1$ elements is bounded above by $\binom{n}{2}$. Thus, for a dissimilarity map δ on X, the size of $\Sigma_i(\delta)$ is bounded above by $\binom{n}{2}$, where $|X| = n$. Using this observation, there is a simple polynomial–time algorithm for reconstructing $\Sigma_i(\delta)$ from δ by adding the elements of X sequentially, along lines similar to the algorithm for finding the splits of X with positive Buneman index. See Bandelt and Dress (1992) for details of this algorithm.

7.5 Ordinal methods

In this section, we consider the existence and uniqueness of tree metric representations of a dissimilarity map δ on X when we are provided with just the relative order of the $\delta(x, y)$ values for all $x, y \in X$. This is relevant to applications for two reasons. First, due to errors associated with estimating distances between species, we may have more confidence in just the order of these values than in their actual numerical values. Second, as we will see in Chapter 8, for certain types of data, the observed distance between two species is, in expectation, described by a non-linear increasing function f of a tree metric. For some models, it may be possible to determine the particular function f and thereby circumvent this problem. However, we may not know the exact form of f for more general and complex models, and so in this case ordinal methods may be particularly useful. Our treatment here mainly follows Bonnot *et al.* (1996) and Guénoche (1998).

Let δ and δ' be non-negative dissimilarity maps on X. We say that δ and δ' are *order equivalent*, and write $\delta \sim \delta'$, if, for some strictly increasing function $f : \mathbb{R} \to \mathbb{R}$, we have

$$f(\delta(x, y)) = \delta'(x, y)$$

for all $x, y \in X$. In other words, for all $w, x, y, z \in X$, f has the property that $\delta(x, y) < \delta(w, z)$ if and only if $f(\delta(x, y)) < f(\delta(w, z))$. It is clear that order equivalence is an equivalence relationship on the set of non-negative dissimilarity maps on X.

A dissimilarity map $\delta : X \times X \to \mathbb{R}^{\geq 0}$ is a *tree ordinal dissimilarity map* if δ is order equivalent to some tree metric $d_{(\mathcal{T};w)}$ on X. In this case, assuming w is strictly positive, we say that $\mathcal{T}$ provides an *ordinal representation* of δ. Unlike the classical case described by Theorem 7.1.8, it is possible for more than one X-tree to provide an ordinal representation of δ. Indeed, it is possible for every phylogenetic X-tree to be such a representation. Furthermore, not every non-negative dissimilarity map is a tree ordinal dissimilarity map. Example 7.5.1 illustrates these last two facts.

Example 7.5.1. Let δ be a non-negative dissimilarity map on $X = \{1, 2, \ldots, n\}$ defined by setting $\delta(i, j) = 2^i + 2^j$, for all distinct $i, j \in X$. We show that every phylogenetic X-tree provides an ordinal representation of δ. For each such phylogenetic tree $\mathcal{T}$, assign, for all $i \in X$, weight 2^i to the edge of $\mathcal{T}$ incident with leaf i, and assign weight $1/n$ to each interior edge of $\mathcal{T}$. If $d_{\mathcal{T}}$ denotes the associated tree metric, then, for all distinct $i, j \in X$, $d_{\mathcal{T}}(i, j) = 2^i + 2^j + c_{\mathcal{T}}(i, j)$ where $0 < c_{\mathcal{T}}(i, j) < 1$. Hence, every phylogenetic X-tree with the above assigned edge weightings provides an ordinal representation of δ.

To show that not every non-negative dissimilarity map is a tree ordinal dissimilarity map, suppose that δ is now a dissimilarity map on $X = \{a, b, c, d\}$ such that $\delta(a, b)$ and $\delta(c, d)$ are both strictly larger than the values $\delta(a, c)$, $\delta(a, d)$, $\delta(b, c)$, and $\delta(b, d)$. If δ is order equivalent to a tree metric d on X, then $d(a, b)$ and $d(c, d)$ are both strictly larger than the values $d(a, c)$, $d(a, d)$, $d(b, c)$, and $d(b, d)$. Consequently, $d(a, b) + d(c, d) > \max\{d(a, c) + d(b, d), d(a, d) + d(b, c)\}$, contradicting the four-point condition. Thus, δ is not a tree ordinal dissimilarity map.

In contrast to Example 7.5.1, an ordinal representation of a tree ordinal dissimilarity map is unique in one natural case.

Proposition 7.5.2. *Let δ and δ' be order equivalent, non-negative dissimilarity maps on X. Suppose that δ' is an ultrametric. Then, δ is an ultrametric. Furthermore, up to isomorphism, there is a unique X-tree that provides an ordinal representation of δ, namely, the unique X-tree that supports a tree metric representation of δ'.*

Proof. Since δ' is an ultrametric on X and δ is order equivalent to δ', it is clear that δ is an ultrametric on X. Thus, by non-negativity, δ is a tree metric, and so, by Theorem 7.1.8, there is a unique X-tree $\mathcal{T}$ that supports a tree metric representation of δ. Similarly, there is a unique X-tree $\mathcal{T}'$ that supports a tree metric representation of δ'. Now, the order equivalence and ultrametricity of δ and δ' imply that, for all subsets S of X of size at most four,

$$\mathcal{T}|S = \mathcal{T}'|S.$$

Hence, by Theorem 6.3.5, $\mathcal{T}$ and $\mathcal{T}'$ are isomorphic. □

Given a non-negative dissimilarity map δ on X, there is an associated map δ_o on X, which is determined only by the relative order of the $\delta(x, y)$ values for all $x, y \in X$. To describe this map, we begin by defining an associated quantity. For all $i, j, x, y \in X$, let

$$D_{ij}(x, y) = \begin{cases} 1, & \text{if } (\delta(x, i) - \delta(x, j))(\delta(y, i) - \delta(y, j)) < 0, \\ \frac{1}{2}, & \text{if exactly one of } \delta(x, i) - \delta(x, j) \text{ and } \delta(y, i) - \delta(y, j) \text{ is zero,} \\ 0, & \text{otherwise.} \end{cases}$$

The map δ_o is defined, for all $x, y \in X$, by setting

$$\delta_o(x, y) = \sum_{\{i,j\} \subseteq X} D_{ij}(x, y).$$

Clearly, if a non-negative dissimilarity map δ' on X is order equivalent to δ, then $\delta_o = (\delta')_o$.

Theorem 7.5.3, due to Bonnot *et al.* (1996), provides a necessary condition for δ to be a tree ordinal dissimilarity map.

Theorem 7.5.3. *If δ is a tree ordinal dissimilarity map on X, then δ_o is a tree metric on X. Furthermore, the unique X-tree that supports a tree metric representation of δ_o is refined by every X-tree that provides an ordinal representation of δ.*

In fact, it is possible to say more about the X-tree that supports a representation of δ_o in the statement of Theorem 7.5.3, but to do so requires one further definition. Let $\mathcal{T} = (T; \phi)$ be an X-tree and let w be a map that assigns real-valued weights to the edges of T. Let x and y be elements of X, and consider the path $\phi(x) = v_0, v_1, \ldots, v_k = \phi(y)$ in T. Then, for some $i \in \{0, 1, \ldots, k-1\}$, the edge $\{v_i, v_{i+1}\}$ *contains the midpoint* of x and y if

$$d_{(\mathcal{T};w)}(\phi(x), v_i) \leq \tfrac{1}{2} d_{(\mathcal{T};w)}(x, y)$$

and

$$d_{(\mathcal{T};w)}(\phi(x), v_{i+1}) \geq \tfrac{1}{2} d_{(\mathcal{T};w)}(x, y).$$

Theorem 7.5.4. *Let δ be a non-negative dissimilarity map on X, and suppose that δ is order equivalent to a tree metric $d_{(\mathcal{T};w)}$ on X. Then, the X-tree that supports a tree metric representation of δ_o has as its set of induced X-splits precisely those X-splits of $\mathcal{T}$ that correspond to an edge that contains a midpoint of some pair of elements in X.*

For a proof of Theorem 7.5.4 as well as for other details, including an explicit description of the edge weighting of the X-tree that supports a tree metric representation of δ_o, see Bonnot *et al.* (1996) and Guénoche (1998).

Bonnot *et al.* (1996) showed by giving an explicit example that if a non-negative dissimilarity map δ is an ultrametric, then δ_o is not necessarily an ultrametric. However, the X-trees that support tree metric representations of δ and δ_o are isomorphic as Corollary 7.5.5 shows.

Corollary 7.5.5. *Let δ be a non-negative ultrametric on X. If $\mathcal{T}$ is the unique X-tree that supports a tree metric representation of δ, then $\mathcal{T}$ is also the unique X-tree that supports a tree metric representation of δ_o.*

Proof. Let $\mathcal{T}'$ denote the unique X-tree that supports a tree metric representation of δ_o. Since δ is an ultrametric with a tree metric representation on $\mathcal{T}$, every edge of $\mathcal{T}$ contains the midpoint of some pair of elements of X. Thus, by Theorem 7.5.4, each X-split of $\mathcal{T}$ is an X-split of $\mathcal{T}'$, and so $\mathcal{T}'$ refines $\mathcal{T}$. However, by Theorem 7.5.3, $\mathcal{T}$ refines $\mathcal{T}'$. Thus, $\mathcal{T}$ and $\mathcal{T}'$ are isomorphic as required. □

A partial converse of the last result is Proposition 7.5.6, whose proof we leave as an exercise.

Proposition 7.5.6. *Let δ be a non-negative dissimilarity map on X. If δ_o is order equivalent to a $\mathcal{T}$-metric, where $\mathcal{T}$ is a binary phylogenetic X-tree, then δ is an ultrametric.*

7.6 Specialist topic: group-valued, non-symmetric dissimilarities

Throughout this chapter, we have restricted ourselves to maps from $X \times X$ into the set of reals and to distances between vertices u and v in a tree that are obtained by summing the real-valued weights of the edges in the path joining u and v. In this section, we consider maps from $X \times X$ into an arbitrary group $\mathcal{G}$, and generalize the existence and uniqueness results of tree metrics given in Sections 7.1 and 7.2 in which distances are now the product of elements of $\mathcal{G}$ corresponding to the edge weights in the directed path from u to v. Apart from its mathematical appeal, we will see in Chapter 8 that there is an underlying biological motivation for considering such maps. This section is based on papers by Böcker and Dress (1998) and Semple and Steel (1999).

To establish these more general results, we first present Theorem 7.6.1, which is a generalization of the equivalence of (i) and (ii) in Theorem 7.2.5.

For a rooted phylogenetic X-tree $\mathcal{T} = (T; \phi)$ with $T = (V, E)$, a map t from V into an arbitrary set M is called a *symbolic dating map*. In addition, t is *discriminating* if $t(u) \neq t(v)$ for all interior edges $\{u, v\}$ of T. Observe that no restrictions are placed on M. Furthermore, we define the map $d_{(\mathcal{T};t)} : X \times X \rightarrow M$ by setting

$$d_{(\mathcal{T};t)}(x, y) = t(\mathrm{lca}_{\mathcal{T}}(\phi(x), \phi(y)))$$

for all $x, y \in X$.

Let δ be a map from $X \times X$ into an arbitrary set M. The pair $(\mathcal{T}; t)$, where $\mathcal{T} = (T; \phi)$ is a rooted phylogenetic X-tree and $t : V(T) \rightarrow M$ is a discriminating symbolic dating map, is a *discriminating symbolic representation* of δ if

$$\delta(x, y) = d_{(\mathcal{T};t)}(x, y)$$

for all $x, y \in X$. Theorem 7.6.1 characterizes when δ has a discriminating symbolic representation.

Let M be an arbitrary set and let δ be a map from $X \times X$ into M. The map δ is said to be a *symbolic ultrametric* (*on* X) if each of the following conditions are satisfied:

(U1) $\delta(i,j) = \delta(j,i)$, for all $i, j \in X$;

(U2) $|\{\delta(i,j), \delta(i,k), \delta(j,k)\}| \leq 2$, for all $i, j, k \in X$; and

(U3) there are no pairwise distinct elements i, j, k, and l of X with

$$\delta(i,j) = \delta(j,k) = \delta(k,l) \neq \delta(j,l) = \delta(l,i) = \delta(i,k).$$

As an example, suppose that δ is an ultrametric on X. Then, δ certainly satisfies (U1) and (U2), and it is straightforward to show that δ satisfies (U3). Hence, δ is a symbolic ultrametric. Böcker and Dress (1998) established Theorem 7.6.1, and we refer the reader to their paper for its proof.

Theorem 7.6.1. *Let M be a finite set, and let δ be a map from $X \times X$ into M. Then, there is a discriminating symbolic representation of δ if and only if δ is a symbolic ultrametric. Furthermore, up to isomorphism, this representation is unique.*

We now describe a polynomial–time algorithm for constructing a discriminating symbolic representation of a symbolic ultrametric δ on X. Let $\mathcal{R}$ denote the set

$$\{ij|k\colon \delta(i,j) \neq \delta(i,k) = \delta(j,k), \text{ where } i, j, k \in X\}$$

of rooted phylogenetic trees, and let $\mathcal{A}(\mathcal{R})$ denote the rooted phylogenetic X-tree that is the output of BUILD when applied to $\mathcal{R}$. Let $(\mathcal{T}; t)$ be a discriminating symbolic representation of δ, where $\mathcal{T} = (T; \phi)$.

We first claim that $\mathcal{A}(\mathcal{R})$ is isomorphic to $\mathcal{T}$. To prove this claim, we argue by induction on the cardinality of X. Let $|X| = n$. Clearly, the claim holds if $n = 1$. Now assume that the result holds for all symbolic ultrametric mappings on X with $|X| \leq n-1$. Let $V_1, V_2, \ldots, V_r$ denote the vertex sets of the maximal rooted subtrees of T lying below the root of T. For all $p \in \{1, 2, \ldots, r\}$, let $X_p = \phi^{-1}(V_p)$. Thus, $X_1, X_2, \ldots, X_r$ are the blocks of a partition of X. Consider the clustering graph $[\mathcal{R}, X]$ on X induced by $\mathcal{R}$. To prove the induction step of the proof, it suffices to show that $X_1, X_2, \ldots, X_r$ are precisely the vertex sets of the connected components of $[\mathcal{R}, X]$. That is, for distinct elements i and j of X, there is a p in $\{1, 2, \ldots, r\}$ so that $i, j \in X_p$ if and only if there is an element k in X such that $ij|k \in \mathcal{R}$.

Suppose that $i, j \in X_p$ for some $p \in \{1, 2, \ldots, r\}$. Let v be the vertex of T corresponding to $\text{lca}(\phi(i), \phi(j))$. Since both i and j are in X_p, it follows that there is an interior vertex u in T, where $u \neq v$, and a k in X such that in T vertices u and v are adjacent, and $u = \text{lca}(\phi(i), \phi(k)) = \text{lca}(\phi(j), \phi(k))$. As u and v are both interior vertices of T and as t is discriminating, it follows that

$$\delta(i,j) \neq \delta(i,k) = \delta(j,k).$$

Thus, $ij|k$ is an element of $\mathcal{R}$.

Now suppose that there is no p in $\{1,2,\ldots,r\}$ such that $i,j \in X_p$. An argument similar to that used in the previous paragraph shows that there is no $k \in X$ such that $ij|k \in \mathcal{R}$. Thus, as claimed, $\mathcal{A}(\mathcal{R})$ is isomorphic to $\mathcal{T}$.

To complete the construction of a discriminating symbolic representation of δ, let V denote the vertex set of $\mathcal{T}$, and let $t : V \to M$ be the map defined, for all $v \in V$, by setting $t(v) = \delta(i,j)$, where i and j are two elements of X such that $v = \mathrm{lca}_{\mathcal{T}}(i,j)$. The pair $(\mathcal{A}(\mathcal{R}); t)$ is a discriminating symbolic representation of δ. Since $\mathcal{A}(\mathcal{R})$ can be constructed in polynomial time and, once $\mathcal{A}(\mathcal{R})$ is constructed, the mapping t can also be constructed in polynomial time, the entire construction of $(\mathcal{A}(\mathcal{R}); t)$ can be done in polynomial time.

With Theorem 7.6.1 and the construction of a discriminating symbolic representation of a symbolic ultrametric in hand, we now turn our attention to maps from $X \times X$ into an arbitrary group $\mathcal{G}$. Unless otherwise stated, the remaining results in this section are due to Semple and Steel (1999), and we refer readers to their paper for full details of omitted proofs. Recall that, for a group $\mathcal{G}$, the identity of $\mathcal{G}$ is denoted by $1_{\mathcal{G}}$ and we multiply elements of $\mathcal{G}$ from left to right.

Given a tree $T = (V,E)$, let $\vec{E}$ denote the set $\{(u,v)\colon \{u,v\} \in E\}$ of arcs. We can regard $\vec{E}$ as the set of pairs in which each member consists of an element of E and an orientation of it. Since T is a tree, it immediately follows that there is a unique directed path between every pair of vertices in the digraph $(V,\vec{E})$.

Let $\mathcal{T} = (T;\phi)$ be an X-tree with $T = (V,E)$ and let w be a mapping from $\vec{E}$ into a group $\mathcal{G}$. We define the map $d_{(\mathcal{T};w)} : X \times X \to \mathcal{G}$, for all $x,y \in X$, as follows:

(i) $d_{(\mathcal{T};w)}(x,y)$ is equal to the ordered product of the weights of the arcs on the directed path in $(V,\vec{E})$ from $\phi(x)$ to $\phi(y)$ if $\phi(x) \neq \phi(y)$; and

(ii) $d_{(\mathcal{T};w)}(x,y)$ is equal to $1_{\mathcal{G}}$ if $\phi(x) = \phi(y)$.

Example 7.6.2. Consider the semi-labelled tree $\mathcal{T} = (T;\phi)$ with $T = (V,E)$ shown in Fig. 7.5 together with the illustrated arc weighting $w : \vec{E} \to (\mathbb{R},+)$. Here, $d_{(\mathcal{T};w)}(d,e) = 7$ and $d_{(\mathcal{T};w)}(e,d) = 3$.

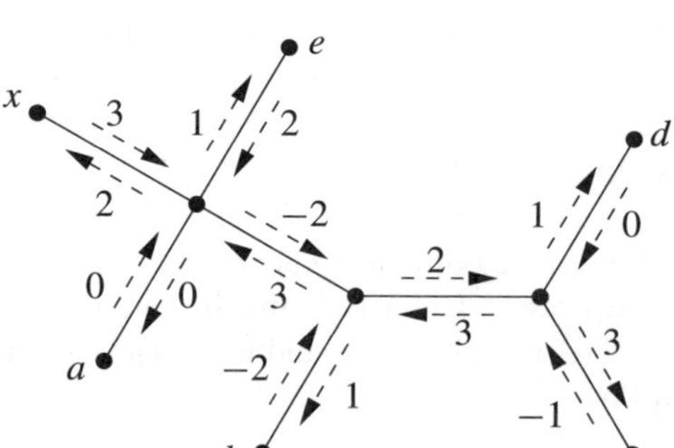

Figure 7.5 A semi-labelled tree with a real-valued arc weighting.

For a group $\mathcal{G}$, a *$\mathcal{G}$-dissimilarity map on X* is a map $\delta : X \times X \to \mathcal{G}$ such that, for all $x \in X$ $\delta(x,x) = 1_{\mathcal{G}}$. Furthermore, we call such a mapping a *tree dissimilarity (on X)* if there exists an X-tree $\mathcal{T} = (T;\phi)$ with $T = (V,E)$ and weight function $w : \vec{E} \to \mathcal{G}$ such that $d_{(\mathcal{T};w)}(x,y) = \delta(x,y)$ for all $x, y \in X$; in which case $(\mathcal{T};w)$ is said to be a *tree representation* of δ. Note that under a $\mathcal{G}$-dissimilarity map δ on X, there may be $x, y \in X$ such that $\delta(x,y) \neq \delta(y,x)$ even when $\mathcal{G}$ is Abelian. We illustrate these notions with an example.

Example 7.6.3. Let $X = \{x,a,b,c,d,e\}$ and consider the $(\mathbb{R},+)$-dissimilarity map δ on X that is illustrated in the table below. It is easily seen that the arc-weighted phylogenetic X-tree shown in Fig. 7.5 is a tree representation of δ, and so δ is a tree dissimilarity map.

	x	a	b	c	d	e
x	0	3	2	6	4	4
a	2	0	−1	3	1	1
b	3	1	0	3	1	2
c	7	5	3	0	0	6
d	8	6	4	3	0	7
e	4	2	1	5	3	0

For a group $\mathcal{G}$, Theorem 7.6.4 essentially determines when a $\mathcal{G}$-dissimilarity map is a tree dissimilarity.

Theorem 7.6.4. *Let $\mathcal{G}$ be a group and let δ be a $\mathcal{G}$-dissimilarity map on X. Let H_δ denote the following (finite) subset of $\mathcal{G}$:*

$$\{\delta(i,k)\delta(j,k)^{-1}\delta(j,l)\delta(i,l)^{-1}:\ i,j,k,l \in X\}.$$

(i) If δ is a tree dissimilarity map on X, then δ satisfies the following conditions:
(G1) *For all distinct elements i, j, and k of X,*

$$\delta(i,j)\delta(k,j)^{-1}\delta(k,i) = \delta(i,k)\delta(j,k)^{-1}\delta(j,i).$$

(G2) *For all four distinct elements of X, we can order these elements as i, j, k, and l so that*

$$\delta(i,k)\delta(j,k)^{-1} = \delta(i,l)\delta(j,l)^{-1}.$$

(ii) If δ satisfies both (G1) *and* (G2), *and H_δ has no elements of order* 2, *then δ is a tree dissimilarity map on X.*

The set H_δ is necessary in the second part of Theorem 7.6.4 as we show in Example 7.6.5. Bandelt and Steel (1995) use this example to show a similar feature of their main result.

Example 7.6.5. Let $\mathcal{G}$ be a group and let a be an element of $\mathcal{G}$ that has order 2; that is, $a^2 = 1$ and $a \neq 1_{\mathcal{G}}$. Let $X = \{1, 2, 3, 4, 5\}$, and define the $\mathcal{G}$-dissimilarity map δ on X by setting, for all $x, y \in X$,

$$\delta(x, y) = \begin{cases} a, & \text{if } \{x, y\} \in \{\{1,4\}, \{2,3\}, \{2,4\}\}, \\ 1_{\mathcal{G}}, & \text{otherwise.} \end{cases}$$

Note that $a \in H_\delta$.

It is straightforward to check that δ satisfies both (G1) and (G2). However, δ is not a tree dissimilarity on X. To see this, consider δ restricted to $\{1, 2, 3, 4\} \times \{1, 2, 3, 4\}$, $\{1, 3, 4, 5\} \times \{1, 3, 4, 5\}$, and $\{1, 2, 3, 5\} \times \{1, 2, 3, 5\}$. Using (G2), a routine check shows that if $(\mathcal{T}; w)$ is a tree representation of δ, then $13|24$, $14|35$, and $15|23$ are elements of $\mathcal{Q}(\mathcal{T})$. But no X-tree can simultaneously display each of these quartets. Thus δ is not a tree dissimilarity on X.

The proof of Theorem 7.6.4 is purely algebraic and it is based on Theorem 7.6.1. Here, we give a detailed outline of this proof.

We begin by describing three further maps, the last two of which are illustrated in Example 7.6.6.

(i) Let δ be a $\mathcal{G}$-dissimilarity map on X, and fix an element x of X. Define the map $\delta_x : X \times X \rightarrow \mathcal{G}$ by setting

$$\delta_x(i, j) = \delta(x, i)\delta(j, i)^{-1}\delta(j, x)$$

for all $i, j \in X$. Note that δ_x is not necessarily a $\mathcal{G}$-dissimilarity map since $\delta_x(i, i)$ may not be equal to $1_{\mathcal{G}}$.

(ii) Let $T = (V, E)$ be a tree. A *discriminating $\mathcal{G}$-dating map* is a function $t : V \rightarrow \mathcal{G}$ with the property that if $\{u, v\}$ is an interior edge of T, then $t(u) \neq t(v)$. Observe the similarity between a discriminating $\mathcal{G}$-dating map and a discriminating symbolic dating map.

(iii) Lastly, let $\mathcal{T} = (T; \phi)$ be a phylogenetic X-tree with $T = (V, E)$, let $t : V \rightarrow \mathcal{G}$ be a discriminating $\mathcal{G}$-dating map, and let x be some fixed element of X. For elements u and v in V, write $u \leq_x v$ if u lies on the path from x to v in T. Define the map $d_x^{(\mathcal{T};t)} : X \times X \rightarrow \mathcal{G}$ by setting, for all $i, j \in X$,

$$d_x^{(\mathcal{T};t)}(i, j) = t(\text{glb}_{\leq_x}\{\phi(i), \phi(j)\})$$

for all $i, j \in X$, where $\text{glb}_{\leq_x}$ denotes the greatest lower bound under the partial order $\leq_x$.

Example 7.6.6. Let $T = (V, E)$ denote the tree shown in Fig. 7.6(a) and let t denote the map from V into the group $(\mathbb{R}, +)$ as shown in Fig. 7.6(a). Then, as $t(u) \neq t(v)$ for all interior edges $\{u, v\}$ of E, the map t is a discriminating $(\mathbb{R}, +)$-dating map. To illustrate the third mapping, let $X = \{x, a, b, c, d, e\}$ and let $\mathcal{T} = (T; \phi)$ denote the phylogenetic

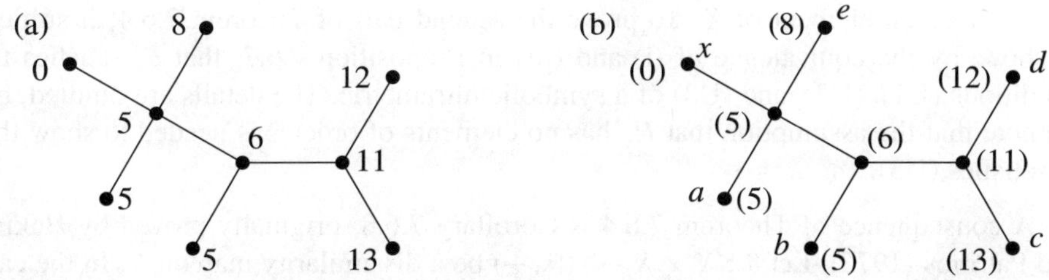

Figure 7.6 (a) A discriminating $(\mathbb{R}, +)$-dating map t. (b) A phylogenetic tree with t shown in brackets.

X-tree shown in Fig. 7.6(b). The map $d_x^{(\mathcal{T};t)} : X \times X \to (\mathbb{R}, +)$ is displayed in the table below:

	x	a	b	c	d	e
x	0	0	0	0	0	0
a	0	5	5	5	5	5
b	0	5	5	6	6	5
c	0	5	6	13	11	5
d	0	5	6	11	12	5
e	0	5	5	5	5	8

Proposition 7.6.7 is crucial to the proof of Theorem 7.6.4. It shows a fundamental relationship between tree dissimilarity maps, the maps (i)–(iii) above, and symbolic ultrametrics.

Proposition 7.6.7. *Let $\mathcal{G}$ be a group, let δ be a $\mathcal{G}$-dissimilarity map on X, and let x be an element of X. The following statements are equivalent for δ:*

(i) δ is a tree dissimilarity on X;

(ii) there is a phylogenetic X-tree $\mathcal{T} = (T; \phi)$ with $T = (V, E)$, and a discriminating $\mathcal{G}$-dating map t on V such that on X

$$\delta_x = d_x^{(\mathcal{T};t)};$$

(iii) δ_x is a symbolic ultrametric on X.

The proof of Proposition 7.6.7 is omitted and instead we make the following remarks. The equivalence of (ii) and (iii) is an immediate consequence of Theorem 7.6.1. To complete the proof of the proposition, one can show the equivalence of (i) and (ii) by utilizing the associated trees in both directions.

Proof of Theorem 7.6.4. If δ is a tree dissimilarity map on X, then it is straightforward to check that (G1) and (G2) must hold by cancelling the products of the appropriate arc weights in $\mathcal{G}$. Thus, the first part of Theorem 7.6.4 holds.

Let x be an element of X. To prove the second part of Theorem 7.6.4, it suffices to show, by the equivalence of (i) and (iii) in Proposition 7.6.7, that δ_x satisfies the conditions (U1), (U2), and (U3) of a symbolic ultrametric. The details are omitted, but we note that the assumption that H_δ has no elements of order 2 is needed to show that δ_x satisfies (U3). □

A consequence of Theorem 7.6.4 is Corollary 7.6.8, originally proved by Hakimi and Patrinos (1972). Let $\delta : X \times X \to (\mathbb{R}, +)$ be a dissimilarity map on X. In the case where δ is a tree dissimilarity map, we say that δ is *tree additive* (*on* X). Note that if δ is tree additive, δ is not necessarily a tree metric, although the converse does hold. We say that δ satisfies the *weak four-point condition* if, for every four not necessarily distinct elements i, j, k, and l of X, at least two of the sums

$$\delta(i,j) + \delta(k,l), \quad \delta(i,k) + \delta(j,l), \quad \text{and} \quad \delta(i,l) + \delta(j,k)$$

are equal.

Corollary 7.6.8. *Let δ be a dissimilarity map on X. Then, δ is tree additive if and only if δ satisfies the weak four-point condition.*

Proof. A routine check similar to that used to prove Lemma 7.1.7 shows that if δ is tree additive, then δ satisfies the weak four-point condition.

To prove the converse, suppose that δ satisfies the weak four-point condition. Since the group $(\mathbb{R}, +)$ has no elements of order 2, it suffices to show, by Theorem 7.6.4, that δ satisfies (G1) and (G2). As $\delta(x, y) = \delta(y, x)$ for all $x, y \in X$, (G1) becomes a simple identity, and so δ trivially satisfies (G1). Furthermore, as δ satisfies the weak four-point condition, it follows that δ satisfies (G2). Hence, δ is tree additive, completing the proof of Corollary 7.6.8. □

If a $\mathcal{G}$-dissimilarity map δ on X is a tree dissimilarity on X, then a tree representation of δ can be constructed in polynomial time. This construction process, which initially uses the process to construct a discriminating symbolic representation of a symbolic ultrametric, is described next.

Fix an element x of X, and let $\mathcal{R}_x$ denote the set

$$\{ij|k\colon \delta_x(i,j) \neq \delta_x(i,k) = \delta_x(j,k), \text{ where } i,j,k \in X - \{x\}\}$$

of rooted phylogenetic trees. Let $\mathcal{A}(\mathcal{R}_x)$ denote the rooted phylogenetic tree on $X - \{x\}$ that is the output of BUILD when applied to $\mathcal{R}_x$. Let $\mathcal{T} = (T; \phi)$ denote the phylogenetic X-tree obtained from $\mathcal{A}(\mathcal{R}_x)$ by adjoining a pendant edge to its root and labelling the new pendant vertex x, and then unrooting the resulting tree. Let $T = (V, E)$ and let t be the map from V into $\mathcal{G}$ defined, for all $v \in V$, by $t(v) = \delta_x(i,j)$, where i and j are two elements of X such that v is the greatest lower bound of i and j under the partial order $\leq_x$. Now t is well defined and, moreover, t is a discriminating $\mathcal{G}$-dating map. This follows from the fact that, by Proposition 7.6.7, δ_x is a symbolic ultrametric on X, and the description of the associated discriminating symbolic representation given earlier in this section. We leave the details to the reader.

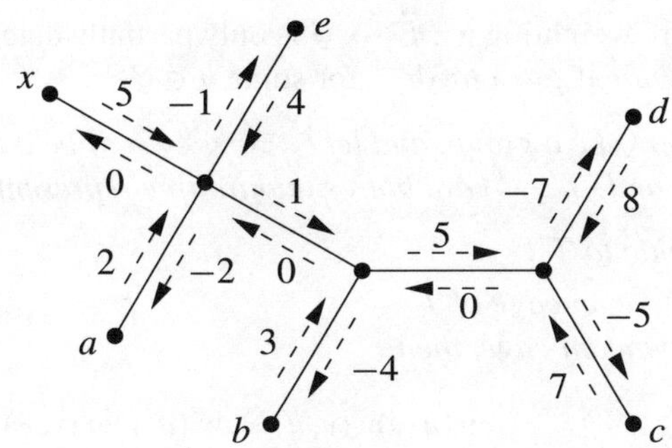

Figure 7.7 A tree representation of δ.

We now define an arc weighting function $w : \vec{E} \to \mathcal{G}$ based on the discriminating $\mathcal{G}$-dating map t and the element x. To each pair of arcs (u, v) and (v, u) in $\vec{E}$, assign the weights $w(u, v)$ and $w(v, u)$, respectively, as follows:

(i) if $\{u, v\}$ is a pendant edge of T with v a pendant vertex and $v \in \phi(X - \{x\})$, set

$$w(u, v) = t(u)^{-1}t(v)\delta(\phi(v), x)^{-1} \quad \text{and} \quad w(v, u) = \delta(\phi^{-1}(v), x);$$

(ii) otherwise, if $u \leq_x v$, set

$$w(u, v) = t(u)^{-1}t(v) \quad \text{and} \quad w(v, u) = 1_{\mathcal{G}}.$$

It can be shown that $(\mathcal{T}; w)$ is a tree representation of δ (see Semple and Steel 1999).

To illustrate the above construction, consider the $(\mathbb{R}, +)$-dissimilarity map δ on $X = \{x, a, b, c, d, e\}$ described in Example 7.6.3. Choosing x as our fixed element of X, the table in Example 7.6.6 describes the map $\delta_x : X \times X \to (\mathbb{R}, +)$. Furthermore, Fig. 7.6(b) shows the phylogenetic X-tree $\mathcal{T}$ obtained from $\mathcal{A}(\mathcal{R}_x)$ and the associated discriminating $(\mathbb{R}, +)$-dating map on the vertex set of $\mathcal{T}$, as described above. Weighting the associated arcs of $\mathcal{T}$ in the way stated in the last paragraph results in the tree representation of δ shown in Fig. 7.7.

Now observe the differences in arc weightings of the tree representations of δ shown in Figs 7.5 and 7.7. We end this section by considering the extent to which a tree representation of a tree dissimilarity is unique following Baake's (1998) approach.

Let $T = (V, E)$ be a tree and let w be a function from $\vec{E}$ into a group $\mathcal{G}$. The *return-trip weights* of an edge $\{u, v\}$ of T are the values $w(u, v)w(v, u)$ and $w(v, u)w(u, v)$. An edge e of T is *properly weighted* if one (or, equivalently, both) of the return-trip weights for e is not equal to $1_{\mathcal{G}}$.

Suppose that δ is a tree dissimilarity map on X. A tree representation $(\mathcal{T}; w)$ of δ is *standard* if $\mathcal{T}$ is a phylogenetic X-tree and every interior edge of T is properly weighted, where $\mathcal{T} = (T; \phi)$. Every tree dissimilarity map has a standard tree representation (Semple and Steel 1999). Theorem 7.6.9 shows that if $(\mathcal{T}; w)$ is a standard tree representation of a tree dissimilarity map δ, then, up to isomorphism, $\mathcal{T}$ is determined by δ. However, the arc weighting $w : \vec{E} \to \mathcal{G}$ is only partially determined. Two elements $g_1, g_2 \in \mathcal{G}$ are *conjugate* if $g_1 = hg_2h^{-1}$ for some $h \in \mathcal{G}$.

Theorem 7.6.9. *Let $\mathcal{G}$ be a group, and let $\delta : X \times X \to \mathcal{G}$ be a tree dissimilarity map. Suppose that $(\mathcal{T}; w)$ and $(\mathcal{T}'; w')$ are both standard tree representations of δ. Then*

(i) $\mathcal{T}$ is isomorphic to $\mathcal{T}'$.

(ii) Let $e = \{u, v\}$ be an edge of $\mathcal{T}$.

(a) If e is a pendant edge, then

$$w(u, v)w(v, u) = w'(u, v)w'(v, u).$$

(b) If e is an interior edge, then $w(u, v)w(v, u)$ and $w'(u, v)w'(v, u)$ are conjugate elements of $\mathcal{G}$.

7.7 Specialist topic: perturbing tree metrics

Earlier in this chapter, we showed how an X-tree $\mathcal{T}$ and an associated positive real-valued edge weighting w can be recovered from an induced tree metric $d = d_{(\mathcal{T};w)}$ on X. In this section, we investigate how far one can perturb d and yet still be able to recover just the X-tree $\mathcal{T}$. This is relevant to applications because dissimilarity maps generated from data cannot be expected to satisfy the four-point condition exactly. Much of this section follows Moulton and Steel (1999).

We will measure the distance be between any two dissimilarity maps δ and δ' on X by the *l_∞-metric* defined as

$$\|\delta - \delta'\|_\infty = \max\{|\delta(x, y) - \delta'(x, y)| : x, y \in X\}.$$

Lemma 7.7.1 connects this metric to the Buneman index.

Lemma 7.7.1. *Let δ and δ' be two dissimilarity maps on X, and let σ be an X-split. Then,*

$$|B_\delta(\sigma) - B_{\delta'}(\sigma)| \leq 2\|\delta - \delta'\|_\infty.$$

Proof. By symmetry, it suffices to show that

$$B_{\delta'}(\sigma) \leq B_\delta(\sigma) + 2\|\delta - \delta'\|_\infty.$$

Writing $\sigma = A|B$, select elements $a, a' \in A$ and $b, b' \in B$ so that

$$B_\delta(\sigma) = \tfrac{1}{2}(\delta(a, b) + \delta(a', b') - \delta(a, a') - \delta(b, b')).$$

Since $|\delta(x, y) - \delta'(x, y)| \leq \|\delta - \delta'\|_\infty$ for all $x, y \in X$, it follows that

$$\begin{aligned} B_{\delta'}(\sigma) &\leq \tfrac{1}{2}(\delta'(a, b) + \delta'(a', b') - \delta'(a, a') - \delta'(b, b')) \\ &\leq B_\delta(\sigma) + 2\|\delta - \delta'\|_\infty \end{aligned}$$

as required. □

We will denote the metric space of tree metrics on X under the l_∞-metric by $\mathcal{T}(X)$. Let $\mathcal{W}(X)$ denote the set of maps $w : \Sigma(X) \rightarrow \mathbb{R}^{\geq 0}$, where we recall that $\Sigma(X)$ denotes the set of all X-splits. We will consider $\mathcal{W}(X)$ as a metric space under either the l_∞-metric or under the l_1-metric, where for all $w, w' \in \mathcal{W}(X)$, these metrics on $\mathcal{W}(X)$ are defined by

$$\|w - w'\|_\infty = \max\{|w(\sigma) - w'(\sigma)|: \sigma \in \Sigma(X)\}$$

and

$$\|w - w'\|_1 = \sum_{\sigma \in \Sigma(X)} |w(\sigma) - w'(\sigma)|,$$

respectively.

We can identify any tree metric $d \in \mathcal{T}(X)$ with a corresponding element in $\mathcal{W}(X)$ as follows. If $(\mathcal{T}; w)$ is a tree metric representation of d, identify d with the element w_d in $\mathcal{W}(X)$ for which

$$w_d(\sigma) = \begin{cases} w(e), & \text{if } e \text{ induces } \sigma \text{ for some edge } e \text{ of } \mathcal{T}, \\ 0, & \text{otherwise,} \end{cases}$$

for all $\sigma \in \Sigma(X)$. Notice that

$$d = \sum_{A|B \in \Sigma(X)} w_d(\sigma)\delta_{A|B}, \tag{7.13}$$

where $\delta_{A|B}$ is the pseudometric defined by setting, for all $x, y \in X$,

$$\delta_{A|B}(x, y) = \begin{cases} 0, & \text{if } x, y \in A \text{ or } x, y \in B, \\ 1, & \text{otherwise.} \end{cases}$$

Theorem 7.7.2 includes results from Farach and Kannan (1999) and Moulton and Steel (1999).

Theorem 7.7.2. *Let d and d' be elements of $\mathcal{T}(X)$. Then*

(i) $\|d - d'\|_\infty \leq \|w_d - w_{d'}\|_1$*; and*
(ii) $\|w_d - w_{d'}\|_\infty \leq 2\|d - d'\|_\infty$.

Proof. Throughout this proof, we adopt the notation $w = w_d$ and $w' = w_{d'}$. We first prove (i). Writing both d and d' in the form of eqn (7.13), we deduce that

$$\begin{aligned}
\|d - d'\|_\infty &= \max_{x,y \in X} \{|d(x,y) - d'(x,y)|\} \\
&= \max_{x,y \in X} \left\{ \Big| \sum_{\sigma \in \Sigma(X)} (w(\sigma) - w'(\sigma)) \cdot \delta_\sigma(x,y) \Big| \right\} \\
&\leq \max_{x,y \in X} \left\{ \sum_{\sigma \in \Sigma(X)} |w(\sigma) - w'(\sigma)| \cdot \delta_\sigma(x,y) \right\} \\
&= \sum_{\sigma \in \Sigma(X)} |w(\sigma) - w'(\sigma)| \cdot \max_{x,y \in X} \{\delta_\sigma(x,y)\} \\
&= \|w - w'\|_1,
\end{aligned}$$

thus establishing (i).

We prove (ii) by showing that, for all $\sigma \in \Sigma(X)$,

$$|w(\sigma) - w'(\sigma)| \leq 2\|d - d'\|_\infty.$$

Let $\mathcal{T}$ and $\mathcal{T}'$ be the X-trees that support tree metric representations of d and d', respectively, and let σ be an X-split. There are three cases to consider.

First, if σ is an X-split induced by both $\mathcal{T}$ and $\mathcal{T}'$, then, by Lemma 7.3.2, $w(\sigma) = B_d(\sigma)$ and $w'(\sigma) = B_{d'}(\sigma)$ and so, by Lemma 7.7.1, we have

$$|w(\sigma) - w'(\sigma)| \leq 2\|d - d'\|_\infty.$$

Second, assume that σ is induced by exactly one of $\mathcal{T}$ and $\mathcal{T}'$. Without loss of generality, assume it is $\mathcal{T}'$. Then, by Lemma 7.3.2, $B_d(\sigma) \leq 0$, and so, by Lemma 7.7.1,

$$B_{d'}(\sigma) \leq B_d(\sigma) + 2\|d - d'\|_\infty \leq 2\|d - d'\|_\infty.$$

Now, by Lemma 7.3.2, $w'(\sigma) = B_{d'}(\sigma)$ and $w(\sigma) = 0$. Thus, by Lemma 7.7.1,

$$|w(\sigma) - w'(\sigma)| = B_{d'}(\sigma) \leq 2\|d - d'\|_\infty.$$

Lastly, if σ is not induced by either $\mathcal{T}$ or $\mathcal{T}'$, then, again by Lemma 7.3.2,

$$|w(\sigma) - w'(\sigma)| = |0 - 0| \leq 2\|d - d'\|_\infty.$$

This completes the proof of (ii). □

The next two examples show that the two inequalities in Theorem 7.7.2 can both be equalities. For the first inequality, let d be the tree metric on X that is induced by the star tree with label set X in which each edge has weight w_1, and let d' be the tree metric on X defined in the same way except that one edge of the star tree has weight w_2 instead of w_1. Then,

$$\|d - d'\|_\infty = |w_1 - w_2| = \|w_d - w_{d'}\|_1.$$

For the second inequality in Theorem 7.7.2, let d be the tree metric on X induced by a quartet tree with label set X in which each edge has weight 2. Let d' be the tree metric on X induced by the same quartet tree, but in this case the interior edge has weight 1 and every other edge has weight $\frac{9}{4}$. Then,

$$\|w_d - w_{d'}\|_\infty = 1 = 2\|d - d'\|_\infty.$$

Theorem 7.7.2(ii) provides a bound on the extent of 'separation' between tree metrics induced by different semi-labelled trees. We make this precise in Corollary 7.7.3.

Corollary 7.7.3. *Suppose that $d = d_{(\mathcal{T};w)}$ and $d' = d_{(\mathcal{T}';w')}$ are tree metrics on X. Let $\omega = \min\{w(e) : e \in E(\mathcal{T})\}$ and let $\omega' = \min\{w(e) : e \in E(\mathcal{T}')\}$.*

(i) If $\|d - d'\|_\infty < \frac{1}{2}\omega$, then $\mathcal{T}'$ refines $\mathcal{T}$.
(ii) If $\|d - d'\|_\infty < \frac{1}{2}\min\{\omega, \omega'\}$, then $\mathcal{T} \cong \mathcal{T}'$.

Proof. To prove (i), suppose that $\mathcal{T}'$ does not refine $\mathcal{T}$. Then, $\mathcal{T}$ has an X-split σ that is not an X-split of $\mathcal{T}'$. Since $\omega = \min\{w(e) : e \in E(\mathcal{T})\}$, it follows that $w_d(\sigma) \geq \omega$ and $w_{d'}(\sigma) = 0$, so $\|w_d - w_{d'}\|_\infty \geq \omega$. By Theorem 7.7.2, this implies that $\|d - d'\|_\infty \geq \frac{1}{2}\omega$; a contradiction.

The second part of the theorem is an immediate consequence of (i). □

Lemma 7.7.4. *Let δ be a non-negative dissimilarity map on X and let $\mathcal{T}_\delta$ be the X-tree returned by the Buneman construction; that is, $\Sigma(\mathcal{T}_\delta) = \Sigma_B(\delta)$. Let $d = d_{(\mathcal{T};w)}$ be a tree metric on X and let $\omega = \min\{w(e) : e \in E(\mathcal{T})\}$. Suppose that*

$$\|d - \delta\|_\infty < \tfrac{1}{2}\omega.$$

Then, $\mathcal{T}_\delta$ refines $\mathcal{T}$. Furthermore, for any X-split σ of $\mathcal{T}_\delta$ that is not an X-split of $\mathcal{T}$, we have $B_\delta(\sigma) < \omega$.

Proof. Suppose that $\|d - \delta\|_\infty < \frac{1}{2}\omega$. Then, for an arbitrary X-split σ, it follows by Lemma 7.7.1 that

$$|B_d(\sigma) - B_\delta(\sigma)| \leq 2\|d - \delta\|_\infty < \omega. \tag{7.14}$$

Let e be an edge of $\mathcal{T}$ and let σ_e denote the X-split of $\mathcal{T}$ corresponding to e. Then, as $B_d(\sigma_e) = w(e) \geq \omega$, we deduce by (7.14) that $B_\delta(\sigma_e) \geq B_d(\sigma_e) - \omega > 0$. Thus, σ_e is an X-split of $\mathcal{T}_\delta$, and therefore $\mathcal{T}_\delta$ refines $\mathcal{T}$. Furthermore, if σ is an X-split of $\mathcal{T}_\delta$ that is not an X-split of $\mathcal{T}$, then, by (7.14),

$$B_\delta(\sigma) < B_d(\sigma) + \omega \leq \omega,$$

which establishes the second claim. □

The next result provides an answer to the question—how far can we perturb a tree metric and still hope to recover the support tree? This is particularly important when investigating the statistical properties of tree reconstruction methods under the types of models that will be discussed in Chapter 8.

Theorem 7.7.5. *Let $d = d_{(\mathcal{T};w)}$ be a tree metric on X such that $\mathcal{T}$ is a binary phylogenetic X-tree. Let $\omega = \min\{w(e)\colon\ e \in E(\mathcal{T})\}$ and let δ be a non-negative dissimilarity map on X.*

(i) If

$$\|d - \delta\|_\infty < \tfrac{1}{2}\omega,$$

then $\mathcal{T}$ is the X-tree returned by both the Buneman construction and the neighbour-joining method when applied to δ.

(ii) The result in (i) is best possible in that, for any real value $\epsilon > 0$, there exists a tree metric $d' = d_{(\mathcal{T}';w')}$ such that $\mathcal{T}'$ is a binary phylogenetic X-tree that is not isomorphic to $\mathcal{T}$ and

$$\|d - d'\|_\infty = \tfrac{1}{2}\omega + \epsilon.$$

Proof. The result for the Buneman construction follows by Lemma 7.7.4. The analogous result for the neighbour-joining method requires a more intricate analysis and the reader is referred to Atteson (1997) for details.

To prove (ii), select an interior edge e_0 of $\mathcal{T}$ for which $w(e_0) = \omega$, and let $\mathcal{T}'$ be one of the two binary phylogenetic X-trees obtained from $\mathcal{T}$ by performing a nearest-neighbour interchange across e_0. Note there is a canonical identification of each edge of $\mathcal{T}'$ with a corresponding edge of $\mathcal{T}$. Under this identification, set, for all $e \in E(\mathcal{T}')$,

$$w'(e) = \begin{cases} \epsilon, & \text{if } e = e_0, \\ w(e) + \tfrac{1}{4}\omega, & \text{if } e \text{ is incident with } e_0, \\ w(e), & \text{otherwise.} \end{cases}$$

It is easily checked that $\|d - d'\|_\infty = \tfrac{1}{2}\omega + \epsilon$ as required. □

Given a dissimilarity map δ on X, we next consider two natural problems:

(i) determine an upper bound on the l_∞ distance to the nearest tree metric; and
(ii) construct or characterize a tree metric, or ultrametric, d that minimizes the distance $\|d - \delta\|_\infty$.

For the first problem, we describe an elegant result due to Gromov (1987). Let hyp(δ) denote the smallest value of $h \geq 0$ such that, for all $i, j, k, l \in X$,

$$\delta(i,j) + \delta(k,l) \leq \max\{\delta(i,k) + \delta(j,l), \delta(i,l) + \delta(j,k)\} + h.$$

The value hyp(δ) is sometimes called the *hyperbolicity* of δ and it is a measure of how far δ deviates from the four-point condition. In particular, by the Tree-Metric Theorem, δ is a tree metric precisely if hyp(δ) $= 0$. Gromov (1987) (see also Bowditch 1991) showed that for any dissimilarity map δ on X satisfying the triangle inequality, there exists a tree metric d on X with

$$\|d - \delta\|_\infty \leq (1 + \log_2 n) \cdot \text{hyp}(\delta),$$

where $n = |X|$. Thus, if δ satisfies the triangle inequality and hyp(δ) is small, then δ is close to a tree metric on X up to a term that grows slowly with n.

We now consider the second problem. For a dissimilarity map δ on X, we call a tree metric or ultrametric d on X that minimizes $\|d - \delta\|_\infty$ an l_∞-optimal tree metric or an l_∞-ultrametric for δ, respectively. The problem of finding such an optimal tree metric on X for δ is NP-hard (Agarwala *et al.* 1999). However, it is easy to construct an l_∞-optimal ultrametric on X for δ using the results in Section 7.2, as we describe next.

For a dissimilarity map δ on X and a positive real number $\epsilon > 0$, let $\delta^{+\epsilon}$ denote the dissimilarity map on X defined, for all $x, y \in X$, by setting

$$\delta^{+\epsilon}(x,y) = \begin{cases} \delta(x,y) + \epsilon, & \text{if } x \neq y, \\ 0, & \text{otherwise.} \end{cases}$$

Note that if δ is an ultrametric, then so too is $\delta^{+\epsilon}$ for all $\epsilon > 0$.

Recall the definition of the sub-dominant ultrametric δ_U that is associated with a dissimilarity map δ from Section 7.2. Since the sub-dominant ultrametric for $\delta^{+\epsilon}$ is $(\delta_U)^{+\epsilon}$, writing $\delta_U^{+\epsilon}$ for either $(\delta^{+\epsilon})_U$ or $(\delta_U)^{+\epsilon}$ is unambiguous. Chepoi and Fichet (2000) established Theorem 7.7.6.

Theorem 7.7.6. *Given any dissimilarity map δ on X, an l_∞-optimal ultrametric for δ is the ultrametric $\delta_U^{+\epsilon}$, where $\epsilon = \frac{1}{2}\|\delta - \delta_U\|_\infty$. Moreover,*

$$\|\delta_U^{+\epsilon} - \delta\|_\infty = \epsilon.$$

Proof. First, it is easily seen that $\|\delta_U^{+\epsilon} - \delta\|_\infty = \epsilon$. Now, by Theorem 7.2.9(ii), $\delta_U(x,y) \leq \delta(x,y)$ for all $x, y \in X$. Therefore, there exists some pair $x_0, y_0 \in X$ for which $\delta_U(x_0,y_0) = \delta(x_0,y_0) - 2\epsilon$. Suppose that δ' is an l_∞-optimal ultrametric for δ. Then, for all $x, y \in X$, $\delta'(x,y) \leq \delta^{+\epsilon}(x,y)$; for otherwise $\delta'(x,y) > \delta(x,y) + \epsilon$, for some $x, y \in X$, which implies that $\delta'(x,y) - \delta(x,y) > \|\delta_U^{+\epsilon} - \delta\|_\infty$, contradicting the minimality of δ'. Consequently, since $\delta_U^{+\epsilon}$ is the sub-dominant ultrametric for $\delta^{+\epsilon}$, it follows by Theorem 7.2.9(iii) that

$$\delta'(x,y) \leq \delta_U^{+\epsilon}(x,y)$$

for all $x, y \in X$. Therefore,

$$\delta'(x_0,y_0) \leq \delta_U^{+\epsilon}(x_0,y_0) = \delta_U(x_0,y_0) + \epsilon = \delta(x_0,y_0) - \epsilon,$$

and hence $\|\delta' - \delta\|_\infty \geq \epsilon$. Since $\|\delta_U^{+\epsilon} - \delta\|_\infty = \epsilon$, this completes the proof of the theorem. □

It follows from Theorem 7.7.6 that an l_∞-optimal ultrametric for a dissimilarity map δ can be constructed in polynomial time. This in turn allows for a polynomial–time calculation, from any dissimilarity map δ on X, of a value that is at most three times the minimum of $\|\delta - d\|_\infty$ over all tree metrics d on X (Agarwala *et al.* 1999).

We complete this section by briefly discussing the continuity of methods for constructing tree metrics. Clearly, the metric space $\mathcal{T}(X)$ of tree metrics sits inside a larger metric space of all dissimilarity maps on X under the same l_∞-metric. Let us denote this latter metric space by $\mathcal{D}(X)$. Any method that constructs tree metrics (or ultrametrics) from arbitrary dissimilarity maps can be regarded as a map from $\mathcal{D}(X)$ to $\mathcal{T}(X)$, and so we can ask if this map is (topologically) continuous. The Buneman construction is clearly continuous and, by Theorem 7.3.3, the mapping fixes each element of $\mathcal{T}(X)$. Regarding the minimum evolution and neighbour-joining methods, since these construct edge-weighted phylogenetic trees, each can be viewed as a map from $\mathcal{D}(X)$ into a proper subspace of $\mathcal{T}(X)$. However, neither of these maps are continuous on all of $\mathcal{D}(X)$. For example, if δ is the dissimilarity map obtained from the shortest path distance in Fig. 7.4(a) by setting $w_{ik|jl} = x$ and setting all other edge weights equal to 1, then neighbour-joining converges to quite different points in $\mathcal{T}(X)$ depending on whether x converges to 1 from above or from below (for details, see Moulton and Steel 1999).

7.8 Exercises

1. Let (X, d) be a metric space with $|X| = 3$. Show that d is a tree metric on X.
2. Let d be a tree metric on X and let $\epsilon > 0$. Let G_ϵ denote the graph that has vertex set X, and an edge joining two vertices x and y precisely if $d(x, y) < \epsilon$. Show that G_ϵ is a chordal graph.
3. Let d be a $\mathcal{T}$-metric on X and let $c : X \to \mathbb{R}^{>0}$ be a function. Let $\delta : X \times X \to \mathbb{R}^{\geq 0}$ be the dissimilarity map on X defined by setting, for all $x, y \in X$,

$$\delta(x, y) = \begin{cases} d(x, y) + c(x) + c(y), & \text{if } x \neq y, \\ 0, & \text{if } x = y. \end{cases}$$

   Show that δ is also a $\mathcal{T}$-metric.
4. (Buneman 1974*b*) Prove that a graph G is a tree if and only if G is connected, has no 3-cycles, and the dissimilarity map on $V(G)$ defined, for all $u, v \in V(G)$, by the graphical distance between u and v satisfies the four-point condition.
5. Prove Lemmas 7.3.2 and 7.4.1.
6. Prove Proposition 7.3.4.
7. Show that, for every tree metric d on X, there is an ultrametric δ on X and a tree metric d' on X for which the tree metric representation is supported by a star tree such that

$$d(x, y) = \delta(x, y) + d'(x, y),$$

   for all $x, y \in X$.
8. (Sankoff 1972) Let δ be an ultrametric on X and let $\Delta = \{\delta(x, y) : x, y \in X\}$. For all $x \in X$ and all $\lambda \in \Delta$, let $c(x, \lambda) = \{y : y \in X \text{ and } \delta(x, y) \leq \lambda\}$. Show that $\{c(x, \lambda) : x \in X, \lambda \in \Delta\}$ is a hierarchy on X.

9. (i) Prove Proposition 7.1.9 and show that this result does not apply to non-binary characters.
   (ii) Construct a sequence $\mathcal{C}$ of binary characters on X so that $\delta_{\mathcal{C}}$ is a tree metric on X but $\mathcal{C}$ is not compatible.
10. Prove Proposition 7.5.6.
11. For a dissimiliarity map $\delta : X \times X \to \mathbb{R}^{\geq 0}$, describe a polynomial–time algorithm to construct the set $\Sigma_B(\delta)$ of splits of X that have positive Buneman index.
12. Let Σ be a set of X-splits and let r be an element of X. Associate with each element in Σ the block that does not contain r. Show that Σ is weakly compatible if and only if the resulting set of subsets of $X - \{r\}$ is a weak hierarchy on $X - \{r\}$.
13. (Bandelt and Dress 1992) Let δ be a dissimilarity map on X. Prove that δ is totally decomposable if and only if, for all $t, u, v, w, x \in X$,

$$i_\delta(\{t,u\}|\{v,w\}) \leq i_\delta(\{t,x\}|\{v,w\}) + i_\delta(\{t,u\}|\{v,x\}),$$

where, for all elements $a, a', b, b' \in X$, $i_\delta(\{a,a'\}|\{b,b'\})$ denotes the isolation index of $\{a,a'\}|\{b,b'\}$ on δ restricted to $\{a,a',b,b'\}$.

14. Let $\mathcal{T} = (T;\phi)$ be a phylogenetic X-tree and $w : E(T) \to \mathbb{R}$ be a real-valued weighting of the edges of T. Let d denote the mapping $d_{(\mathcal{T};w)}$ on X and let $W = \sum_{e\in E(T)} w(e)$. In this exercise, we describe two quite different ways to write W as a linear combination of the $d(x,y)$ values.
   (i) Let $(x_1, x_2, \ldots, x_n)$ be a cyclic permutation of X that provides a circular ordering for $\mathcal{T}$. Show that

$$W = d(x_n, x_1) + \frac{1}{2}\sum_{i=1}^{n-1} d(x_i, x_{i+1}).$$

   (ii) (Pauplin 2000) For every pair of distinct elements $x, y \in X$, let $I(x,y)$ denote the set of interior vertices of T in the path connecting $\phi(x)$ and $\phi(y)$. Let $\lambda : X \times X \to \mathbb{R}^{\geq 0}$ be the dissimilarity map on X defined, for all $x, y \in X$, in terms of the degrees of the vertices of T as follows:

$$\lambda(x,y) = \begin{cases} \prod_{v\in I(x,y)} (d(v)-1)^{-1}, & \text{if } x \neq y, \\ 0, & \text{if } x = y. \end{cases}$$

   Show that

$$W = \sum_{\{x,y\}\subseteq X} \lambda(x,y)d(x,y).$$

15. (Bryant 1997) Let $\mathcal{Q}$ be a set of quartet trees such that, for each subset X' of $\mathcal{L}(\mathcal{Q})$ of size four, there is exactly one quartet tree in $\mathcal{Q}$ that has label set X'. For all $q = ab|cd \in \mathcal{Q}$, let $\mathcal{D}_q$ denote the set of dissimilarity maps δ on X such that

$$\delta(a,c) + \delta(b,d) = \delta(a,d) + \delta(b,c).$$

Let $\mathcal{T}$ be the X-tree for which $\Sigma(\mathcal{T}) = \Sigma(\mathcal{Q})$ as implied by Theorem 6.3.7(i). Prove that a dissimilarity map δ on X is an element of $\bigcap_{q \in \mathcal{Q}} \mathcal{D}_q$ if and only if $\delta = d_{(\mathcal{T};w)}$ for some real-valued weighting w of the edges of $\mathcal{T}$.

16. Let $\mathcal{T}$ be a binary phylogenetic X-tree and let $w : E(\mathcal{T}) \rightarrow (\mathbb{R} - \{0\})$ be a function on $E(\mathcal{T})$. Let d denote the mapping $d_{(\mathcal{T};w)}$ on X and let $\mathcal{Q}$ denote the set of all quartet trees $aa'|bb'$ such that $a, a', b, b' \in X$ and

$$d(a,a') + d(b,b') \notin \{d(a,b) + d(a',b'), d(a,b') + d(a',b)\}.$$

Show that

$$\mathrm{scl}_2(\mathcal{Q}) = \mathcal{Q}(\mathcal{T}).$$

8

Markov models on trees

In previous chapters, we have regarded data such as characters or dissimilarity maps on X as being pre-given, and we have then discussed various ways that such data can be described by an X-tree. This, in turn, led to methods for reconstructing X-trees from characters or dissimilarity maps.

In this chapter, we consider how such characters might 'evolve' on some underlying phylogenetic tree according to various stochastic models. An obvious question is whether the techniques we have described earlier will recover this underlying phylogenetic tree when applied to data that is generated by such a process. A related question is how to use these stochastic models to devise more appropriate methods for tree reconstruction. This is a large and expanding area of research, and in this chapter we only touch on some of the main developments. The results described here involve extensive use of probability theory, and readers less familiar with this area may wish to consult the background material in the Preliminaries.

Unless otherwise stated, throughout this chapter a character will always mean a full character.

8.1 Matrix algebra

In this chapter, we will make frequent use of particular operations on matrices and we review these here. Throughout, we will write I to denote an identity matrix and 0 to denote a zero matrix. Furthermore, if the rows and columns of a matrix A are indexed by a set S, we write $A = [A_{ij}]$ to indicate that, for all $i, j \in S$, A_{ij} is the entry of A having row and column index i and j, respectively.

Let $A = [A_{ij}]$ be an $n \times n$ matrix. The *trace* of A, denoted $\text{tr}(A)$, is defined to be $\sum_{i=1}^{n} A_{ii}$ and the *determinant* of A, denoted $\det(A)$, is defined to be

$$\sum_{\pi} (-1)^{\text{sgn}(\pi)} \prod_{i=1}^{n} A_{i\pi(i)},$$

where the summation is over all permutations π of $\{1, \ldots, n\}$ and $\text{sgn}(\pi)$ denotes the parity of the permutation π.

Let A and B be square matrices of the same order. Then,

$$\det(AB) = \det(A)\det(B)$$

and

$$\det(A^{\mathrm{t}}) = \det(A),$$

where A^{t} denotes the transpose of A. Also, if A has rank 1, then

$$\det(I + A) = 1 + \mathrm{tr}(A). \tag{8.1}$$

The *exponential* of a square matrix A, denoted $\exp(A)$, is defined to be $\exp(A) = I + \sum_{k>0}(1/k!)A^k$. If $AB = BA$, then

$$\exp(A + B) = \exp(A)\exp(B).$$

Jacobi's identity, relating the determinant, trace, and exponential of A, states

$$\det(\exp(A)) = e^{\mathrm{tr}(A)}. \tag{8.2}$$

The matrix function exp has a left multiplicative inverse, the *logarithm* function log; that is, $\log(\exp(A)) = A$ for all square matrices A. Furthermore,

$$\log(A) = \log(I - (I - A)) = -\sum_{k\geq 1} \frac{1}{k}(I - A)^k$$

provided this series converges.

Later in this chapter, we will apply the notation exp and log to a vector. If v is a vector of rank n with i-th component v_i, then $\exp(v)$ and $\log(v)$ denote the vector of rank n whose i-th component is e^{v_i} and $\log(v_i)$, respectively.

We pause to introduce some further definitions and conventions. Given a vector π, we let $D(\pi)$ denote the diagonal matrix defined by

$$D(\pi)_{ij} = \begin{cases} \pi_i, & \text{if } i = j, \\ 0, & \text{otherwise.} \end{cases}$$

A *transition matrix* is a square matrix with all its entries non-negative and with each of its rows summing to 1. A *permutation matrix* is any matrix that can be obtained from an identity matrix by permuting rows or, equivalently, by permuting columns. Note that a permutation matrix is a particular type of transition matrix. In Lemma 8.1.1 (whose proof is omitted) for a vector π, we write $\pi > 0$ as shorthand for $\pi_i > 0$ for all i.

Lemma 8.1.1. *Let P be an $n \times n$ transition matrix. Then*

(i) $|\det(P)| \leq 1$. *Furthermore,* $\det(P) = 1$ *if and only if P is a permutation matrix.*

(ii) Let π be a row vector of order n with $\pi > 0$ and let $\pi' = \pi P$. Suppose that $\det(P) \neq 0$. *Then, $\pi' > 0$ which implies that $D(\pi')^{-1}$ exists, and so*

$$Q = D(\pi')^{-1} P^t D(\pi)$$

is well defined. Furthermore, the following hold: Q is a transition matrix; $0 < \det(PQ) \leq 1$*; and* $\det(PQ) = 1$ *if and only if P, or equivalently Q, is a permutation matrix.*

8.2 Markov processes on trees and graphs

In this section, we describe a framework for modelling the stochastic evolution of characters on a phylogenetic tree. This concept of a 'Markov process' on a phylogenetic tree generalizes the more familiar notion of a 'Markov chain.'

In this section, we adopt the following convention. Let $\xi_1, \xi_2, \ldots, \xi_k$ be random variables on a sample space S taking values in a set U. Let $A = \{1, 2, \ldots, k\}$. For a subset B of A and an event E in S, we will write the equation $\mathbb{P}(E| \bigwedge_{i\in A} \xi_i) = \mathbb{P}(E| \bigwedge_{i\in B} \xi_i)$ as short-hand for the statement that, for all selections of elements u_i from U, we have $\mathbb{P}(E| \bigcap_{i\in A}\{\xi_i = u_i\}) = \mathbb{P}(E| \bigcap_{i\in B}\{\xi_i = u_i\})$.

Definition 8.2.1. Let T be a rooted tree with vertex set V. Viewing the edges of T as arcs directed away from the root, let $\leq$ be any total ordering of V such that, whenever (u, v) is an arc of T, we have $u < v$. A *Markov process on* T with state set C is a family $\{\xi_v : v \in V\}$ of random variables such that, whenever (u, v) is an arc of T and $\alpha \in C$,

$$\mathbb{P}\left(\xi_v = \alpha | \bigwedge_{w<v} \xi_w\right) = \mathbb{P}(\xi_v = \alpha|\xi_u). \tag{8.3}$$

We refer to condition (8.3) as the *Markov property*. Informally, it states that, for each arc (u, v) of T, the value of ξ_v, conditional on ξ_u, is independent of the ξ-values at all other 'earlier' vertices. This captures the notion that the stochastic evolution of the state depends only on the immediately ancestral state and not on interactions from other 'earlier' species in the tree.

Markov processes on trees are important in biological classification for a variety of reasons, including the following:

- analysing how existing tree reconstruction methods perform if characters evolve according to Markov-based models;
- developing tree-reconstruction methods appropriate to particular models;
- investigating theoretical questions, such as how much 'phylogenetic information' is carried by a sequence of characters, each of which evolves under a Markov process.

Let T be a rooted tree. For each arc $e = (u, v)$ of T, a Markov process on T with state set C induces an associated transition matrix, denoted $P(e) = [P(e)_{\alpha\beta}]$, with rows and columns indexed over C, and defined by setting

$$P(e)_{\alpha\beta} = \mathbb{P}(\xi_v = \beta|\xi_u = \alpha),$$

for all $\alpha, \beta \in C$. The matrix $P(e)$ is called the *transition matrix* for e and the probability $\mathbb{P}(\xi_u \neq \xi_v)$, denoted $p(e)$, is called the *substitution probability* for $e = (u, v)$. For each vertex v of T, let π^v denote the row vector indexed by C and defined, for all $\alpha \in C$, by

$$\pi^v_\alpha = \mathbb{P}(\xi_v = \alpha).$$

If (u, v) is an arc of T, then, by the law of total probability,

$$\pi^v = \pi^u P(e). \tag{8.4}$$

Example 8.2.2. A familiar example of a Markov process on a rooted tree is a finite *Markov chain*. This is a sequence $\xi_0, \xi_1, \xi_2, \ldots, \xi_n, \ldots$, of random variables taking values in some finite state set C and that satisfies

$$\mathbb{P}\left(\xi_n = \alpha | \bigwedge_{j<n} \xi_j\right) = \mathbb{P}(\xi_n = \alpha | \xi_{n-1}),$$

for all $n > 0$ and $\alpha \in C$.

A Markov chain can be regarded as a Markov process on the rooted tree that has vertex set $0, 1, 2, \ldots, n, \ldots$, with vertex 0 as the root and with an arc from i to $i+1$ for each $i \geq 0$. Let $P(n)$ be the (transition) matrix defined by $P(n)_{\alpha\beta} = \mathbb{P}(\xi_n = \beta | \xi_{n-1} = \alpha)$. By eqn (8.4), $\pi^n = \pi^{n-1}P(n)$, and so $\pi^n = \pi^0 P(1)P(2)\cdots P(n)$.

For a Markov process on a rooted tree $T = (V, E)$ with state set C, let $\overline{\chi} : V \to C$ be a function and let

$$p(\overline{\chi}) = \mathbb{P}\left(\bigcap_{v \in V} \{\xi_v = \overline{\chi}(v)\}\right),$$

which is the joint probability that each vertex v in T is in state $\overline{\chi}(v)$. We can express this joint probability in a more useful form as follows. Let $v_1, \ldots, v_p$ be a total ordering $\leq$ of V such that, whenever (u, v) is an arc of T, we have $u < v$. Then, by the product rule,

$$p(\overline{\chi}) = \mathbb{P}\left(\xi_{v_1} = \overline{\chi}(v_1)\right) \prod_{i=2}^{p} \mathbb{P}\left(\xi_{v_i} = \overline{\chi}(v_i) | \bigcap_{j<i} \{\xi_{v_j} = \overline{\chi}(v_j)\}\right).$$

Furthermore, since this total order extends the partial order $\leq_T$, v_1 is the root of T and so, by the Markov property, we obtain the equality

$$p(\overline{\chi}) = \pi^{\rho}_{\overline{\chi}(\rho)} \prod_{e=(u,v)\in E(T)} P(e)_{\overline{\chi}(u)\overline{\chi}(v)}. \tag{8.5}$$

It is convenient at this stage to impose the following two mild technical conditions on a Markov process on a tree $T = (V, E)$ with root vertex ρ and state set C:

(M1) $\pi^{\rho}_{\alpha} > 0$, for all $\alpha \in C$, and
(M2) $\det(P(e)) \neq 0$, for each $e \in E$.

Unless otherwise stated, we will assume that these two conditions are satisfied throughout the rest of this chapter. However, we will often mention them explicitly either for emphasis or to indicate exactly how they are required. Later in this chapter, we will impose a variety of further constraints on a Markov process on a tree and, in these cases, we will usually refer to a 'Markov model' rather than a Markov process.

By Lemma 8.1.1 and eqn (8.4), conditions (M1) and (M2) imply the following weaker condition:

$$\pi_\alpha^v > 0, \quad \text{for all } v \in V \text{and all } \alpha \in C. \tag{8.6}$$

This weaker condition allows us to 're-root' the Markov process on T at any vertex of T. When we select a new vertex as a root, this may reverse the orientation of some arcs. For any such arc $e = (u, v)$ of T, the corresponding transition matrix, denoted $\overline{P}(e)$, is related to $P(e)$ as follows. For all $\alpha, \beta \in C$,

$$\overline{P}(e)_{\alpha\beta} = \mathbb{P}(\xi_u = \beta | \xi_v = \alpha) = \mathbb{P}(\xi_v = \alpha | \xi_u = \beta)\frac{\pi_\beta^u}{\pi_\alpha^v} = P(e)_{\beta\alpha}\frac{\pi_\beta^u}{\pi_\alpha^v},$$

where the second equality is derived from Bayes' rule. By Lemma 8.1.1, (8.6) ensures that $\overline{P}(e)$ is defined for all arcs, and we can express $\overline{P}(e)$ in matrix notation as

$$\overline{P}(e) = D(\pi^v)^{-1} P^{\mathrm{t}} D(\pi^u).$$

The Markov process resulting from this re-rooting induces the same distribution on ξ as the original Markov process, and it satisfies (8.6), as well as (M1) and (M2) provided the original Markov process does so. This ability to re-root the tree means that we can select a leaf of T as the root if we wish.

Markov random fields. A Markov process on a rooted tree is a special case of a more general stochastic process that can be defined on any graph. Given a graph $G = (V, E)$, let $N(v)$ denote the set of vertices of G adjacent to v. Then, a collection $\{\xi_v : v \in V\}$ of random variables taking values in state set C is called a *Markov random field on G* provided the following local (conditional) independence condition holds:

$$\mathbb{P}\left(\xi_v = \alpha | \bigwedge_{u \in V - \{v\}} \xi_u\right) = \mathbb{P}\left(\xi_v = \alpha | \bigwedge_{u \in N(v)} \xi_u\right),$$

for each $v \in V$ and each $\alpha \in C$. That is, for all $v \in V$, the value of ξ_v given the values ξ_u for all $u \in N(v)$ is conditionally independent of the remaining ξ-values.

It is not immediately clear that, when we specialize this definition to the case where the underlying graph is a tree T, the resulting process corresponds to a Markov process on T relative to some root vertex. The complication is that a Markov random field involves conditioning on all of the neighbours of a vertex, not just the one that comes immediately 'earlier' under some rooting of T. It is relatively easy to show by eqn (8.5) that a Markov process on a rooted tree T is a Markov random field on T. Establishing the converse [assuming (8.6)], requires a more delicate argument (Chang, personal communication).

8.3 Phylogenetic models of character evolution

In this section, we link Markov processes on trees with the central aspects of phylogenetics, namely phylogenetic trees and characters. We begin by introducing a slight generalization of the concept of a rooted phylogenetic tree.

Let $\mathcal{T} = (T; \phi)$ be a phylogenetic tree. A *rooting* of $\mathcal{T}$ is rooted tree T_ρ obtained by either distinguishing any vertex of T (including possibly a leaf) as a root, or subdividing some edge of T and regarding the resulting degree-two vertex as a root. In either case, we let ρ denote the root vertex of T_ρ.

Suppose that $\mathcal{T} = (T; \phi)$ is a phylogenetic tree. A *Markov process on* $\mathcal{T}$ is a Markov process on T_ρ for some rooting of $\mathcal{T}$. If $\mathcal{T} = (T; \phi)$ is a rooted phylogenetic tree, then a *Markov process on* $\mathcal{T}$ is Markov process on T.

For a Markov process on a phylogenetic X-tree $\mathcal{T} = (T; \phi)$ and a full character $\chi : X \to C$, let

$$p(\chi) = \mathbb{P}\left(\bigcap_{x \in X} \{\xi_{\phi(x)} = \chi(x)\}\right).$$

In other words, $p(\chi)$ is the probability that, for all $x \in X$, the leaf of T_ρ labelled $\phi(x)$ takes the state specified by the character χ. The value of $p(\chi)$ can be expressed as the sum of the probabilities $p(\overline{\chi})$ over all extensions $\overline{\chi}$ of χ to T_ρ. In particular,

$$p(\chi) = \sum_{\overline{\chi}: V(T_\rho) \to C, \overline{\chi} \circ \phi = \chi} p(\overline{\chi}). \tag{8.7}$$

Definition 8.3.1. Let $\mathcal{T} = (T; \phi)$ be a phylogenetic X-tree and suppose we have a Markov process on $\mathcal{T}$. The map $\chi \mapsto p(\chi)$ is the *character distribution* on X induced by the Markov process on $\mathcal{T}$. Furthermore, for all $x, y \in X$, let

$$p(x, y) = \mathbb{P}(\xi_{\phi(x)} \neq \xi_{\phi(y)});$$

that is, $p(x, y)$ is the probability that vertex $\phi(x)$ is assigned a different state to the vertex $\phi(y)$. From (8.7), we can write

$$p(x, y) = \sum_{\chi : \chi(x) \neq \chi(y)} p(\chi).$$

The map $(x, y) \mapsto p(x, y)$ is the *dissimilarity distribution* on X.

Example 8.3.2. Let $\mathcal{T} = (T; \phi)$ be the rooted phylogenetic tree shown in Fig. 8.1 with label set $\{x_1, x_2, x_3\}$. Suppose that we have a Markov process on $\mathcal{T}$ with state set $\{0, 1\}$ so that $\pi^\rho = [\pi_0^\rho, \pi_1^\rho]$ and, for each $i \in \{1, 2, 3, 4\}$,

$$P(e_i) = \begin{bmatrix} 1 - p_i & p_i \\ q_i & 1 - q_i \end{bmatrix},$$

where $0 \leq p_i \leq 1$ and $0 \leq q_i \leq 1$. For the character χ defined by $\chi(x_1) = \chi(x_2) = 0$ and $\chi(x_3) = 1$,

$$\begin{aligned} p(\chi) = \pi_0^\rho p_3[(1 - p_1)(1 - p_2)(1 - p_4) + q_1 q_2 p_4] \\ + \pi_1^\rho (1 - q_3)[(1 - p_1)(1 - p_2) q_4 + q_1 q_2 (1 - q_4)]. \end{aligned}$$

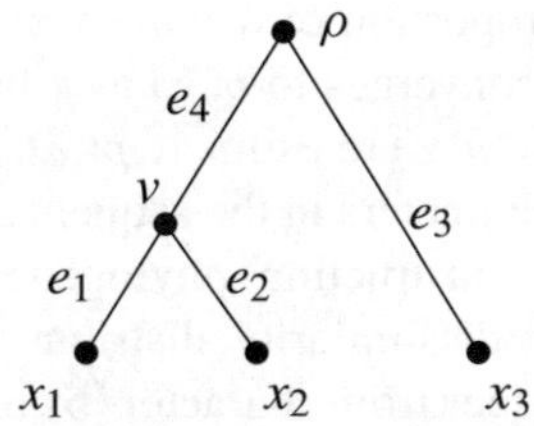

Figure 8.1

Furthermore, letting $\pi^v = \pi^\rho P(e_4)$, we have

$$p(x_1, x_2) = \pi_0^v[(1 - p_1)p_2 + p_1(1 - p_2)] + \pi_1^v[q_1(1 - q_2) + q_2(1 - q_1)].$$

A fundamental question is the following: what does the character distribution of a Markov process on a phylogenetic X-tree $\mathcal{T}$ tell us about $\mathcal{T}$? We shall see in Section 8.4 that, except for degenerate cases, it is possible to reconstruct $\mathcal{T}$, up to the position of the root, from the character distribution. This reconstruction only requires the 'marginal' distribution on pairs of elements of X. Furthermore, under much stronger assumptions, one can reconstruct $\mathcal{T}$ from the dissimilarity distribution on X.

We end this section by discussing three further aspects of the character distribution of a Markov process on a phylogenetic X-tree. First we describe some notation and observations that are useful in specifying character distributions. Then, we briefly outline the relevance of character distributions to tree reconstruction in biology. Finally, we will describe how the character distribution can be calculated in polynomial time in $|X|$.

Parameterizing the character distribution. Given a Markov process on a phylogenetic X-tree $\mathcal{T} = (T; \phi)$, the induced character distribution is determined by a pair $(\mathcal{T}, \theta)$ where, relative to some rooting of $\mathcal{T}$, θ comprises the vector π^ρ and the transition matrices $P(e)$ assigned to each arc e of T_ρ. We will let θ denote these parameters additional to $\mathcal{T}$, and write $p(\chi|\mathcal{T}, \theta)$, instead of just $p(\chi)$, for the character distribution when we wish to make explicit $\mathcal{T}$ and the associated parameters that comprise θ. Later, we will see that, for certain Markov processes, the choice of any particular vertex in a phylogenetic tree $\mathcal{T}$ as a root vertex ρ has no effect on the transition matrices. In this case, we may refer to the transition matrix associated with an edge of $\mathcal{T}$.

Suppose we have a character distribution induced by a Markov process on a phylogenetic X-tree $\mathcal{T} = (T; \phi)$. Then, provided (M1) and (M2) hold, we can choose the rooting of $\mathcal{T}$ to be based on any vertex of T and obtain the same character distribution, once the transition matrices and vector π^ρ are adjusted appropriately. In particular, although the root vertex ρ in T_ρ for the original Markov process may have degree 2, we can obtain the same character distribution by rooting $\mathcal{T}$ with an existing vertex of T.

Relevance of the character distribution. Suppose that we have a sequence of k characters on X that have been generated independently and identically according to a Markov process on a phylogenetic tree (aligned DNA/RNA sequences are often modelled in this way). Let $\hat{p}(\chi)$ denote the proportion of occurrences of character χ in the data. By the law of large numbers, $\hat{p}(\chi)$ converges to $p(\chi)$ as k becomes large. Thus, provided k is sufficiently large, one can use $\hat{p}(\chi)$ to estimate $p(\chi)$. Similarly, one can estimate $p(x, y)$ by the proportion $\hat{p}(x, y)$ of characters in the sequence that assign x a different state to y.

Thus, techniques for reconstructing phylogenetic trees from either the character distribution or from the dissimilarity distribution lead naturally to methods for reconstructing phylogenetic trees from characters by using the 'plug-in' technique of substituting $\hat{p}(\chi)$ in place of $p(\chi)$ or $\hat{p}(x, y)$ in place of $p(x, y)$, respectively. One caution with this approach is that in applications the number k of characters in the data is often much less than the total number of possible characters taking values in the given state set, and this forces $\hat{p}(\chi)$ to equal zero for most choices of χ. Consequently, it is helpful to develop techniques for reconstructing trees from marginal distributions of the character distribution, such as the $p(x, y)$ values, rather than requiring all of the individual $p(\chi)$ values.

A further rationale for establishing the reconstructability of a phylogenetic tree from its character distribution is that it guarantees the statistical consistency of other approaches to tree reconstruction such as maximum likelihood. We discuss this further in Section 8.9.

Computing the character distribution. Suppose that we have a Markov process on a phylogenetic X-tree $\mathcal{T} = (T; \phi)$ with state set C and a rooting of $\mathcal{T}$ by an interior vertex. Let $T = (V, E)$ and let $\chi : X \to C$ be a character. Combining eqns (8.5) and (8.7), we can express $p(\chi)$ as

$$p(\chi) = \sum_{\overline{\chi}:V\to C, \overline{\chi}\circ\phi=\chi} \pi^{\rho}_{\overline{\chi}(\rho)} \prod_{e=(u,v)\in E} P(e)_{\overline{\chi}(u)\overline{\chi}(v)}. \tag{8.8}$$

Notice that if $\mathcal{T}$ has ν interior vertices, then the summation in (8.8) involves $|C|^{\nu}$ terms, which is problematic for a binary phylogenetic X-tree with $|X|$ large. However, there is an alternative way to compute $p(\chi)$ that runs in linear time in $|X|$.

This alternative approach uses dynamic programming based on the fundamental decomposition of a rooted phylogenetic tree into rooted phylogenetic subtrees, illustrated in Fig. 2.5. To describe the required recursion, we need to introduce a further definition. Let $p(\chi|\mathcal{T}, \theta, \alpha)$ denote the conditional probability that $\xi_{\phi(x)} = \chi(x)$ for all $x \in X$, given that $\xi_{\rho} = \alpha$. By the law of total probability,

$$p(\chi|\mathcal{T}, \theta) = \sum_{\alpha\in C} p(\chi|\mathcal{T}, \theta, \alpha)\pi^{\rho}_{\alpha}. \tag{8.9}$$

The desired recursion applies to the quantity $p(\chi|\mathcal{T}, \theta, \alpha)$. For all $v \in V$, let

$$X[v] = \{x \in X : \phi(x) \leq_T v\} \quad \text{and} \quad \mathcal{T}[v] = \mathcal{T}|X[v].$$

Informally, $\mathcal{T}[v]$ is the phylogenetic subtree of $\mathcal{T}$ rooted at v. Suppose $v_1, \ldots, v_k$ are the vertices of $\mathcal{T}[v]$ incident with the root v. Let $\theta[v]$ denote the restriction of θ to $\mathcal{T}[v]$.

Let χ_v denote the restriction of χ to $X[v]$ and, for all $i \in \{1, 2, \ldots, k\}$, let χ_i denote the restriction of χ to $X[v_i]$. Then, by the Markov property,

$$p(\chi_v|\mathcal{T}[v], \theta[v], \alpha) = \prod_{i=1}^{k} \left[\sum_{\beta \in C} p(\chi_i|\mathcal{T}[v_i], \theta[v_i], \beta) P((v, v_i))_{\alpha\beta} \right]. \tag{8.10}$$

Now, $p(\chi)$ can be efficiently computed by starting at the leaves of $\mathcal{T}$ and working up to the root, using (8.10) to compute the values $p(\chi_v|\mathcal{T}[v], \theta[v], \beta)$ for each $v \in V$ and $\beta \in C$. Once these values have been determined for $v = \rho$, apply eqn (8.9) to obtain $p(\chi|\mathcal{T}, \theta)$.

8.4 Tree reconstruction for the general Markov process

In this section, we show that the character distribution of a Markov process on a phylogenetic tree $\mathcal{T}$ suffices to determine $\mathcal{T}$, subject to (M1), (M2), and one other mild condition. To illustrate the necessity of these conditions, we begin with an example.

Example 8.4.1. For each quartet tree $\mathcal{T}$ on $X = \{1, 2, 3, 4\}$, consider a Markov process on $\mathcal{T}$ with any interior vertex of $\mathcal{T}$ as the root vertex ρ, and with state set $C = \{0, 1\}$. Take $\pi^\rho = [0.5, 0.5]$ and, for a fixed value of $p \in [0, 0.5]$, assign to each edge of $\mathcal{T}$ the transition matrix

$$\begin{bmatrix} 1-p & p \\ p & 1-p \end{bmatrix}.$$

If $p = 1/2$, then each of the three quartet trees induce the same character distribution on X, namely the distribution that independently assigns each element of X state 0 or 1 with equal probability. Furthermore, if $p = 0$, then again all quartet trees induce the same character distribution on X. In this case, it is the distribution that assigns all elements of X the same state (0 or 1) with equal probability. In both cases, we have different phylogenetic trees inducing the same character distribution on X.

In Example 8.4.1, the case $p = 1/2$ violates condition (M2), while the case $p = 0$ gives a transition matrix that is a permutation matrix. For a Markov process on a rooted phylogenetic X-tree $\mathcal{T}$, the requirement than none of the transition matrices are a permutation matrix is equivalent, by Lemma 8.1.1(i), to the following condition:

(M3) $\det(P(e)) \neq \pm 1$ for all edges e of $\mathcal{T}$.

We will see shortly, that conditions (M1)–(M3) allow us to reconstruct $\mathcal{T}$ (up to the position of the root vertex) from the character distribution on X induced by the Markov process. This is based on the following construction of a dissimilarity map on X from this distribution.

Suppose that we have a Markov process on a phylogenetic X-tree $\mathcal{T} = (T; \phi)$ with state set C satisfying (M1) and (M2). Then, for all $x, y \in X$ and $\alpha, \beta \in C$, the conditional probability

$$P^{xy}_{\alpha\beta} = \mathbb{P}(\xi_{\phi(y)} = \beta | \xi_{\phi(x)} = \alpha), \tag{8.11}$$

is well defined. Let $P^{xy} = [P^{xy}_{\alpha\beta}]$ denote the corresponding $|C| \times |C|$ matrix. Thus, for all $x, y \in X$, P^{xy} is the matrix of conditional probabilities of observing particular states at leaf $\phi(y)$ given the state at leaf $\phi(x)$. Lemma 8.4.2 relates P^{xy} and P^{yx} to the transition matrices associated to the arcs of T_ρ once a root vertex ρ of T specified. Its straightforward proof is omitted.

Lemma 8.4.2. *Suppose we have a Markov process on a phylogenetic* X*-tree* $\mathcal{T} = (T; \phi)$ *satisfying* (M1) *and* (M2). *Select a rooting* T_ρ *of* $\mathcal{T}$. *Let* $x, y \in X$, *and let* v *be the last vertex of* T *shared in the paths from* ρ *to* $\phi(x)$ *and from* ρ *to* $\phi(y)$. *Let* $e_1, \ldots, e_r$ *and* $e'_1, \ldots, e'_s$ *denote the arcs in* T_ρ *from* v *to* $\phi(x)$ *and from* v *to* $\phi(y)$, *respectively. Then,*

$$P^{xy} = \overline{P}(e_r) \cdots \overline{P}(e_1) P(e'_1) \cdots P(e'_s)$$

and

$$P^{yx} = \overline{P}(e'_s) \cdots \overline{P}(e'_1) P(e_1) \cdots P(e_r).$$

It follows from Lemmas 8.4.2 and8.1.1 that, for all $x, y \in X$, $\det(P^{xy}P^{yx}) > 0$. Thus, for all $x, y \in X$, we can let

$$\delta(x, y) = -\tfrac{1}{2} \log(\det(P^{xy}P^{yx})). \tag{8.12}$$

As $\delta(x, x) = 0$ and $\delta(x, y) = \delta(y, x)$, for all $x, y \in X$, δ is a dissimilarity map on X.

Lake (1994) described an alternative way to express $\delta(x, y)$. For all $x, y \in X$ and $\alpha, \beta \in C$, let $J^{xy}_{\alpha\beta}$ be the joint probability

$$\mathbb{P}(\{\xi_{\phi(x)} = \alpha\} \cap \{\xi_{\phi(y)} = \beta\}),$$

and let $J^{xy} = [J^{xy}_{\alpha\beta}]$ denote the corresponding $|C| \times |C|$ matrix. Since $P^{xy} = (J^{xx})^{-1} J^{xy}$, we can write, for all $x, y \in X$,

$$\delta(x, y) = -\log\left(\frac{|\det(J^{xy})|}{\sqrt{\det(J^{xx})\det(J^{yy})}}\right). \tag{8.13}$$

The dissimilarity map on X we have described by (8.12) and (8.13) has become known as a *logdet transform* or as a *paralinear distance.*

Theorem 8.4.3. *Suppose we have a Markov process on a phylogenetic* X*-tree* $\mathcal{T} = (T; \phi)$ *satisfying assumptions* (M1) *and* (M2). *Then, the dissimilarity map* δ *on* X *defined by* (8.12) *is a tree metric on* X. *In particular, let* $w : E(T) \to \mathbb{R}$ *be the edge weighting of* T *defined, for all* $e \in E(T)$ *by*

$$w(e) = -\tfrac{1}{2} \log(\det(P(e)\overline{P}(e))) \geq 0.$$

Then, for all $x, y \in X$,

$$\delta(x, y) = d_{(\mathcal{T}, w)}(x, y).$$

Furthermore, if the Markov process $(\mathcal{T}, \theta)$ *satisfies* (M3), *then* $w(e) > 0$ *for all edges of* T.

Proof. First note that, by Lemma 8.1.1, $w(e) \geq 0$ for each edge e of $\mathcal{T}$. It suffices to show that $\delta(x, y) = d_{(\mathcal{T},w)}(x, y)$ for all distinct elements x and y in X. Select a rooting T_ρ of $\mathcal{T}$ using any vertex of T as the root and let $e_1, \ldots, e_r$ and $e'_1, \ldots, e'_s$ be the arcs of T_ρ as described in the statement of Lemma 8.4.2. Then, by Lemma 8.4.2,

$$\log(\det(P^{xy}P^{yx})) = \sum_{i=1}^{r} \log(\det(P(e_i)\overline{P}(e_i))) + \sum_{j=1}^{s} \log(\det(P(e'_j)\overline{P}(e'_j))).$$

Consequently, $\delta(x, y) = d_{(\mathcal{T},w)}(x, y)$ as claimed. Furthermore, if the Markov process also satisfies (M3), then, by Lemma 8.1.1, we have $0 < \det(P(e)\overline{P}(e)) < 1$ for all $e \in E(T)$ and so $w(e) > 0$ for all $e \in E(T)$. □

Combining Theorem 8.4.3 with Theorem 7.1.8, we see that the logdet transform δ uniquely determines $\mathcal{T}$. Furthermore, $\mathcal{T}$ can be reconstructed from δ in polynomial time by the techniques described in Chapter 7. Thus, we have the following result.

Corollary 8.4.4. *For a Markov process on a phylogenetic X-tree $\mathcal{T}$ satisfying* (M1)–(M3)*, $\mathcal{T}$ can be reconstructed from the character distribution induced by $(\mathcal{T}, \theta)$ in polynomial time.*

In Corollary 8.4.4, we do not require the entire character distribution to reconstruct $\mathcal{T}$, only the (marginal) distribution on pairs of elements of the label set X of $\mathcal{T}$. The logdet transform given by (8.13) has been applied to DNA sequence data in conjunction with the methods described in Chapter 7 for reconstructing trees from dissimilarity maps. Theorem 8.4.3 and Corollary 8.4.4 are also significant because they guarantee the statistical consistency of methods such as maximum likelihood (described in Section 8.9) when applied to data generated by a Markov process on a tree satisfying (M1)–(M3).

Once $\mathcal{T} = (T; \phi)$ has been recovered from the character distribution, it is natural to ask whether one can also recover the other parameters that underly the Markov process. If we specify a root vertex ρ of T, then the parameters of interest are the vector π^ρ and the transition matrices $P(e)$ associated to the arcs of the rooted tree T_ρ. One might hope that the induced character distribution on pairs of elements of X—given by the joint probability matrices J^{xy} for all $x, y \in X$—might suffice to determine these parameters, since these are sufficient to reconstruct $\mathcal{T}$. Indeed, if T is rooted on a leaf labelled x, then we can determine π^ρ as the row sums of J^{xy}. However, even when a root vertex for T has been specified the transition matrices on the arcs of T are not necessarily uniquely determined by the J^{xy} matrices. Nevertheless, Baake (1998) established the following result, which also follows immediately from Theorem 7.6.9.

Proposition 8.4.5. *Suppose we have a Markov process on a phylogenetic X-tree $\mathcal{T} = (T; \phi)$ satisfying* (M1)–(M3)*. Select any vertex of T as a root. Then, the matrices J^{xy}, for all $x, y \in X$, determine the product $P(e)\overline{P}(e)$ for each pendant edge e of T, and $P(e)\overline{P}(e)$ is determined up to conjugacy for each interior edge e of T.*

In relation to Proposition 8.4.5, notice that J^{xy} does not depend on the choice of the root vertex for T. So, although the transition matrices for the Markov process on T

depend on the choice of the root vertex, the quantity $P(e)\overline{P}(e)$ is invariant to such choices for pendant edges, and is invariant up to conjugation for interior edges.

Chang (1996*a*) showed that under slightly stronger conditions than (M1)–(M3), the joint probability distribution on *triples* of elements of X suffices to determine $P(e)$.

8.5 Stationary and reversible models

In biology, one often views the evolution of character states in a state set C on a rooted tree T as evolving according to some continuous-time Markov process. Under such a process, there is some rate at which each state will randomly mutate into another state; for each arc e of T, we let Q_e denote the rate matrix associated with e. This matrix is indexed by the character states in C and, for each distinct pair $\alpha, \beta \in C$, $(Q_e)_{\alpha\beta}$ is the instantaneous rate of substitution from state α to state β on arc e. By conservation, each row of Q_e sums to 0. Each time a state change occurs in this process, we say that a *substitution* has occurred.

For each arc e of T, we assume that the rate matrix Q_e operates at intensity λ_e and for duration t_e on e. The relationship between the triple (Q_e, λ_e, t_e) and the transition matrix $P(e)$ of the associated Markov process is given by the fundamental relationship

$$P(e) = \exp(Q_e\mu_e),$$

where $\mu_e = \lambda_e t_e$. Furthermore, by eqn (8.2),

$$\det(P(e)) = e^{\mathrm{tr}(Q_e)\mu_e}.$$

Consequently, condition (M2) applies to the corresponding Markov process and condition (M3) is equivalent to the condition that $\mu_e > 0$ for all e.

For the remainder of this section, we consider a particular type of continuous-time Markov process that has become widely used in phylogenetic analysis.

Definition 8.5.1. Let $\mathcal{T} = (T; \phi)$ be a phylogenetic tree and consider a rooting of $\mathcal{T}$. A Markov process on $\mathcal{T}$ is said to be a *stationary process* with *equilibrium vector* π if

(i) there exists a fixed matrix Q so that, for each arc e of T_ρ, $P(e) = \exp(Q\mu_e)$,
(ii) $\pi Q = 0$ and $\pi = \pi^\rho$.

A stationary Markov process is *reversible* if the matrix $D(\pi)Q$ is symmetric.

Reversible stationary processes have the feature that if they are represented on a phylogenetic tree $\mathcal{T}$, then changing the choice of a root vertex has no effect on the resulting transition matrices since $\overline{P}(e) = P(e)$ for each arc e of T_ρ. Consequently, a reversible stationary process can be represented by a pair $(\mathcal{T}, \theta)$, where $\mathcal{T}$ is a (unrooted) phylogenetic tree, and θ consists of Q together with the weight function $e \mapsto \lambda_e$ defined on all edges e of $\mathcal{T}$. For the remainder of this section we will work with reversible stationary processes.

Suppose we have a stationary process on $\mathcal{T}$. It is immediate from Definition 8.5.1 that, for all vertices v of $\mathcal{T}$, we have $\pi^v = \pi$. Furthermore, if all the off-diagonal entries of Q are strictly positive, then, for each edge e of $\mathcal{T}$, the substitution probability $p(e)$ converges to the limiting value

$$p_\infty = 1 - \sum_{\alpha \in C} \pi_\alpha^2 \tag{8.14}$$

as $\mu_e \to \infty$.

For a reversible stationary process on $\mathcal{T} = (T; \phi)$, let $\kappa(e)$ denote the expected number of substitutions that occur on edge e assuming these occur according to a continuous-time process. Since $\sum_{\alpha,\beta \in C: \alpha \neq \beta} \pi_\alpha Q_{\alpha\beta} = -\sum_\alpha \pi_\alpha Q_{\alpha\alpha}$ is the instantaneous rate of substitution on any edge, and since μ_e is the scaled duration of this process on e, $\kappa(e)$ can be written conveniently as

$$\kappa(e) = -\text{tr}(D(\pi) Q_e)\mu_e, \tag{8.15}$$

for all $e \in E(\mathcal{T})$. If each edge e of $\mathcal{T}$ is weighted by $\kappa(e)$, then the value of the induced tree metric on X for a pair of elements, x and y say, is referred to as the *evolutionary distance* between x and y, and is denoted $\kappa(x, y)$. It equals the expected total number of substitutions that occur on the path in T between $\phi(x)$ and $\phi(y)$.

If for a reversible stationary process, λ_e is the same for all edges e of $\mathcal{T}$, then the rate of substitutions per unit time is constant throughout the tree. In this case, a *molecular clock* is said to apply and the expected number of substitutions along the path from the root of $\mathcal{T}$ to any leaf is the same. Consequently, since the t_e values induce an ultrametric on X, the evolutionary distances between elements of X also induce an ultrametric on X. We now describe a further subclass of stationary processes.

For a reversible stationary process on a phylogenetic X-tree $\mathcal{T}$, the matrix P^{xy}, defined by eqn (8.11), satisfies

$$P^{xy} = P^{yx} = \exp(Q\mu(x, y)), \tag{8.16}$$

where

$$\mu(x, y) = \sum_{e \in P(T; \phi(x), \phi(y))} \mu_e.$$

Therefore, by (8.15), the evolutionary distance between any pair $x, y \in X$ is given by

$$\kappa(x, y) = -\text{tr}(D(\pi) \log(P^{xy})). \tag{8.17}$$

The rate matrices corresponding to reversible stationary processes have some elegant properties. For example, the eigenvalues of the rate matrices are real, and a useful spectral representation exists for describing the transition matrices in terms of the eigenvectors and eigenvalues of Q.

Many reversible stationary processes have been proposed for DNA sequences, such as the Kimura 3ST model, which we discuss in Section 8.10. We next describe a particular class (and subclass) of reversible stationary processes.

The equal input model. Let π be any positive probability distribution on a state set C. A canonical reversible stationary process that has π as its equilibrium distribution is obtained by using the rate matrix $Q = [Q_{\alpha\beta}]$ defined, for all $\alpha, \beta \in C$ by

$$Q_{\alpha\beta} = \begin{cases} \pi_\beta, & \text{if } \alpha \neq \beta, \\ \pi_\beta - 1, & \text{if } \alpha = \beta. \end{cases}$$

It is easily checked that $\pi Q = 0$; that is, π is the equilibrium vector for Q. Furthermore, $D(\pi)Q$ is symmetric and thus the model is also reversible. For DNA sequences, where $|C| = 4$, this model is known as the 'Tajima–Nei equal input model' or the 'Felsenstein's 1981 model' (see Swofford *et al.* 1996).

The equal input model is easy to analyse, as Proposition 8.5.2 demonstrates. Recall from eqn (8.14) that $p_\infty = 1 - \sum_{\alpha \in C} \pi_\alpha^2$.

Proposition 8.5.2. *Suppose we have an equal input model on a phylogenetic tree $\mathcal{T}$ with state set C. Then, for all edges e of $\mathcal{T}$, $\alpha, \beta \in C$, and $x, y \in X$,*

(i) $P(e)_{\alpha\beta} = \begin{cases} \pi_\alpha + (1 - \pi_\alpha)e^{-\mu_e}, & \text{if } \alpha = \beta, \\ \pi_\beta(1 - e^{-\mu_e}), & \text{if } \alpha \neq \beta. \end{cases}$

(ii) $\kappa(e) = p_\infty \mu_e$.

(iii) $p(e) = p_\infty(1 - e^{-\kappa_e / p_\infty})$.

(iv) $p(x, y) = p_\infty(1 - e^{-\kappa(x,y)/p_\infty})$, *and so* $\kappa(x, y) = -p_\infty \log(1 - p(x, y)/p_\infty)$.

Proof. We first prove (i). By taking $A = \pi\pi^{\mathrm{t}}$ in eqn (8.1), it can be shown that, for all $e \in E(\mathcal{T})$, the matrix $Q\mu_e$ has eigenvalues $-\mu_e$ and 0 with multiplicity r and 1, respectively. Thus, for all $\alpha, \beta \in C$, $P(e)_{\alpha\beta} = a_{\alpha\beta} + b_{\alpha\beta}e^{-\mu_e}$ for constants $a_{\alpha\beta}$ and $b_{\alpha\beta}$. Furthermore, when $\mu_e = 0$ we have

$$P(e)_{\alpha\beta} = \begin{cases} 1, & \text{if } \alpha = \beta, \\ 0, & \text{if } \alpha \neq \beta. \end{cases}$$

As $\mu_e \to \infty$, we have $P(e)_{\alpha\beta} = p_\infty$. These two boundary conditions suffice to determine the constants $a_{\alpha\beta}$ and $b_{\alpha\beta}$ stated in part (i) of the proposition.

We next prove (ii). By eqn (8.15), for all $e \in E(\mathcal{T})$,

$$\kappa(e) = -\mathrm{tr}(D(\pi)Q)\mu_e = -\sum_{\alpha \in C} \pi_{\alpha \in C}(\pi_\alpha - 1)\mu_e = p_\infty \mu_e,$$

as required. For the proof of (iii) it follows by (i) that

$$p(e) = \sum_{\alpha \in C} \sum_{\beta \in C : \alpha \neq \beta} \pi_\alpha P(e)_{\alpha\beta} = p_\infty(1 - e^{-\mu_e}).$$

Part (iii) now follows by part (ii).

Finally, we establish part (iv). The first equation in part (iv) follows from part (iii) and eqn (8.16). The second equation is an immediate consequence. □

In Proposition 8.5.2, note that for $\kappa(e)$ small, $p(e)$ is approximately equal to $\kappa(e)$ but, as $\kappa(e)$ grows, $p(e)$ underestimates $\kappa(e)$.

The r-state symmetric (N_r) model. We now discuss a particular case of the equal input model on a phylogenetic tree $\mathcal{T}$, which was introduced by Neyman (1971). In this simple model on a state set C of size r underlying $r \times r$ rate matrix Q has the property that all the off-diagonal entries are equal. It follows that the equilibrium vector π for this model induces the uniform distribution on $|C|$ and whenever there is a substitution from a particular state, each of the $r-1$ remaining states is equally likely to be the new state. Thus, for all $e \in E(\mathcal{T})$ the transition matrix associated to e is described, for all $\alpha, \beta \in C$, by

$$P(e)_{\alpha\beta} = \begin{cases} 1 - p(e), & \text{if } \alpha = \beta, \\ p(e)/(r-1), & \text{if } \alpha \neq \beta. \end{cases}$$

where $p(e)$ is the substitution probability associated with e. It is a simple exercise to show that

$$\det(P(e)) = \left(1 - \frac{r}{(r-1)}p(e)\right)^{r-1}. \tag{8.18}$$

Thus, the only constraints on the N_r-model required for the model to satisfy (M1)–(M3) (or, equivalently, for the continuous time process to act with positive duration on each edge of $\mathcal{T}$) is that

$$0 < p(e) < \frac{r-1}{r}, \tag{8.19}$$

In case $r = 4$, this model is known in biology as the 'Jukes–Cantor' model. The N_r-model also arises in statistical physics, where it is sometimes referred to as the 'r-state Potts model', and the N_2-model is closely related to the 'Ising model' of ferromagnetism.

An alternative description of the N_r-model is known in statistical physics as the 'Fortuin–Kasteleyn random cluster' representation. In this model, each edge e of $\mathcal{T}$ is treated independently and is deleted with probability $[r/(r-1)]p(e)$; otherwise, it is not deleted. The subsets of X that label leaves that are connected in the resulting random graph form a partition of X. Each block of the partition is then independently assigned the same state from C, and according to the uniform probability distribution on C. It can be shown that the probability distribution on characters on X that results from this stochastic process is identical to that under the N_r-model, with corresponding transition probabilities $p(e)$. Since π is uniform on $|C|$, it follows by eqn (8.14) that $p_\infty = (r-1)/r$. Combining this with Proposition 8.5.2 we obtain the following result.

Proposition 8.5.3. *Suppose we have the N_r-model on a phylogenetic tree $\mathcal{T}$. Then,*

(i) $p(e) = ((r-1)/r)(1 - e^{-(r/(r-1))\kappa(e)})$, *and*

(ii) $\kappa(x, y) = -((r-1)/r)\log(1 - (r/(r-1))p(x, y))$.

The next result (Tuffley and Steel 1997), sets an upper bound on the probability of generating a particular character χ on X under the N_r-model on a phylogenetic X-tree $\mathcal{T}$. This upper bound involves just the parsimony score of χ on $\mathcal{T}$.

Theorem 8.5.4. *Given a character χ on X, a phylogenetic X-tree $\mathcal{T}$, and the N_r-model on $\mathcal{T}$, then*

$$\sup_{\theta}\{p(\chi|\mathcal{T},\theta)\} = r^{-l(\chi,\mathcal{T})},$$

where $l(\chi,\mathcal{T})$ denotes the parsimony score of χ on $\mathcal{T}$ and the supremum is over all choices of θ that satisfy (8.19).

8.6 The Hadamard representation

In this section, we describe a surprisingly compact representation of the character distribution for the N_2-model. In fact, we describe this representation for a slight generalization of the N_2-model and in the specialist topic of this chapter we indicate how it extends further to Markov models that correspond to a random walk on an Abelian group.

Let $\mathcal{T} = (T;\phi)$ be a phylogenetic X-tree and let C be the state set $\{0,1\}$. Select a vertex ρ of T as a root and let π^ρ be an arbitrary vector of probabilities [possibly in violation of (M1)]. To each arc e of T_ρ assign the transition matrix

$$\begin{bmatrix} 1-p(e) & p(e) \\ p(e) & 1-p(e) \end{bmatrix},$$

where $0 \leq p(e) < \frac{1}{2}$. We call the resulting Markov process on $\mathcal{T}$ the *extended N_2-model*. It produces the same character distribution as the N_2–model, provided $\pi^\rho = [0.5, 0.5]$.

For the remainder of this section it will be convenient to take $X = \{1,\ldots,n\}$, where $n \geq 2$, and to let X' denote the set $\{1,\ldots,n-1\}$. There is a natural way to encode pairs of (complementary) binary characters on X in terms of subsets of X'. For a subset A of X' and an element $i \in \{0,1\}$, let χ_A^i be the character that assigns state i precisely to those elements of X that are not in A. For the extended N_2 model, let

$$\overline{p}_A = p(\chi_A^0) + p(\chi_A^1).$$

Note that $\overline{p}_A$ is the probability of generating a binary character on X under the extended N_2-model for which the induced split of X is $A|(X-A)$. The proof of Lemma 8.6.1 is straightforward and is left as an exercise.

Lemma 8.6.1. *Let $\mathcal{T}$ be a phylogenetic tree and let ρ be a root vertex of $\mathcal{T}$.*

(i) For the extended N_2-model on $\mathcal{T}$, $\overline{p}_A$ is independent of the value taken by the vector π^ρ and the choice of the vertex ρ.
(ii) For the N_2-model on $\mathcal{T}$, $p(\chi_A^0) = p(\chi_A^1) = \frac{1}{2}\overline{p}_A$.

Note that, although the N_2-model on $\mathcal{T}$ is a reversible stationary process, and therefore the transition matrices do not change when we re-root $\mathcal{T}$, this is no longer true for the extended N_2-model. However, by part (i) of Lemma 8.6.1, if we are only interested in the values $\overline{p}_A$ for $A \subseteq \{1,\ldots,n-1\}$, then we may again return to a parameterization that involves a (unrooted) phylogenetic X-tree $\mathcal{T}$, together with the substitution probability $p(e)$ associated to each edge e of $\mathcal{T}$.

The representation of the extended N_2-model indicated earlier is based on a particular class of orthogonal matrices, which we now describe. A *Hadamard matrix* of rank k is a $k \times k$ matrix H in which each entry is $+1$ or -1, and $HH^t = kI$. The latter property implies $H^{-1} = (1/k)H$.

There is a particular construction for such a matrix when $k = 2^n$ for all $n \geq 1$. Under any fixed ordering of the subsets of $\{1, \ldots, n\}$, let H_n denote the matrix indexed by this ordering, and defined by

$$H_n = [H_{AA'}] = \left[(-1)^{|A \cap A'|}\right].$$

Lemma 8.6.2. *The matrix H_n is a Hadamard matrix. Furthermore, $H_n^t = H_n$ and $H_n^{-1} = 2^{-n}H_n$.*

It is possible to order the subsets of $\{1, \ldots, n\}$ so that the Hadamard matrix H_{n+1} has the block representation

$$H_{n+1} = \begin{pmatrix} H_n & H_n \\ H_n & -H_n \end{pmatrix}.$$

Hadamard matrices of this form are sometimes called *Sylvester matrices*.

Now suppose we have an extended N_2-model on a phylogenetic X-tree $\mathcal{T}$. For $e \in E(\mathcal{T})$, let

$$\theta(e) = 1 - 2p(e) \quad \text{and} \quad \gamma(e) = -\tfrac{1}{2}\log(\theta(e)).$$

If we are dealing with the standard N_2-model, then, by Proposition 8.5.3(i), $\gamma(e)$ is precisely the expected number of substitutions on edge e. Let γ denote the vector, indexed by the 2^{n-1} subsets of X', and defined by setting, for all $A \subseteq \{1, \ldots, n-1\}$,

$$\gamma_A = \begin{cases} 0, & \text{if } A|(X - A) \notin \Sigma(\mathcal{T}) \cup \{\emptyset|X\}, \\ \gamma(e), & \text{if } A|(X - A) \text{ corresponds to edge } e \text{ of } \mathcal{T}, \\ -\sum_{e \in E(\mathcal{T})} \gamma(e), & \text{if } A = \emptyset. \end{cases}$$

Let $\overline{p} = [\overline{p}_A]$ denote the vector of $\overline{p}_A$ values. A fundamental result that has proved useful for both theoretical investigations and applications to data is the following remarkable theorem due to Hendy (1989).

Theorem 8.6.3. *For the extended N_2-model on a phylogenetic tree X-tree $\mathcal{T}$,*

$$\overline{p} = H_{n-1}^{-1} \exp(H_{n-1}\gamma)$$

and so, equivalently,

$$\gamma = H_{n-1}^{-1} \log(H_{n-1}\overline{p}),$$

where exp *and* log *are applied component-wise.*

Theorem 8.6.3 expresses $\overline{p}$ via a summation indexed in a way that is entirely independent of $\mathcal{T}$, a feature we will make even more explicit in Corollary 8.6.6. Although

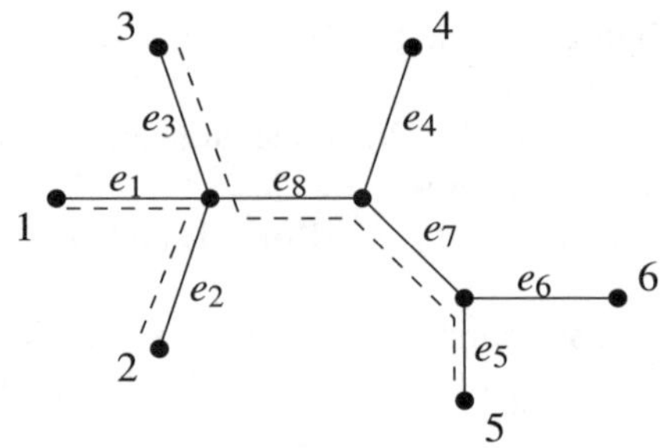

Figure 8.2 The path set induced by $\{1, 2, 3, 5\}$ is $\{e_1, e_2, e_3, e_5, e_8\}$.

Theorem 8.6.3 is restricted to the extended N_2-model, we describe how its scope can be broadened in Section 8.10.

Theorem 8.6.3 has been applied directly to reconstruct phylogenetic trees from genetic data. It has been a useful tool in calculating the character distribution as well as a valuable analytical tool for addressing some theoretical questions concerning the properties of tree reconstruction procedures.

In the rest of this section, we provide further background that is useful in interpreting and applying Theorem 8.6.3. In particular, we present two lemmas, and show how they furnish a proof of Theorem 8.6.3. We begin by showing how to express $H_{n-1}\gamma$ in Theorem 8.6.3 in terms of collections of paths in $\mathcal{T}$.

Suppose that $\mathcal{T} = (T; \phi)$ is a phylogenetic X-tree and let Y be an even-sized subset of X. Construct a set of $\frac{1}{2}|Y|$ edge-disjoint paths in T each of which connects a pair of leaves labelled by elements from Y. It can be shown that if $\mathcal{T}$ is binary, then this set of paths is uniquely determined by Y. More generally, for any phylogenetic X-tree $\mathcal{T}$, the set $P(\mathcal{T}, Y)$ of edges used in the paths is uniquely determined and is called the *path set* induced by Y on $\mathcal{T}$. If $Y = \emptyset$, we set $P(\mathcal{T}, Y) = \emptyset$. Note that if $Y = \{x_1, x_2\}$, then $P(\mathcal{T}, Y)$ is simply the set of edges in the path connecting $\phi(x_1)$ and $\phi(x_2)$. An example of a path set is shown in Fig. 8.2.

Returning to the proof of Theorem 8.6.3, let

$$f : 2^{X'} \to \{Y \subseteq X \colon\ |Y| \equiv 0 \ (\mathrm{mod}\ 2)\}$$

be the map defined, for all $A \subseteq X' = \{1, \ldots, n-1\}$ by

$$f(A) = \begin{cases} A, & \text{if } |A| \equiv 0 \ (\mathrm{mod}\ 2), \\ A \cup \{n\}, & \text{otherwise.} \end{cases}$$

The inverse f^{-1} of this mapping is defined by $Y \mapsto Y \cap X'$.

Lemma 8.6.4. *Let $\mathcal{T}$ be a phylogenetic X-tree and let γ be the vector defined above and induced by $\mathcal{T}$. Then, for all $A \subseteq \{1, \ldots, n-1\}$,*

$$(H_{n-1}\gamma)_A = -2 \sum_{e \in P(\mathcal{T}, f(A))} \gamma(e)$$

and

$$\exp\,(H_{n-1}\gamma)_A = \prod_{e\in P(\mathcal{T},f(A))} \theta(e). \tag{8.20}$$

For an even-sized subset Y of X, let $\mathrm{sgn}(Y) = (-1)^Z$, where Z is a random variable corresponding to the total number of edges $e = \{u, v\}$ in $P(\mathcal{T}, Y)$ for which $\xi_u \neq \xi_v$. Informally, the event $\mathrm{sgn}(Y) = -1$ corresponds to the existence of an odd number of substitution events on the edges in $P(\mathcal{T}, Y)$. The following lemma expresses the probability of this event in two quite different ways and it is the crucial connection for the proof of Theorem 8.6.3.

Lemma 8.6.5. *Suppose that we have the extended N_2-model on a phylogenetic X-tree $\mathcal{T}$. Then, for all even-sized subset Y of X,*

(i) $\mathbb{P}(\mathrm{sgn}(Y) = -1) = \frac{1}{2}(1 - \prod_{e\in P(\mathcal{T},Y)} \theta(e)$, *and*

(ii) $\mathbb{P}(\mathrm{sgn}(f(A)) = -1) = \frac{1}{2}(1 - (H_{n-1}\overline{p})_A)$.

Proof of Theorem 8.6.3. Setting $Y = f(A)$ in the two parts of Lemma 8.6.5, we get

$$(H_{n-1}\overline{p})_A = \prod_{e\in P(\mathcal{T},f(A))} \theta(e)$$

for all subsets of A of X'. Furthermore, by Lemma 8.6.4, the right-hand side of the last equation is equal to $\exp\,(H_{n-1}\gamma)_A$. Since this holds for all subsets A of X',

$$H_{n-1}\overline{p} = \exp\,(H_{n-1}\gamma),$$

and so

$$\overline{p} = H_{n-1}^{-1} \exp\,(H_{n-1}\gamma),$$

completing the proof of Theorem 8.6.3. □

The argument used in the proof of Theorem 8.6.3 provides an additional way to express the vector $\overline{p}$.

Corollary 8.6.6. *Suppose that we have the extended N_2-model on a phylogenetic X-tree $\mathcal{T}$. Then, for all subsets A of $\{1, \ldots, n-1\}$,*

$$\overline{p}_A = \frac{1}{2^{n-1}} \sum_{Y\subseteq X:\, |Y|\equiv 0 (\mathrm{mod}\ 2)} \left[(-1)^{|Y\cap A|} \prod_{e\in P(\mathcal{T},Y)} \theta(e) \right].$$

An interesting aspect of the representation in Corollary 8.6.6 is that it applies without any restriction on the substitution probabilities of the edges of $\mathcal{T}$.

Example 8.6.7. Consider the extended N_2-model on the phylogenetic X-tree shown in Fig. 8.3. For all $i \in \{1, 2, 3, 4, 5\}$, let $\theta_i = 1 - p(e_i)$. Then, by Corollary 8.6.6,

$$\overline{p}_\emptyset = \tfrac{1}{8}\,(1 + \theta_1\theta_2 + \theta_3\theta_4 + \theta_5(\theta_1\theta_3 + \theta_2\theta_3 + \theta_1\theta_4 + \theta_2\theta_4) + \theta_1\theta_2\theta_3\theta_4)\,.$$

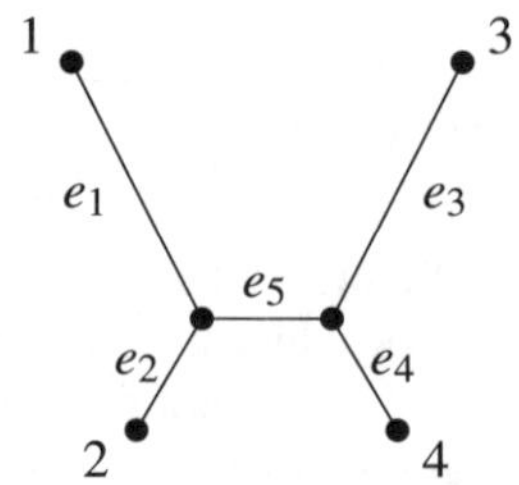

Figure 8.3 The Felsenstein zone.

To obtain $\overline{p}_{\{1,2\}}$ we can simply replace θ_5 by $-\theta_5$ in this last expression or, equivalently, replace θ_1, θ_2 by $-\theta_1, -\theta_2$, or replace θ_i by $-\theta_i$ for all i, to obtain

$$\overline{p}_{\{1,2\}} = \tfrac{1}{8}\left(1 + \theta_1\theta_2 + \theta_3\theta_4 - \theta_5(\theta_1\theta_3 + \theta_2\theta_3 + \theta_1\theta_4 + \theta_2\theta_4) + \theta_1\theta_2\theta_3\theta_4\right).$$

Notice that the substitutions (in the signs of the θ_i's) that convert $\overline{p}_{\emptyset}$ to $\overline{p}_{\{1,2\}}$ correspond precisely to the ways of extending a full character on X that generates the X-split $\{1,2\}|\{3,4\}$ to all the vertices of the tree. For example, the substitution $\theta_1, \theta_2 \mapsto -\theta_1, -\theta_2$ corresponds to a transition precisely on each of the two edges incident with leaves 1 and 2. This is a general and sometimes useful feature of the Hadamard representation.

8.7 The 'Felsenstein zone'

Suppose we generate a sequence of k characters independently and identically, according to a Markov process on a binary phylogenetic tree $\mathcal{T}$. If we apply the maximum parsimony method to the resulting sequence of characters, are we likely to recover the underlying tree $\mathcal{T}$ up to the position of the root? In particular, as k grows, does the probability of selecting $\mathcal{T}$ as the unique maximum parsimony tree converge to 1 or to 0, or to some other value? In a landmark paper, Felsenstein (1978*b*) showed that, even for a binary phylogenetic tree with four leaves under the simple N_2-model, the substitution probabilities can be chosen so that the probability of selecting the original quartet tree by the maximum parsimony method converges to 0 as k grows.

In this section, we describe Felsenstein's example, and our approach will also illustrate the use of the Hadamard representation. In Section 8.9, we will consider the issue of 'statistical consistency' indicated above in more detail.

Consider the extended N_2-model on the phylogenetic X-tree $\mathcal{T}$ shown in Fig. 8.3. Suppose that $p(e_1) = p(e_3) = p$ and $p(e_2) = p(e_4) = p(e_5) = q$, and consider a full character on X randomly generated under this model. Then, the expected parsimony score of this randomly generated character on $\mathcal{T}$ can be written as

$$1 + \overline{p}_{\{1,3\}} + \overline{p}_{\{2,3\}}$$

since any such character has parsimony score 1 on $\mathcal{T}$, except for the two pairs of characters on X that induce the X-splits $13|24$ and $14|23$, which have parsimony score 2 on $\mathcal{T}$.

Similarly, the expected parsimony score of a randomly generated character on X on the rival quartet tree $\mathcal{T}' = 13|24$ is

$$1 + \overline{p}_{\{1,2\}} + \overline{p}_{\{2,3\}}.$$

Let Δ denote the difference between the expected parsimony score of a randomly generated character on $\mathcal{T}'$ and on $\mathcal{T}$; that is, $\Delta = \overline{p}_{\{1,2\}} - \overline{p}_{\{1,3\}}$.

Now suppose we generate a sequence of k random characters independently under this model. If $\Delta < 0$, then, by standard statistical theory, the probability that the parsimony score of the sequence of characters on $\mathcal{T}$ exceeds the parsimony score of the characters on $\mathcal{T}'$ tends to 1 as k grows. Consequently, if $\Delta < 0$, then maximum parsimony will select an incorrect tree with probability tending to 1 as the number of characters grows. Thus, we now determine precisely when $\Delta < 0$.

Let us order the subsets of $\{1, 2, 3\}$ as

$$\emptyset, \{1\}, \{2\}, \{3\}, \{1, 2\}, \{1, 3\}, \{2, 3\}, \{1, 2, 3\}.$$

Then, $\Delta = w\overline{p}$ where $w = [0, 0, 0, 0, 1, -1, 0, 0]$. Therefore, by Theorem 8.6.3,

$$\Delta = wH_3^{-1} \exp{(H_3\gamma)} = \tfrac{1}{8}wH_3 \exp{(H_3\gamma)} = u \exp{(H_3\gamma)},$$

where $u = \frac{1}{4}[0, 0, -1, 1, 1, -1, 0, 0]$ since $u_A = \frac{1}{8}[(-1)^{|\{1,2\}\cap A|} - (-1)^{|\{1,3\}\cap A|}]$ for all $A \subseteq \{1, 2, 3\}$. Let $x = 1 - 2p$ and $y = 1 - 2q$. Then, by eqn (8.20),

$$\Delta = \tfrac{1}{4}(-y^3 + 2xy - x^2y) = \tfrac{1}{4}y(-y^2 - x^2 + 2x).$$

Thus, $\Delta < 0$ if and only if $-x^2 - y^2 + 2x < 0$. Replacing x and y by their expressions in terms of p and q, the latter becomes

$$p^2 > q(1 - q).$$

Observe that the substitution probabilities can be arbitrarily small and yet maximum parsimony may still favour the wrong tree. When $|X| = 4$, it is possible to characterize exactly the conditions on the substitution probabilities for the statistical consistency of maximum parsimony. Letting $\theta_i = (1 - 2p(e_i))$ for all $i \in \{1, 2, 3, 4, 5\}$, an extension of the above argument establishes the following result.

Theorem 8.7.1. *For sequences of characters generated by the phylogenetic tree $\mathcal{T}$ shown in Fig.* 8.3 *under the extended N_2-model the probability that maximum parsimony selects $\mathcal{T}$ converges to* 1 *as the number of characters grows if and only if*

$$\theta_5 < \min\left\{\frac{\theta_1\theta_2 + \theta_3\theta_4}{\theta_1\theta_3 + \theta_2\theta_4}, \frac{\theta_1\theta_2 + \theta_3\theta_4}{\theta_1\theta_4 + \theta_2\theta_3}\right\}.$$

If we realize the N_2-model as a stationary process, it can be shown that the condition described in the last theorem always holds under a molecular clock. However, when the number of leaves in a phylogenetic tree $\mathcal{T}$ exceeds four, maximum parsimony can fail to converge on $\mathcal{T}$ for the N_2-model, even when a molecular clock applies (for details see Hendy and Penny 1989).

For phylogenetic trees with more than four leaves, two sufficient conditions for maximum parsimony to converge on the correct tree have been described in the literature. One condition, which applies under the N_r-model, is that the substitution probability on each edge is sufficiently small, and in addition each interior edge also has a substitution probability that is bounded away from 0 (Steel 2000). A second sufficient condition, which applies to the N_r-model and to some extensions, is that the state set is sufficiently large and the substitution probabilities are bounded away from 0 and 1 (Steel and Penny 2000).

8.8 Rate variation across characters

So far in this chapter we have assumed that each character is generated according to the same Markov process on a phylogenetic tree. However, it is useful to allow different characters to evolve at different 'rates'.

To make this idea precise, suppose we have a reversible stationary Markov model with underlying rate matrix Q and edge parameters μ_e. Under a distribution of rates-across-sites, the μ_e values are multiplied by a positive random value ν drawn from some distribution $\mathcal{D}$. The character then evolves as according to a reversible stationary model with rate matrix Q and edge parameters $\nu\mu_e$. For a full character χ on X let $p(\chi\,|\,\nu)$ denote the probability of generating character χ conditional on the selection of the rate ν. If $p_{\mathcal{D}}(\chi)$ denotes the probability of generating χ when ν is selected according to $\mathcal{D}$, then $p_{\mathcal{D}}(\chi)$ is the expected value of $p(\chi\,|\,\nu)$. Thus, if $\mathcal{D}$ has a probability density function $f(\nu)$, then

$$p_{\mathcal{D}}(\chi) = \int_0^\infty p(\chi\,|\,\nu)f(\nu)\,d\nu.$$

Such 'mixture' models have become widely used in molecular phylogenetics, and a common choice for the distribution $\mathcal{D}$ is the gamma distribution, where a positive probability is sometimes also given to the value $\nu = 0$.

Given a distribution $\mathcal{D}$, let $M_{\mathcal{D}}(x)$ denote the 'moment generating function' defined as the expected value of $e^{\nu x}$ if ν is randomly selected according to $\mathcal{D}$. Thus, if $\mathcal{D}$ has a probability density function $f(\mu)$, then

$$M_{\mathcal{D}}(x) = \int_0^\infty e^{\nu x}f(\mu)\,d\nu.$$

Let $M_{\mathcal{D}}^{-1}(x)$ denote the inverse of this function. As an example, for the gamma distribution defined by the density function,

$$f(\lambda) = \frac{e^{-k\lambda}x^{k-1}k^k}{\Gamma(k)}$$

where $\lambda > 0$, one has $M_{\mathcal{D}}(x) = (k/(k-x))^k$ and $M_{\mathcal{D}}^{-1}(x) = k(1 - x^{-1/k})$. Note that, although $M_{\mathcal{D}}$ and $M_{\mathcal{D}}^{-1}$ are defined on scalars, it is straightforward to extend their domain to matrices. For example, one may diagonalize the matrices and apply the function componentwise to the entries in the diagonal matrix.

A mixture model may no longer satisfy the Markov property (8.3) as shown by Chang (1996*b*). Also, there appears to be no recursive way to calculate $p_{\mathcal{D}}(\chi)$ in general. However, an approximation scheme, based on discretizing a continuous distribution, has been developed by Yang (1994).

Many of the results concerning stationary processes can be extended to allow rate variation provided one replaces the functions exp and log by the functions $M_{\mathcal{D}}$ and $M_{\mathcal{D}}^{-1}$, respectively. We illustrate this with two examples.

In the case of the N_2-model on a phylogenetic X-tree $\mathcal{T}$ with a rate distribution $\mathcal{D}$, one can derive the following analogue of Theorem8.6.3. For $X = \{1, \ldots n\}$ let $\overline{p}_{\mathcal{D}}$ denote the vector, indexed by subsets A of $\{1, \ldots, n-1\}$ and for which $(\overline{p}_{\mathcal{D}})_A$ is the sum of the probabilities (under this model) of the two binary characters on X that induce the split $X|(X-A)$.

Theorem 8.8.1. *For the N_2-model on a phylogenetic tree X-tree $\mathcal{T}$, with a rate distribution $\mathcal{D}$,*

$$\overline{p}_{\mathcal{D}} = H_{n-1}^{-1} M_{\mathcal{D}}(H_{n-1}\gamma)$$

and so, equivalently,

$$\gamma = H_{n-1}^{-1} M_{\mathcal{D}}^{-1}(H_{n-1}\overline{p}_{\mathcal{D}}),$$

where $M_{\mathcal{D}}$ and $M_{\mathcal{D}}^{-1}$ are applied componentwise.

As a second example, consider the N_r-model on a phylogenetic tree with an associated rate distribution $\mathcal{D}$. We can extend Proposition 8.5.3(ii) to give the evolutionary distance between any pair of elements $x, y \in X$ as follows:

$$\kappa(x, y) = -\frac{(r-1)}{r} M_{\mathcal{D}}^{-1}\left(1 - \frac{r}{(r-1)} p(x, y)\right).$$

More generally, given any reversible stationary process the evolutionary distance between any pair of elements $x, y \in X$ can be expressed in terms of the joint probability matrix J^{xy} by

$$\kappa(x, y) = -\mathrm{tr}(D(\pi) M_{\mathcal{D}}^{-1}(D(\pi)^{-1} J^{xy})). \tag{8.21}$$

Equation (8.21) was independently established by Gu and Li (1996), and Waddell and Steel (1997).

For the remainder of this section, we comment on the implications of these extensions for tree reconstruction. By eqn (8.21), we see that under a reversible stationary process on a phylogenetic tree $\mathcal{T}$ with a rate distribution $\mathcal{D}$, we can reconstruct $\mathcal{T}$ from its induced character distribution provided the function $M_{\mathcal{D}}$ is known. In fact, this reconstruction of $\mathcal{T}$ requires just the joint distributions J^{xy}, since π can be determined from these matrices. A natural question is whether one can extend this to allow the reconstruction of $\mathcal{T}$ from its induced character distribution, without knowledge of the distribution $\mathcal{D}$. Theorem 8.8.2 shows that this is possible under certain conditions; for example, when the underlying μ_e values obey a molecular clock. However, in general it is not possible, even when the underlying model is the simple N_2-model, and so no method can always recover the underlying phylogenetic tree from a rate-variation model without some knowledge of the distribution $\mathcal{D}$. The proof of the first part of this theorem is left as an exercise. The second part is more involved and relies on Theorem 8.8.1 (see Steel *et al.* 1994 for details).

Theorem 8.8.2.

(i) Let $\mathcal{T}$ be a binary phylogenetic X-tree. For any reversible stationary process on $\mathcal{T}$ with rate distribution $\mathcal{D}$, the two dissimilarity maps on X given by $(x, y) \mapsto p(x, y)$ and $(x, y) \mapsto \kappa(x, y)$ (for all $x, y \in X$) are ordinally equivalent. Consequently, if a molecular clock applies, then $\mathcal{T}$ can be reconstructed from the dissimilarity distribution without knowledge of $\mathcal{D}$.

(ii) For each binary phylogenetic X-tree $\mathcal{T}$, there exists an associated set $\mu_{\mathcal{T}}$ of μ_e values for the N_2-model on $\mathcal{T}$ together with an associated rate distribution $\mathcal{D}_{\mathcal{T}}$ so that the character distribution induced by the triple $(\mathcal{T}, \mu_{\mathcal{T}}, \mathcal{D}_{\mathcal{T}})$ takes the same value for all choices of $\mathcal{T}$.

Other types of rate variation have also been explored in molecular phylogenetics. For example, one may know which particular sites are evolving quickly and which are evolving slowly, in which case the appropriate analysis would incorporate this knowledge. Note also that, although this section has allowed characters to evolve at different rates, the rate for each character is fixed throughout the tree. Alternative models that allow the rate of a character to vary throughout the tree have been investigated. In the simplest case, each character may have rate 0 or 1, which may vary (perhaps randomly) throughout the tree. Such a model reflects a mechanism of molecular evolution called the 'covarion model' that Walter Fitch and colleagues first put forward around 1970. For a mathematical analysis of this type of model, see Tuffley and Steel (1998). Recently, such models have been successfully applied to study molecular evolution, as for example, in the study by Galtier (2001).

8.9 Model-based reconstruction methods

In this section, we describe techniques for reconstructing phylogenetic trees from character data that are explicitly based on Markov processes of character evolution.

One simple and fast technique for reconstructing a phylogenetic tree $\mathcal{T}$ is to estimate a dissimilarity map on X for which $\mathcal{T}$ supports a tree metric representation. For example, we may estimate the logdet transform, or for stationary processes, the evolutionary distance. Since the true values of these dissimilarity maps are tree metrics and $\mathcal{T}$ supports a tree metric representation, one can then apply the methods from the previous chapter to reconstruct $\mathcal{T}$. The usual way to estimate these dissimilarity maps is simply to use the 'plug-in' approach of replacing probabilities by observed frequencies. Thus, for the logdet transform, the entries in J^{xy}, for each $x, y \in X$, are estimated by the proportion of characters at which x and y are in the various pairs of prescribed states. Similarly, to estimate the values $p(x, y)$ for each $x, y \in X$ (in order to compute evolutionary distances), one uses the observed proportion of characters where x and y take different states. This plug-in approach has been widely used in biology and has sound statistical properties provided the number of characters is reasonably large. However, other approaches that we describe below are also widely used, and in certain situations may be more informative.

Maximum likelihood. The method of *maximum likelihood* is a common procedure in statistical inference, but we will describe it here just in the context of phylogenetics.

Let $\mathcal{T}$ be a phylogenetic X-tree and let $\mathcal{C} = (\chi_1, \ldots, \chi_k)$ be a sequence of characters on X. For a Markov process on $\mathcal{T}$ and with parameters described by θ the probability that a sequence of k independently and identically generated characters yields $\mathcal{C} = (\chi_1, \ldots, \chi_k)$ is

$$\mathbb{P}(\mathcal{C}|\mathcal{T}, \theta) = \prod_{i=1}^{k} p(\chi_i|\mathcal{T}, \theta).$$

Now any model under consideration allows θ to take values in some permitted set Θ. The maximum likelihood method for such a model selects the phylogenetic tree(s) that maximize $\sup_{\theta \in \Theta}\{\mathbb{P}(\mathcal{C}|\mathcal{T}, \theta)\}$.

As with the maximum parsimony method, maximum likelihood involves two optimization steps; first maximizing $\mathbb{P}(\mathcal{C}|\mathcal{T}, \theta)$ for a fixed phylogenetic X-tree $\mathcal{T}$, and then maximizing this value over the set of all phylogenetic X-trees. However, in contrast to the maximum parsimony method, no general procedure is known that guarantees to solve the first optimization step. Analytic solutions have been restricted to phylogenetic trees with at most four leaves under the N_2-model (Chor *et al.* 2000, Yang 2000). Consequently, various heuristics are usually applied to the fixed-tree optimization problem. Furthermore, the function $\mathbb{P}(\mathcal{C}|\mathcal{T}, \theta)$ can have multiple optima as θ varies, even for simple models such as the N_2-model. Despite the lack of a general procedure, heuristic methods appear to perform well in finding approximate solutions to this first optimization step, and maximum likelihood is now widely used in phylogenetic analysis. It is often applied in statistical tests involving likelihood ratios or (parametric and non-parametric) bootstrap. For a statistical perspective, the interested reader is referred to Goldman (1990).

Models for which θ is allowed to vary from character to character have also been considered. Usually, this is modelled by selecting some parameter in θ according to a distribution. For example, for a model that allows rate variation across sites as in Section 8.8, it is the rate parameter ν associated with the character that randomly varies, and the maximum likelihood procedure is applied by taking $\mathbb{P}(\mathcal{C}|\mathcal{T}, \theta) = \prod_{i=1}^{k} p_{\mathcal{D}}(\chi_i|\mathcal{T}, \theta)$.

Bayesian methods. Suppose we have a prior joint probability distribution on the set of pairs $(\mathcal{T}, \theta)$, where $\mathcal{T}$ is a binary phylogenetic X-tree and θ is the parameters in some Markov process on $\mathcal{T}$. Such probability distributions are often based on simple models for random speciation and extinction, and these generally provide a marginal distribution on binary phylogenetic trees that is identical to the Yule distribution. Given such a joint prior distribution, one can express the posterior distribution of a binary phylogenetic X-tree $\mathcal{T}$, conditional on a sequence $\mathcal{C}$ of characters on X, as

$$\mathbb{P}(\mathcal{T}|\mathcal{C}) = \mathbb{P}(\mathcal{T}) \int \frac{1}{N} \mathbb{P}(\mathcal{C}|\mathcal{T}, \theta) f(\theta|\mathcal{T})\, d\theta$$

where N is a normalizing constant fixed over all binary phylogenetic X-trees $\mathcal{T}$, $\mathbb{P}(\mathcal{T})$ is the prior (marginal) probability of $\mathcal{T}$, and $f(\theta|\mathcal{T})$ is the density of the conditional distribution of θ given $\mathcal{T}$.

Such approaches to phylogeny have been considered by several investigators, using Markov chain monte carlo techniques to estimate quantities in the calculations. For details, see Yang and Rannala (1997) and Mau *et al.* (1999). However, this 'Bayesian' approach to phylogenetics still has limited acceptance, largely due to concerns about the underlying prior distributions.

Consistency and convergence. In Section 8.7, we saw that under certain conditions, the maximum parsimony method will tend to select, with increasing probability, an incorrect tree when applied to k characters generated independently and identically under the N_2-model. We now consider this issue of statistical consistency in more detail, along with the related issue of the rate at which a statistically consistent method converges on the true phylogenetic tree as the number of characters grows.

Much of this discussion can be phrased in a general framework as follows. Suppose we have a finite set A and, for each element $a \in A$, an associated parameter set $\Theta(a)$. Furthermore, suppose that for each pair (a, θ), where $a \in A$ and $\theta \in \Theta(a)$, we have a random variable $Y(a, \theta)$ taking values in a finite set U. Now, independently generate k elements of U, say $u_1, u_2, \ldots, u_k$, by the random variable $Y(a, \theta)$, and let Y^k be the random variable that assigns to the pair (a, θ) the k-tuple $(u_1, \ldots, u_k) \in U^k$. Given a reconstruction method m that assigns an element from A to each element of U^k, let

$$r^{(k)}(a, \theta; m) = \mathbb{P}(m(Y^k) = a).$$

We may regard $r^{(k)}(a, \theta; m)$ as the probability of correctly reconstructing a from the random variable Y^k that a induces. The method m is said be a *statistically consistent* estimator of a if $r^{(k)}(a, \theta; m)$ converges to 1 as k tends to infinity, for each $a \in A$ and $\theta \in \Theta(a)$.

Now if a and b are distinct elements of A, and (a, θ) and (b, θ') induce exactly the same probability distribution on U, then it is easily checked that no reconstruction method can simultaneously be a statistically consistent estimator of both a and b. To state a converse result, we need to introduce a metric on random variables W and Z taking values in the finite set U. Let

$$d(W, Z) = \sum_{u \in U} |\mathbb{P}(W = u) - \mathbb{P}(Z = u)|.$$

Suppose (a, θ) has the property that, for some constant $d_0 > 0$,

$$d(Y(a, \theta), Y(b, \theta')) \geq d_0, \quad \text{for all } b \neq a, \theta' \in \Theta(b). \tag{8.22}$$

Then, under this separation condition, it can be shown that the maximum likelihood method is a consistent estimator of a.

We now specialize this general result to the phylogenetic context, where the set A corresponds to the set of binary phylogenetic X-trees, and the set $\Theta(a)$ corresponds to the parameters associated with a Markov model. The random variable $Y(a, \theta)$ is the associated randomly generated character χ on X.

Condition (8.22) asserts that the induced character distribution $p(\chi|\mathcal{T}, \theta)$ cannot be approximated arbitrarily closely by the induced character distribution of a different phylogenetic X-tree $\mathcal{T}'$ by selecting appropriate associated parameters θ'. This applies for the general Markov process on a binary phylogenetic X-tree [under assumptions (M1)–(M3)] by Theorem 8.4.3, and for reversible stationary processes on a binary phylogenetic X-tree with a known rate distribution by eqn (8.21).

A more careful analysis is required to determine the number of characters required for a method to correctly reconstruct the underlying tree. However, using results from Section 7.7, one can derive upper and lower bounds on the number of characters k required to accurately reconstruct binary phylogenetic X-trees. For example, under the N_2-model and when $|X| = 4$, suppose that the interior edge of the three binary phylogenetic X-trees has substitution probability p. Then as p converges to zero, the value of k required to accurately reconstruct each tree grows at a rate proportional to $1/p^2$. Furthermore, a simple counting argument shows that under any model of sequence evolution and any method of tree reconstruction, the sequence length k required to accurately reconstruct just the class of all caterpillar trees on X must grow at least as fast as a constant times $\log(n)$, where $n = |X|$. Surprisingly, these lower bounds on k are at least close to optimal, as the upper bound results of Erdös *et al.* (1999) suggest.

8.10 Specialist topic: algebraic properties of Markov processes on trees

In this section, we investigate two algebraic properties of Markov processes on trees. First, we describe how some models have an alternative group-theoretic description, which allows for an extension of the Hadamard representation described in Section 8.6 to certain models on a larger state set. Second, we discuss some polynomial-based properties of the character distribution induced by Markov processes, known as 'phylogenetic invariants'.

Group-based models. For a set C of character states, suppose that we have a binary operation $*$ on C such that $\mathcal{G} = (C, *)$ is a group. Let $\mathcal{T}$ be a phylogenetic X-tree and let T_ρ be a rooting of $\mathcal{T}$. Now, assign a random element ξ_ρ of C to the root ρ of T_ρ according to a probability distribution π^ρ and, for each arc $e = (u, v)$ of T_ρ, randomly and independently assign an element $\xi(u, v)$ of C. For each arc e of T_ρ, let h_e be the probability distribution on C defined, for each $\alpha \in C$, by

$$h_e(\alpha) = \mathbb{P}(\xi(u, v) = \alpha).$$

Once elements of C have been randomly assigned in this way to all the arcs of T_ρ, one obtains a family of C-valued random variables $\{\xi_v \colon v \in V(T)\}$ as follows. If

$\rho = v_1, v_2, \ldots, v_k = v$ is the (unique) path from ρ to v, then set

$$\xi(v) = \xi_\rho * \xi(v_1, v_2) * \cdots * \xi(v_{k-1}, v_k).$$

It is easily checked that $\{\xi_v\colon\ v \in V(T)\}$ is a Markov process on T_ρ, and that, for each arc e of T_ρ, the associated transition matrix $P(e) = [P(e)_{\alpha\beta}]$ is given by

$$P(e)_{\alpha\beta} = h_e(\alpha^{-1}\beta)$$

for all $\alpha, \beta \in C$.

A Markov process on a rooted tree with state set C is *group based* if it can be represented in this way for some binary group operation $*$ on C and some appropriate choices of h_e, in which case we say that the model can be *realized* by the group $(C, *)$. Similarly, a Markov process on a phylogenetic tree $\mathcal{T}$ is said to be *group based* if the associated Markov process on some rooting of $\mathcal{T}$ is group based.

Lemma 8.10.1 shows that not every Markov process is group based since their transition matrices have the special property of being *doubly stochastic*; that is, each row and each column sums to 1.

Lemma 8.10.1. *In a group-based Markov process, each transition matrix is doubly stochastic.*

Proof. Suppose we have a group-based Markov process on a rooted tree T with state space C. Then, for each arc e of T and all $\alpha, \beta \in C$,

$$\sum_{\alpha \in C} P(e)_{\alpha\beta} = \sum_{\alpha \in C} h_e(\alpha^{-1}\beta) = \sum_{\gamma \in C} h_e(\gamma) = 1$$

and

$$\sum_{\beta \in C} P(e)_{\alpha\beta} = \sum_{\beta \in C} h_e(\alpha^{-1}\beta) = \sum_{\gamma \in C} h_e(\gamma) = 1.$$

□

Group-based models have many appealing properties. For example, given a group-based Markov process on a rooted tree T, and any two edge-disjoint paths $P(T; u, v)$ and $P(T; u', v')$, the events

$$E = \{\xi_u \neq \xi_v\} \quad \text{and} \quad E' = \{\xi_{u'} \neq \xi_{v'}\}$$

are independent.

Several of the models we have discussed in this chapter are group based. For example, the extended N_2-model on a phylogenetic tree can be realized by the cyclic group $\mathbb{Z}_2$, which consists of the elements 0 and 1 under addition mod 2. In this case, for each edge e of T, we choose $h_e(1) = p(e)$ and $h_e(0) = 1 - p(e)$. Indeed, for all $r \geq 2$, the N_r-model on a phylogenetic tree may be realized using a group of order r. Evans and Speed (1993) made an important observation that a model of DNA substitution called the Kimura 3ST model could be realized by the 'Klein group' $\mathbb{Z}_2 \times \mathbb{Z}_2$ of order 4. Several authors, beginning with Evans and Speed (1993), applied this observation to analyse properties of this model.

Much of the analysis described in Section 8.6 on the Hadamard representation carries over directly to models based on Abelian groups with natural analogues of Lemma 8.6.1 and Theorem 8.6.3. The underlying theoretical tool for deriving these analogues is the discrete Fourier inversion formula over finite Abelian groups, which we outline next (for further details the interested reader should consult Diaconis 1988).

Let $\mathcal{G}$ be a finite Abelian group, let $\mathbb{C}$ denote the field of complex numbers, and let $\mathbb{C}^1$ denote the Abelian group of complex numbers of modulus 1 under multiplication. Given an Abelian group $\mathcal{G}$, let $\hat{\mathcal{G}}$ denote the group of homomorphisms from $\mathcal{G}$ into $\mathbb{C}^1$ under the binary operation $*$ defined by $(\zeta_1 * \zeta_2)(g) = \zeta_1(g) \cdot \zeta_2(g)$ for all $g \in \mathcal{G}$. The group $\hat{\mathcal{G}}$ is called the *character group* for $\mathcal{G}$ and $\hat{\mathcal{G}}$ is isomorphic to $\mathcal{G}$. The character group plays a central role in the following Fourier inversion theorem.

Proposition 8.10.2. *Let $\mathcal{G}$ be a finite Abelian group and let f be a function from $\mathcal{G}$ into $\mathbb{C}$. Let $\hat{f} : \hat{\mathcal{G}} \to \mathbb{C}$ be defined, for all $\zeta \in \hat{\mathcal{G}}$, by*

$$\hat{f}(\zeta) = \sum_{g \in \mathcal{G}} \zeta(g) f(g).$$

Then, for all $g \in \mathcal{G}$,

$$f(g) = \frac{1}{|\mathcal{G}|} \sum_{\zeta \in \hat{\mathcal{G}}} \overline{\zeta(g)} \hat{f}(\zeta),$$

where $\bar{z}$ denotes the complex conjugate of the complex number z.

For example, when $\mathcal{G} = \mathbb{Z}_2$, the elements of $\hat{\mathcal{G}}$ are the homomorphisms ζ_1, ζ_{-1} defined by $\zeta_{\pm 1}(1) = \pm 1$. Taking $\mathcal{G} = \mathbb{Z}_2^k$, the elements of $\hat{\mathcal{G}}$ are all the k-tuples $(\zeta^1, \ldots, \zeta^k)$ where, for all $i \in \{1, \ldots, k\}$, $\zeta^i \in \{\zeta_{+1}, \zeta_{-1}\}$ and

$$(\zeta^1, \ldots, \zeta^k)(g_1, \ldots, g_k) = \prod_{i=1}^{k} \zeta_i(g_i).$$

Thus, an element of $\hat{\mathbb{Z}_2^k}$ maps each element of $\mathbb{Z}^k$ to $+1$ or -1. Furthermore, by the orthogonality of group characters, the matrix $[\zeta(g)]$ is a Hadamard matrix, where ζ ranges over the elements of $\hat{\mathbb{Z}_2^k}$ and g ranges over the elements of $\mathbb{Z}_2^k$. When $k = n - 1$, this is precisely the Hadamard matrix H_{n-1} encountered in Section 8.6. Proposition 8.10.2 provides an alternative algebraic approach to the inversion formula described by Theorem 8.6.3, and its extension to the Kimura 3ST model, which is realized by the group $\mathbb{Z}_2 \times \mathbb{Z}_2$. Evans and Speed (1993) also applied Proposition 8.10.2 to analyse the Kimura 3ST model. More generally, for any finite Abelian group $\mathcal{G}$, the matrix $[\zeta(g)]$ (where g ranges over $\mathcal{G}$ and ζ ranges over $\hat{\mathcal{G}}$) is a unitary matrix. In this more general setting, Proposition 8.10.2 allows an an algebraic proof of the representation theory for Abelian group-based models, which extends the results described in Section 8.6 (for details, see Székely *et al.* 1993).

Phylogenetic invariants. The construction and classification of 'phylogenetic invariants' has been an active area of mathematical research in phylogenetics. In the second half of this specialist topic, we provide an overview of some of the concepts and results in this area. One motivation for investigating phylogenetic invariants is that they provide techniques for developing new methods for tree reconstruction. Phylogenetic invariants can also be used to improve existing tree reconstruction methods such as maximum likelihood (see Chor *et al.* 2000). In addition, phylogenetic invariants also have some attractive mathematical properties, and their algebraic geometry has been investigated by Hagedorn (2000), and, for the Kimura 3ST model and submodels, by Evans and Zhou (1998).

We first recall some properties of rings of polynomials. Given (formal) variables $w_1, w_2, \ldots, w_k$, let $\mathbb{R}[w_1, w_2, \ldots, w_k]$ denote the ring of all polynomials in the variables $w_1, w_2, \ldots, w_k$ with real coefficients. The proof of Proposition 8.10.3 involves a straightforward application of concepts from linear algebra and it is left as an exercise.

Proposition 8.10.3. *Let $\{p_1, p_2, \ldots, p_r\} \subset \mathbb{R}[y_1, y_2, \ldots y_s]$ and suppose that $r > s$. Then, there exists a non-zero real polynomial $q \in \mathbb{R}[z_1, z_2, \ldots, z_r]$ such that, when $q(p_1, p_2, \ldots, p_r)$ is evaluated as an element of $\mathbb{R}[y_1, y_2, \ldots, y_s]$, it is the zero element of $\mathbb{R}[y_1, y_2, \ldots, y_s]$.*

As a simple illustration of Proposition 8.10.3, suppose that $p_1 = y_1 + y_2, p_2 = y_1y_2$, and $p_3 = y_1^2$. Let

$$q(z_1, z_2, z_3) = z_1^2 z_3 - 2z_2z_3 - z_2^2 - z_3^2.$$

A routine check shows that $q(p_1, p_2, p_3)$ is the zero polynomial in $\mathbb{R}[y_1, y_2, y_3]$.

Let $P = \{p_1, p_2, \ldots, p_r\}$ be a subset of $\mathbb{R}[y_1, y_2, \ldots, y_s]$ such that $r > s$. Then, any element q as described by Proposition 8.10.3 is an *invariant* for P. The set of all invariants of P together with the zero element of $\mathbb{R}[z_1, \ldots, z_r]$ is denoted by inv(P). Clearly, if $q_1, \ldots, q_t \in$ inv(P), then, for all choices of elements $a_1, a_2, \ldots, a_t \in \mathbb{R}[z_1, z_2, \ldots z_r]$, the element

$$q = a_1q_1 + \cdots + a_tq_t \tag{8.23}$$

lies in inv(P). This property implies that, in the language of commutative algebra, inv(P) forms an 'ideal' of the polynomial ring $\mathbb{R}[z_1, z_2, \ldots, z_r]$. Consequently, by Hilbert's basis theorem (see Jacobson 1980), inv(P) has a finite basis. In other words, there exists a fixed set of elements $q_1, \ldots, q_t$ of inv(P) such that, for all $q \in$ inv(P), there exist elements $a_1, \ldots, a_t \in \mathbb{R}[z_1, z_2, \ldots z_r]$ such that q can be expressed in the form shown in eqn (8.23). The computationally versatile theory of Gröbner bases is often useful for constructing such a basis.

We now explain the relevance of these abstract algebraic considerations to the phylogenetic setting. Let X and C be fixed. Suppose we have a Markov process on a phylogenetic X-tree $\mathcal{T}$ and a character $\chi : X \to C$. Then, by eqns (8.5) and (8.7), the probability $p(\chi|\mathcal{T}, \theta)$ is a polynomial in the entries of the transition matrices $P(e)$ and the entries of the vector π^ρ that are specified when a rooting of $\mathcal{T}$ is chosen. Indeed, some of the entries in these matrices can be eliminated since each row of $P(e)$ and the entries in π^ρ sum to 1. Further elimination of variables may also be possible if additional constraints are applied to the model, as in the case of the N_r-model. Thus, for a particular model, we will let $Y = \{y_1, y_2, \ldots, y_s\}$ denote a minimal set of variables for which $p(\chi\,|\mathcal{T},\theta)$ can be written as a polynomial, denoted $p_{(\mathcal{T},\chi)}(y_1,\ldots,y_s)$, in $y_1,\ldots,y_s$. In this way we may view the character distribution as a collection $P = \{p_{(\mathcal{T},\chi)}(y_1,\ldots,y_s)\}$ of real polynomials (one for each full character $\chi : X \to C$) in a fixed set variables $(y_1,\ldots,y_s)$. An element of inv(P) is called a *phylogenetic invariant* for $\mathcal{T}$. Thus, a phylogenetic invariant for $\mathcal{T}$ is a polynomial q in variables indexed by characters on X with the following property: q vanishes on the character distribution induced by the pair $(\mathcal{T},\theta)$ for all choices of θ that are possible under the model being considered.

A polynomial q that is a phylogenetic invariant for all phylogenetic X-trees under some model is called a *model invariant*. In contrast, if q is a phylogenetic invariant for some—but not all—phylogenetic X-trees, then q is said to be a *phylogenetically informative* invariant. Since $\sum_\chi p_{(\mathcal{T},\chi)} = 1$, the linear polynomial $q = \sum_\chi z_\chi - 1$ is a model invariant for all models and is called the *trivial* invariant.

Proposition 8.10.3 implies the existence of non-trivial phylogenetic invariants for a general Markov process on a phylogenetic X-tree $\mathcal{T}$, with state set C whenever $n = |X| \geq 3, c = |C| \geq 2$, and $n + c \geq 6$. This follows from the following counting argument. There are c^n characters $\chi : X \to C$ and each of these corresponds to a polynomial whose variable correspond to parameters in the Markov process. If we eliminate any one of these polynomials (so as to exclude the trivial invariant), we obtain $c^n - 1$ polynomials. Now, $\mathcal{T}$ has at most $2n-3$ edges and each edge e has c^2-c non-redundant entries in $P(e)$, while π^ρ has $c-1$ non-redundant entries. Applying Proposition 8.10.3 with $r = c^n-1$ and $s \leq ((2n-3)c+1)(c-1)$, we deduce the existence of such an invariant.

In fact, it is possible to give explicit descriptions of phylogenetic invariants for a general Markov process. To do so, we first describe some functional identities that arise from the group-based dissimilarity results of Section 7.6.

Theorem 8.10.4. *Suppose we have a Markov process with parameters $(\mathcal{T},\theta)$. Let J^{xy} be the joint probability matrix of states at x and y.*

(i) For all elements $x, y, z \in X$, let $J^{xyz} = J^{xy}(J^{zy})^{-1}J^{zx}$. For all $\alpha,\beta \in C$, let

$$r_{\alpha\beta} = J^{xyz}_{\alpha\beta} - J^{xyz}_{\beta\alpha}.$$

Then, for all choices of θ, $r_{\alpha\beta} = 0$.

(ii) If $\mathcal{T}|\{x,y,w,z\}$ is either the quartet tree $xy|wz$ or a star tree, then, for all $\alpha,\beta \in C$, let

$$r'_{\alpha\beta} = (J^{xw}(J^{yw})^{-1})_{\alpha\beta} - (J^{xz}(J^{yz})^{-1})_{\alpha\beta}.$$

Then, for all choices of θ, $r'_{\alpha\beta} = 0$.

Proof. We may assume that (M1)–(M3) hold, since if a polynomial vanishes everywhere on an open subset of Euclidean space then it vanishes on the closure of this set. Therefore, for all $x, y \in X$, P^{xy} exists, and $P^{xy} = (J^{xx})^{-1}J^{xy}$. Now, let us take the viewpoint of Section 7.6 whereby elements of a group $\mathcal{G}$, in this case the group of real invertible $|C| \times |C|$ matrices under matrix multiplication, are assigned to each orientation of an edge e of $\mathcal{T}$. Then, the mapping $\delta : X \times X \to \mathcal{G}$ defined, for all $x, y \in X$, by

$\delta(x, y) = P^{xy}$ is a $\mathcal{G}$-dissimilarity map. Therefore, by Theorem 7.6.4, δ satisfies (G1) and (G2). This translates into the conditions that the matrices $[r_{\alpha\beta}]$ and $[r'_{\alpha\beta}]$ both equal the zero matrix, as required. □

We now explain the significance of Theorem 8.10.4 to phylogenetic invariants. Clearly, for all $\alpha, \beta \in C$, the functions $r_{\alpha\beta}$ and $r'_{\alpha\beta}$ can be written as a ratio of two polynomials in the entries of the J^{xy}, and thereby of the character distribution. Apart from certain degenerate cases ($\alpha = \beta$ or $|C| = 2$ in the case of $r_{\alpha\beta}$) the numerators of these ratios are not identically the zero polynomial in the entries of the character distribution, in which case the numerator of $r_{\alpha\beta}$ is a model invariant (and valid for any Markov model), and the numerator of $r'_{\alpha\beta}$ is a phylogenetic invariant for $\mathcal{T}$ under any Markov process on $\mathcal{T}$.

The Hadamard representation, and its extension to other group-based models, provides a particularly explicit description for the invariants of the extended N_2-model and, more generally, of group-based models. Referring to the notation of Section 8.6 these invariants arise directly from the equation $\gamma_A = 0$ for any subset A of X' for which $A|(X - A)$ is not a split of $\mathcal{T}$.

Example 8.10.5. Consider the extended N_2-model on the binary phylogenetic tree $\mathcal{T}$ shown in Fig. 8.4. In the notation of Section 8.6, $\gamma_{\{1,3\}} = \gamma_{\{1,4\}} = 0$. Using the equality $\gamma = H_3^{-1} \log(H_3\overline{p})$ in the statement of Theorem 8.6.3, these translate into linear functions involving logarithms of linear combinations of the $\overline{p}_A$ values, and so they can be re-expressed as a pair of polynomial equations in the $\overline{p}_A$ values, and thereby the $p(\chi)$ values. However, Lemma 8.6.4 provides a more direct route to a pair of quadratic phylogenetic invariants for $\mathcal{T}$. Define an equivalence relation $\sim$ on pairs of path sets on $\mathcal{T}$ by writing $(P, Q) \sim (P', Q')$ if, for each edge e of $\mathcal{T}$,

$$|\{R \in \{P, Q\}\colon\ e \in R\}| = |\{R \in \{P', Q'\}\colon\ e \in R\}|.$$

For example, for the phylogenetic tree $\mathcal{T}$ in Fig 8.4, it is easily seen that

$$(P(\mathcal{T}, \{1, 2, 3, 4\}), P(\mathcal{T}, \emptyset)) \sim (P(\mathcal{T}, \{1, 2\}), P(\mathcal{T}, \{3, 4\}))$$

and

$$(P(\mathcal{T}, \{1, 3\}), P(\mathcal{T}, \{2, 4\})) \sim (P(\mathcal{T}, \{1, 4\}), P(\mathcal{T}, \{2, 3\})).$$

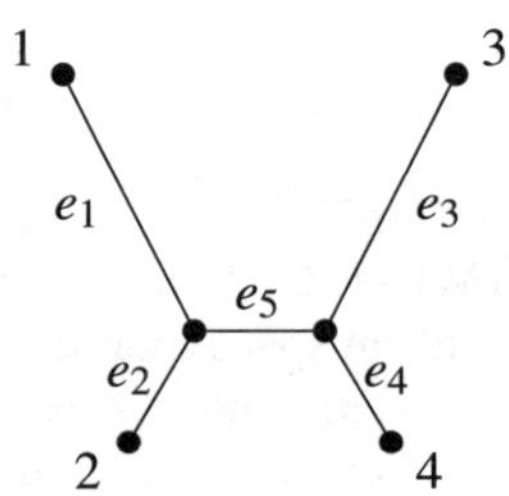

Figure 8.4

Notice that if $(P, Q) \sim (P', Q')$, then, by eqn (8.20), we have

$$\exp(H_3\gamma)_P \cdot \exp(H_3\gamma)_Q = \exp(H_3\gamma)_{P'} \cdot \exp(H_3\gamma)_{Q'}. \tag{8.24}$$

Applying eqn (8.24) to the two relationships just described for $\mathcal{T}$, and noting that $\exp(H_3\gamma) = H\overline{p}$ by eqn (8.24), we obtain the quadratic invariants

$$q = (H_3\overline{p})_{\{1,2\}}(H_3\overline{p})_{\{3,4\}} - (H_3\overline{p})_{\{1,2,3,4\}}$$

and

$$q' = (H_3\overline{p})_{\{1,3\}}(H_3\overline{p})_{\{2,4\}} - (H_3\overline{p})_{\{1,4\}}(H_3\overline{p})_{\{2,3\}}.$$

Cavender and Felsenstein (1987) first determined the existence of such a pair of quadratic phylogenetic invariants for the N_2-model, but by using a slightly different approach.

All the phylogenetic invariants that arise under the extended N_2-model by virtue of the identities $\gamma_A = 0$ are phylogenetically informative. However, the invariants that arise analogously under other group-based models may also include model invariants. For example, the Kimura 3ST model, which is realized by the group $\mathbb{Z}_2 \times \mathbb{Z}_2$, has both model invariants and phylogenetically informative invariants. Any invariants for the Kimura 3ST model automatically are valid for any submodel, in particular for the N_4-model. However, there are further invariants that arise for the N_4-model, and some of these are linear polynomials in the entries of the character distribution. Such invariants are called *linear phylogenetic invariants* and they have certain desirable statistical properties. In particular, they are still valid even if each character evolves at a different rate. Since the set of linear phylogenetic invariants forms a real vector space, it is interesting to consider the dimension of this space. In the case of the N_4-model, the dimension of this space has been determined explicitly by a formula that involves the Fibonacci numbers (for details see Steel and Fu 1995).

8.11 Exercises

1. Show that, for any Markov process on a rooted tree T, the substitution probability $p(e)$ for an arc $e = (u, v)$ of T can be written

$$p(e) = 1 - \mathrm{tr}(D(\pi^u)P(e)).$$

2. Suppose we have a Markov process on a rooted phylogenetic X-tree $\mathcal{T}$, with state set $C = \{0, 1\}$, and that $\pi_0^\rho = 1$.
   (i) If ρ lies on the path connecting $\phi(x)$ and $\phi(y)$, where $x, y \in X$, show that

$$\det(J^{xy}) = 0.$$

   (ii) Suppose that the transition event $1 \to 0$ has zero probability on each arc of $\mathcal{T}$, and the transition event $0 \to 1$ has strictly positive probability on each arc of

$\mathcal{T}$. Show that $\mathcal{T}$ can be reconstructed from the induced character distribution by using the dissimilarity map δ on X defined, for all distinct elements $x, y \in X$, by

$$\delta(x,y) = -2\log\left(\mathbb{P}(\{\phi(x)=0\}\cap\{\phi(y)=0\})\right) + \log\left(\mathbb{P}(\phi(x)=0)\right) + \log\left(\mathbb{P}(\phi(y)=0)\right).$$

3. Show that for any Markov process on a rooted phylogenetic X-tree $\mathcal{T}$, we may write, for all $x, y \in X$,
$$J^{xy} = (P^{ux})^t D(\pi^u) P^{uy},$$
where vertex $u = \mathrm{lca}_{\mathcal{T}}(\phi(x), \phi(y))$.
4. (Steel 1994) Using Exercise 3 from Chapter 7 and Theorem 8.4.3 show that, for a Markov process on a phylogenetic X-tree $\mathcal{T}$ satisfying (M1)–(M3), the following modified logdet transform defined, for all $x, y \in X$, as
$$\delta(x,y) = \begin{cases} -\log\left(\det(J^{xy})\right), & \text{if } x \neq y,\\ 0, & \text{otherwise,}\end{cases}$$
is a $\mathcal{T}$-metric.
5. We have seen that Corollary 8.4.4 can fail without condition (M2) or condition (M3). Show that it can also fail without condition (M1), even for the case $|C| = 2$.
6. For the N_2-model, and the dissimilarity map δ on X defined by eqn (8.12), show that, for all $x, y \in X$, $\delta(x,y) = -\log(1 - 2p(x,y))$.
7. Consider the N_2-model with state set C on a phylogenetic X-tree, and a full character $\chi : X \to C$. Assuming condition (8.19) show that
$$\mathbb{P}(\chi|\mathcal{T},\theta) \leq \tfrac{1}{2}p^{l(\chi,\mathcal{T})},$$
where
$$p = \max\{p(x,y)\colon\ x, y \in X, \chi(x) \neq \chi(y)\}$$
and where $l(\chi, \mathcal{T})$ is the parsimony score of χ on $\mathcal{T}$.
8. For a reversible stationary process on a phylogenetic X-tree, show that $p(x,y)$ is related to the evolutionary distance $\kappa(x,y)$ by the equation
$$p(x,y) = 1 - \mathrm{tr}\left(D(\pi)\exp\left(Q\frac{-\kappa(x,y)}{\mathrm{tr}(D(\pi)Q)}\right)\right),$$
for all $x, y \in X$.
9. (Tuffley and Steel 1997) Let $\mathcal{C}$ be a sequence of characters on X each of which is generated independently under the N_r-model, and satisfying8.19, though possibly with different parameters across the characters. Using Theorem 8.5.4, show that the maximum likelihood tree(s) for $\mathcal{C}$ under this unconstrained model is precisely the maximum parsimony tree(s) for $\mathcal{C}$.

10. Prove Theorem 8.7.1.
11. Given two vectors u and v of the same rank, their *inner product*, denoted $\langle u, v\rangle$, is defined as $\langle u, v\rangle = \sum_i u_i v_i$. Referring to Theorem 8.6.3, let $\overline{p} = \overline{p}(\gamma)$, and $\overline{p}' = \overline{p}(\gamma')$, where the vectors γ and γ' may correspond to the same or different phylogenetic X-trees. Show that

$$\langle \overline{p}, \overline{p}' \rangle = \overline{p}_{\emptyset}.$$

12. Consider the Hadamard representation for the extended N_2-model on a phylogenetic X-tree, where $X = \{1, 2, \ldots, n\}$. Let $X' = \{1, \ldots, n-1\}$.
    (i) Show that

$$\overline{p}_{\emptyset} = \max\{\overline{p}_A : A \subseteq X'\}.$$

    (ii) Show that for this model there are no *linear* phylogenetic invariants. That is, if

$$\sum_{A \subseteq X'} c_A \overline{p}_A = 0$$

    for some collection of constants c_A, show that $c_A = 0$ for all $A \subseteq X'$.

References

Adams, E. N. III (1986). *N*-trees as nestings: complexity, similarity and consensus. *Journal of Classification*, **3**, 299–317.

Agarwala, R. and Fernández-Baca, D. (1994). A polynomial–time algorithm for the phylogeny problem when the number of character states is fixed. *SIAM Journal on Computing*, **23**, 1216–1224.

Agarwala, R., Bafna, V., Farach, M., Narayanan, B., Paterson, M., and Thorup, M. (1999). On the approximability of numerical taxonomy: fitting distances by tree metrics. *SIAM Journal on Computing*, **28**, 1073–1085.

Aho, A. V., Sagiv, Y., Szymanski, T. G., and Ullman, J. D. (1981). Inferring a tree from lowest common ancestors with an application to the optimization of relational expressions. *SIAM Journal on Computing*, **10**, 405–421.

Aldous, D. (1996). Probability distributions on cladograms. In: *Random Discrete Structures*, IMA Volumes in Mathematics and its Applications, Vol. 76 (ed. D. Aldous and R. Pemantle). Springer-Verlag, New York, pp. 1–18.

Allen, B. L. and Steel, M. (2001). Subtree transfer operations and their induced metrics on evolutionary trees. *Annals of Combinatorics*, **5**, 1–13.

Amir, A. and Keselman, D. (1997). Maximum agreement subtree in a set of evolutionary trees: metrics and efficient algorithms. *SIAM Journal on Computing*, **26**, 1656–1669.

Archie, J. W. and Felsenstein, J. (1993). The number of evolutionary steps on random and minimum length trees for random evolutionary data. *Theoretical Population Biology*, **43**, 52–79.

Atteson, K. (1997). The performance of the NJ method of phylogeny reconstruction. In: *Mathematical Hierarchies and Biology*, DIMACS Series in Discrete Mathematics and Theoretical Computer Science, Vol. 37 (ed. B. Mirkin, F. R. McMorris, F. S. Roberts, and A. Rzhetsky). American Mathematical Society, Providence, Rhode Island, pp. 133–147.

Baake, E. (1998). What can and what cannot be inferred from pairwise sequence comparisons? *Mathematical Biosciences*, **154**, 1–21.

Balasubramanian, R., Fellows, M. R., and Raman, V. (1998). An improved fixed-parameter algorithm for vertex cover. *Information Processing Letters*, **65**, 163–168.

Bandelt, H.-J. (1984). Retracts of hypercubes. *Journal of Graph Theory*, **8**, 501–510.

Bandelt, H.-J. and Dress, A. W. M. (1986). Reconstructing the shape of a tree from observed dissimilarity data. *Advances in Applied Mathematics*, **7**, 309–343.

Bandelt, H.-J. and Dress, A. W. M. (1989). Weak hierarchies associated with similarity measures—an additive clustering technique. *Bulletin of Mathematical Biology*, **51**, 133–166.

Bandelt, H.-J. and Dress, A. W. M. (1992). A canonical decomposition theory for metrics on a finite set. *Advances in Mathematics*, **92**, 47–105.

Bandelt, H.-J. and Steel, M. A. (1995). Symmetric matrices representable by weighted trees over a cancellative abelian monoid. *SIAM Journal on Discrete Mathematics*, **8**, 517–525.

Bandelt, H.-J., Forster, P., Sykes, B. C., and Richards, M. B. (1995). Mitochondrial portraits of human populations using median networks. *Genetics*, **141**, 743–753.

Barthélemy, J.-P. (1989). From copair hypergraphs to median graphs with latent vertices. *Discrete Mathematics*, **76**, 9–28.

Barthélemy, J.-P. and Guénoche, A. (1991). *Trees and Proximity Representations*. Wiley, London.

Barthélemy, J.-P. and McMorris, F. R. (1986). The median procedure for *n*-trees. *Journal of Classification*, **3**, 329–334.

Benham, C., Kannan, S., Paterson, M., and Warnow, T. (1995). Hen's teeth and whale's feet: generalized characters and their compatibility. *Journal of Computational Biology*, **2**, 515–525.

Benzer, S. (1959). On the topology of the genetic fine structure. *Proceedings of the National Academy of Sciences USA*, **45**, 1607–1620.

Berry, V. and Gascuel, O. (2000). Inferring evolutionary trees with strong combinatorial evidence. *Journal of Theoretical Computer Science*, **240**, 271–298.

Bininda-Emonds, O. R. P., Gittleman, J. L., and Steel, M. A. (2002). The (super) tree of life: procedures, problems and prospects. *Annual Reviews of Ecology and Systematics*, **33**, 265–289.

Böcker, S. (1999). From subtrees to supertrees. Unpublished PhD thesis, Fakultät für Mathematik, Universität Bielefeld, Bielefeld.

Böcker, S. and Dress, A. W. M. (1998). Recovering symbolically dated, rooted trees from symbolic ultrametrics. *Advances in Mathematics*, **138**, 105–125.

Böcker, S. and Dress, A. W. M. (2001). Patchworks. *Advances in Mathematics*, **157**, 1–21.

Böcker, S., Dress, A. W. M., and Steel, M. (1999). Patching up *X*-trees. *Annals of Combinatorics*, **3**, 1–12.

Böcker, S., Bryant, D., Dress, A. W. M., and Steel, M. A. (2000). Algorithmic aspects of tree amalgamation. *Journal of Algorithms*, **37**, 522–537.

Bodlaender, H. L., Fellows, M. R., and Warnow, T. J. (1992). Two strikes against perfect phylogeny. In: *Proceedings of the International Colloquium on Automata, Languages and Programming*, Lecture Notes in Computer Science, Vol. 623. Springer-Verlag, Berlin, pp. 273–283.

Bonet, M., Steel, M., Warnow, T., and Yooseph, S. (1998). Better methods for solving parsimony and compatibility. *Journal of Computational Biology*, **5**, 391–407.

Bonnot, F., Guénoche, A., and Perrier, X. (1996). Properties of an order distance associated with a tree distance. In: *Ordinal and Symbolic Data Analyis* (ed. E. Diday *et al.*). Springer-Verlag, Berlin, pp. 252–261.

Bowditch, B. (1991). Notes on Gromov's hyperbolicity criterion for path-metric spaces. In: *Group Theory from a Geometrical Viewpoint* (ed. E. Ghys, A. Haefliger, and A. Verjovsky). World Scientific, Singapore, pp. 64–167.

Brown, J. K. M. (1994). Probabilities of evolutionary trees. *Systematic Biology*, **43**, 78–91.

Bryant, D. (1996). Hunting for trees in binary character sets: efficient algorithms for extraction, enumeration and optimization. *Journal of Computational Biology*, **3**, 275–288.

Bryant, D. (1997). Building trees, hunting for trees, and comparing trees: theory and methods in phylogenetic analysis. Unpublished PhD thesis, University of Canterbury.

Bryant, D. and Steel, M. (1995). Extension operations on sets of leaf-labelled trees. *Advances in Applied Mathematics*, **16**, 425–453.

Buneman, P. (1971). The recovery of trees from measures of dissimilarity. In: *Mathematics in the Archaeological and Historical Sciences* (ed. F. R. Hodson, D. G. Kendall, and P. Tautu), Edinburgh University Press, Edinburgh, pp. 387–395.

Buneman, P. (1974*a*). A characterisation of rigid circuit graphs. *Discrete Mathematics*, **9**, 205–212.

Buneman, P. (1974*b*). A note on the metric property of trees. *Journal of Combinatorial Theory, Series B*, **17**, 48–50.

Carter, M., Hendy, M. D., Penny, D., Székely, L. A., and Wormald, N. C. (1990). On the distribution of lengths of evolutionary trees. *SIAM Journal on Discrete Mathematics*, **3**, 38–47.

Cavender, J. A. and Felsenstein, J. (1987). Invariants of phylogenies in a simple case with discrete states. *Journal of Classification*, **4**, 57–71.

Chang, J. T. (1996*a*). Full reconstruction of Markov models on evolutionary trees: identifiability and consistency. *Mathematical Biosciences*, **137**, 51–73.

Chang, J. T. (1996*b*). Inconsistency of evolutionary tree topology reconstruction methods when substitution rates vary across characters. *Mathematical Biosciences*, **134**, 189–216.

Charleston, M. A. (1998). Jungles: a new solution to the host/parasite phylogeny reconciliation problem. *Mathematical Biosciences*, **149**, 191–223.

Chepoi, V. and Fichet, B. (2000). l_∞-Approximation via subdominants. *Journal of Mathematical Psychology*, **44**, 600–616.

Chor, B., Hendy, M. D., Holland, B. R., and Penny, D. (2000). Multiple maxima of likelihood in phylogenetic trees: an analytic approach. *Molecular Biology and Evolution*, **17**, 1529–1541.

Colonius, H. and Schulze, H. H. (1981). Tree structures for proximity data. *British Journal of Mathematical and Statistical Psychology*, **34**, 167–180.

Constantinescu, M. and Sankoff, D. (1995). An efficient algorithm for supertrees. *Journal of Classification*, **12**, 101–112.

Dahlhaus, E., Johnson, D. S., Papadimitriou, C. H., Seymour, P. D., and Yannakakis, M. (1994). The complexity of multiterminal cuts. *SIAM Journal on Computing*, **23**, 864–894.

Darwin, C. (1872). *On the Origin of Species, 6th edn*. John Murray, London.

Day, W. H. E. (1987). Computational complexity of inferring phylogenies from dissimilarity matrices. *Bulletin of Mathematical Biology*, **49**, 461–67.

Day, W. H. E. and Sankoff, D. (1986). Computational complexity of inferring phylogeny by compatibility. *Systematic Zoology*, **35**, 224–229.

Dekker, M. C. H. (1986). Reconstruction methods for derivation trees. Unpublished Masters thesis, Vrije Universiteit, Amsterdam, Netherlands.

Denis, F. and Gascuel, O. (2002). On the consistency of the minimum evolution principle of phylogenetic inference. *Discrete Applied Mathematics* (in press).

Diaconis, P. (1988). *Group Representations in Probability and Statistics*, Institute of Mathematical Statistics, Lecture Notes-Monograph Series, Vol. 11 (ed. S. S. Gupta). Hayward, California.

Diaconis, P. W. and Holmes, S. P. (1998). Matchings and phylogenetic trees. *Proceedings of the National Academy of Sciences*, **95**, 14600–14602.

Dirac, G. A. (1961). On rigid circuit graphs. *Abhandlungen aus dem Mathematischen Seminar der Universität Hamburg*, **25**, 71–76.

Dobson, A. J. (1974). Unrooted trees for numerical taxonomy. *Journal of Applied Probability*, **11**, 32–42.

Downey, R. G. and Fellows, M. R. (1999). *Parameterized Complexity*. Springer-Verlag, New York.

Dress, A. W. M., Huson, D., and Moulton, V. (1996). Analysing and visualizing sequence and distance data using SPLITSTREE. *Discrete Applied Mathematics*, **71**, 95–109.

Dress, A., Hendy, M., Huber, K., and Moulton, V. (1997*a*). On the number of vertices and edges of the Buneman graph. *Annals of Combinatorics*, **1**, 329–337.

Dress, A., Moulton, V., and Steel, M. (1997*b*). Trees, taxonomy and strongly compatible multi-state characters. *Advances in Applied Mathematics*, **19**, 1–30.

Erdös, P. L. and Székely, L. A. (1989). Applications of antilexicographic order I: an enumerative theory of trees. *Advances in Applied Mathematics*, **10**, 488–496.

Erdös, P. L. and Székely, L. A. (1992). Evolutionary trees: an integer multicommodity max-flow–min-cut theorem. *Advances in Applied Mathematics*, **13**, 375–389.

Erdös, P. L. and Székely, L. A. (1993). Counting bichromatic evolutionary trees. *Discrete Applied Mathematics*, **47**, 1–8.

Erdös, P. L. and Székely, L. A. (1994). On weighted multiway cuts in trees. *Mathematical Programming*, **65**, 93–105.

Erdös, P. L., Székely, L. A., Steel, M., and Warnow, T. (1999). A few logs suffice to build (almost) all trees (I). *Random Structures and Algorithms*, **14**, 153–184.

Estabrook, G. F. and McMorris, F. R. (1977). When are two qualitative taxonomic characters compatible? *Journal of Mathematical Biology*, **4**, 195–200.

Evans, S. N. and Speed, T. P. (1993). Invariants of some probability models used in phylogenetic inference. *Annals of Statistics*, **21**, 355–377.

Evans, S. N. and Zhou, X. (1998). Constructing and counting phylogenetic invariants. *Journal of Computational Biology*, **5**, 713–724.

Faith, D. P. and Trueman, J. W. H. (2001). Towards an inclusive philosophy for phylogenetic inference. *Systematic Biology*, **50**, 331–350.

Farach, M. and Kannan, S. (1999). Efficient algorithms for inverting evolution. *Journal of the Association for Computing Machinery*, **46**, 437–449.

Farach, M., Kannan, S., and Warnow, T. (1995). A robust model for finding optimal evolutionary trees. *Algorithmica*, **13**, 155–179.

Felsenstein, J. (1978*a*). The number of evolutionary trees. *Systematic Zoology*, **27**, 27–33.

Felsenstein, J. (1978*b*). Cases in which parsimony or compatibility methods will be positively misleading. *Systematic Zoology*, **27**, 401–410.

Foulds, L. R. (1984). Maximum savings in the Steiner problem in phylogeny. *Journal of Theoretical Biology*, **107**, 471–474.

Foulds, L. R. and Graham, R. L. (1982). The Steiner problem in phylogeny is NP-complete. *Advances in Applied Mathematics*, **3**, 43–49.

Fulkerson, D. R. and Gross O. A. (1965). Incidence matrices and interval graphs. *Pacific Journal of Mathematics*, **15**, 835–855.

Galtier, N. (2001). Maximum-likelihood phylogenetic analysis under a covarion-like model. *Molecular Biology and Evolution*, **18**, 866–873.

Garey, M. R. and Johnson, D. S. (1979). *Computers and Intractability. A Guide to the Theory of NP-completeness.* Freeman, San Francisco.

Gascuel, O. (1997). Concerning the NJ algorithm and its unweighted version, UNJ. In: *Mathematical Hierarchies and Biology*, DIMACS Series in Discrete Mathematics and Theoretical Computer Science, Vol. 37 (ed. B. Mirkin, F. R. McMorris, F. S. Roberts, and A. Rzhetsky). American Mathematical Society, pp. 149–170.

Gavril, F. (1974). The intersection graphs of subtrees in trees are exactly the chordal graphs. *Journal of Combinatorial Theory, Series B*, **16**, 47–56.

Goldman, N. (1990). Maximum likelihood inference of phylogenetic trees, with special reference to a Poisson process model of DNA substitution and to parsimony analysis. *Systematic Zoology*, **39**, 345–361.

Golumbic, M. C. (1980). *Algorithmic Graph Theory and Perfect Graphs.* Academic Press, New York.

Goulden, I. P. and Jackson, D. M. (1983). *Combinatorial Enumeration*. Wiley, New York.

Gromov, M. (1987). Hyperbolic groups. In: *Essays in Group Theory*, Mathematical Sciences Research Institute Publication 8 (ed. S. Gerstin). Springer-Verlag, New York, pp. 75–263.

Gu, X. and Li, W.-H. (1996). A general additive distance with time-reversibility and rate variation among nucleotide sites. *Proceedings of the National Academy of Sciences USA*, **93**, 4671–4676.

Guénoche, A. (1998). Ordinal properties of tree distances. *Discrete Mathematics*, **192**, 103–117.

Gusfield, D. (1991). Efficient algorithms for inferring evolutionary trees. *Networks*, **21**, 19–28.

Gusfield, D. (1997). *Algorithms on Strings, Trees and Sequences*. Cambridge University Press, Cambridge.

Hagedorn, T. R. (2000). Determining the number and structure of phylogenetic invariants. *Advances in Applied Mathematics*, **24**, 1–21.

Hakimi, S. L. and Patrinos, A. N. (1972). The distance matrix of a graph and its tree realization. *Quarterly of Applied Mathematics*, **30**, 255–269.

Hamel, A. M. and Steel, M. A. (1997). The length of a random leaf coloration on a random tree. *SIAM Journal on Discrete Mathematics*, **10**, 359–372.

Harding, E. F. (1971). The probabilities of rooted tree shapes generated by random bifurcation. *Advances in Applied Probability*, **3**, 44–77.

Hartigan, J. A. (1973). Minimum mutation fits to a given tree. *Biometrics*, **29**, 53–65.

Hendy, M. D. (1989). The relationship between simple evolutionary tree models and observable sequence data. *Systematic Zoology*, **38**, 310–321.

Hendy, M. D. and Penny, D. (1989). A framework for the quantitative study of evolutionary trees. *Systematic Zoology*, **38**, 297–309.

Hendy, M. D., Little, C. H. C., and Penny, D. (1984). Comparing trees with pendant vertices labelled. *SIAM Journal on Applied Mathematics*, **44**, 1054–1065.

Hendy, M. D., Penny, D., and Steel, M. A. (1994). A discrete fourier spectral analysis for evolutionary trees. *Proceedings of the National Academy of Sciences, USA*, **91**, 3339–3343.

Henzinger, M. R., King, V., and Warnow, T. (1999). Constructing a tree from homeomorphic subtrees, with applications to computional evolutionary biology. *Algorithmica*, **24**, 1–13.

Jacobson, N. (1980). *Basic Algebra (II)*. W. H. Freeman, San Fransisco.

Jiang, T., Kearney, P. E., and Li, M. (1998). Orchestrating quartets: approximation and data correction. *Proceedings of the 39th IEEE Symposium on the Foundations of Computer Science (FOCS)*, 416–425.

Kannan, S. and Warnow, T. (1997). A fast algorithm for the computation and enumeration of perfect phylogenies. *SIAM Journal on Computing*, **26**, 1749–1763.

Kannan, S. K., Lawler, E. L., and Warnow, T. J. (1996). Determining the evolutionary tree using experiments. *Journal of Algorithms*, **21**, 26–50.

Knuth, D. E. (1997). *The Art of Computer Programming*, Vol. 3. Addison-Wesley, Reading, MA.

Kruskal, J. B. (1956). On the shortest spanning tree of a graph and the traveling salesman problem. *Proceedings of the American Mathematical Society*, **7**, 48–50.

Kubicka, E., Kubicki, G., and McMorris, F. R. (1992). On agreement subtrees of two binary trees. *Congressus Numerantium*, **88**, 217–224.

Lake, J. A. (1994). Reconstructing evolutionary trees from DNA and protein sequences: paralinear distances. *Proceedings of the National Academy of Sciences USA*, **91**, 1455–1459.

LeQuesne, W. J. (1982). Compatibility analysis and its applications. *Zoological Journal of the Linnean Society*, **74**, 267–275.

Li, M., Tromp, J., and Zhang, L. (1996). On the nearest neighbour interchange distance between evolutionary trees. *Journal of Theoretical Biology*, **182**, 463–467.

Maddison, W. and Slatkin, M. (1991). Null models for the number of evolutionary steps in a character on a phylogenetic tree. *Evolution*, **45**, 1184–1197.

Marczewski, E. (1945). Sur deux propriétés des classes d'ensembles. *Fundamenta Mathematica*, **33**, 303–307.

Margush, T. and McMorris, F. R. (1981). Consensus *n*-trees. *Bulletin of Mathematical Biology*, **43**, 239–244.

Mau, B., Newton, M. A., and Larget, B. (1999). Bayesian phylogenetic inference via Markov chain Monte Carlo methods. *Biometrics*, **55**, 1–12.

McKee, T. A. and McMorris, F. R. (1999). *Topics in Intersection Graph Theory*. Society for Industrial and Applied Mathematics, Philadelphia.

McMorris, F. R. (1985). Axioms for consensus functions on undirected phylogenetic trees. *Mathematical Biosciences*, **74**, 17–21.

McMorris, F. R. and Meacham, C. A. (1983). Partition intersection graphs. *ARS Combinatoria*, **16B**, 135–138.

McMorris, F. R., Warnow, T. J., and Wimer, T. (1994). Triangulating vertex-colored graphs. *SIAM Journal on Discrete Mathematics*, **7**, 296–306.

Meacham, C. A. (1983). Theoretical and computational considerations of the compatibility of qualitative taxonomic characters. In: *Numerical Taxonomy*, NATO ASI Series, Vol. G1 (ed. J. Felsenstein). Springer-Verlag, Berlin, pp. 304–314.

Menger, K. (1927). Zur allgemeinen Kurventheorie. *Fundamenta Mathematica*, **10**, 96–115.

Moulton, V. and Steel, M. (1999). Retractions of finite distance functions onto tree metrics. *Discrete Applied Mathematics*, **91**, 215–233.

Mulder, M. (1978). The structure of median graphs. *Discrete Mathematics*, **24**, 197–204.

Mulder, M. and Schrijver, A. (1979). Median graphs and Helly hypergraphs. *Discrete Mathematics*, **25**, 41–50.

Murtagh, F. (1984). Counting dendograms: a survey. *Discrete Applied Mathematics*, **7**, 191–199.

Narushima, H. and Hanazawa, M. (1997). A more efficient algorithm for MPR problems in phylogeny. *Discrete Applied Mathematics*, **80**, 231–238.

Neyman, J. (1971). Molecular studies of evolution: a source of novel statistical problems. In: *Statistical Decision Theory and Related Topics* (ed. S. Gupta and J. Yackel). Academic Press, New York, pp. 1–27.

Ng, M. P. and Wormald, N. C. (1996). Reconstruction of rooted trees from subtrees. *Discrete Applied Mathematics*, **69**, 19–31.

Ng, M. P., Steel, M., and Wormald, N. C. (2000). The difficulty of constructing a leaf-labelled tree including or avoiding given subtree. *Discrete Applied Mathematics*, **98**, 227–235.

Pauplin, Y. (2000). Direct calculation of a tree length using a distance matrix. *Journal of Molecular Evolution*, **51**, 41–47.

Pearl, J. and Tarsi, M. (1986). Structuring causal trees. *Journal of Complexity*, **2**, 60–77.

Pe'er, I., Shamir, R., and Sharan, R. (2000). Incomplete directed perfect phylogeny. In: *Proceedings of the Eleventh Symposium on Combinatorial Pattern Matching (CPM)*, Lecture Notes in Computer Science, Vol. 1848 (ed. D. Sankoff). Springer-Verlag, New York, pp. 143–153.

Penny, D., Foulds, L. R., and Hendy, M. D. (1982). Testing the theory of evolution by comparing phylogenetic trees constructed from five different protein sequences. *Nature*, **297**, 197–200.

Prömel, H. J. and Steger, A. (2000). A new approximation algorithm for the Steiner tree problem with performance ratio 5/3. *Journal of Algorithms*, **36**, 89–101.

Rinsma, I., Hendy, M. D., and Penny, D. (1990). Minimally coloured trees. *Mathematical Biosciences*, **98**, 201–210.

Robinson, D. F. (1971). Comparison of labeled trees with valency three. *Journal of Combinatorial Theory*, **11**, 105–119.

Robinson, D. F. (1973). Extending a function on a graph. *Discrete Mathematics*, **6**, 89–99.

Robinson, D. F. and Foulds, L. R. (1981). Comparison of phylogenetic trees. *Mathematical Biosciences*, **53**, 131–147.

Rose, D. J. (1970). Triangulated graphs and the elimination process. *Journal of Mathematical Analysis and Applications*, **32**, 597–609.

Rzhetsky, A. and Nei, M. (1993). Theoretical foundation of the minimum-evolution method of phylogenetic inference. *Molecular Biology and Evolution*, **10**, 1073–1095.

Sankoff, D. (1972). Reconstructing the history and geography of an evolutionary tree. *American Mathematical Monthly*, **79**, 596–603.

Sankoff, D. (1975). Minimal mutation trees of sequences. *SIAM Journal on Applied Mathematics*, **28**, 35–42.

Schröder, E. (1870). Vier combinatorische probleme. *Zeitschrift für Mathematik und Physik*, **15**, 361–376.

Semple, C. (2002). Reconstructing minimal rooted trees. *Discrete Applied Mathematics* (in press).

Semple, C. and Steel, M. (1999). Tree representations of non-symmetric group-valued proximities. *Advances in Applied Mathematics*, **23**, 300–321.

Semple, C. and Steel, M. (2000). A supertree method for rooted trees. *Discrete Applied Mathematics*, **105**, 147–158.

Semple, C. and Steel, M. (2001). Tree reconstruction via a closure operation on partial splits. In: *Proceedings of Journées Ouvertes: Biologie, Informatique et Mathématique*, Lecture Notes in Computer Science (ed. O. Gascuel and M.-F. Sagot). Springer-Verlag, Berlin, pp. 126–134.

Semple, C. and Steel, M. (2002*a*). A characterization for a set of partial partitions to define an X-tree. *Discrete Mathematics*, **247**, 169–186.

Semple, C. and Steel, M. (2002*b*). Tree reconstruction from multi-state characters. *Advances in Applied Mathematics*, **28**, 169–184.

Simões-Pereira, J. M. S. (1969). A note on the tree realizability of a distance matrix. *Journal of Combinatorial Theory*, **6**, 303–310.

Slowinski, J. B. (1990). Probabilities of n-trees under two models: a demonstration that asymmetrical interior nodes are not improbable. *Systematic Zoololgy*, **39**, 89–94.

Stanley, R. P. (1997). *Enumerative Combinatorics* 1, Cambridge Studies in Advanced Mathematics 49 (ed. W. Fulton, D. J. H. Garling, K. Ribet, and P. Walters). Cambridge University Press, Cambridge.

Steel, M. (1992). The complexity of reconstructing trees from qualitative characters and subtrees. *Journal of Classification*, **9**, 91–116.

Steel, M. A. (1993). Distributions on bicoloured binary trees arising from the principle of parsimony. *Discrete Applied Mathematics*, **41**, 245–261.

Steel, M. (1994). Recovering a tree from the leaf colourations it generates under Markov model. *Applied Mathematical Letters*, **7**, 19–23.

Steel, M. (2000). Sufficient conditions for two tree reconstruction techniques to succeed on sufficiently long sequences. *SIAM Journal on Discrete Mathematics*, **14**, 36–48.

Steel, M. A. and Fu, Y. X. (1995). Classifying and counting linear phylogenetic invariants for the Jukes–Cantor model. *Journal of Computational Biology*, **2**, 39–47.

Steel, M. A. and Penny, D. (1993). Distribution of tree comparison metrics—some new results. *Systematic Biology*, **42**, 126–141.

Steel, M. and Penny, D. (2000). Parsimony, likelihood and the role of models in molecular phylogenetics. *Molecular Biology and Evolution*, **17**, 839–850.

Steel, M. and Warnow, T. (1993). Kaikoura tree theorems: computing the maximum agreement subtree. *Information Processing Letters*, **48**, 77–82.

Steel, M. A., Lockhart, P., and Penny, D. (1993). Confidence in evolutionary trees from biological sequence data. *Nature*, **364**, 440–442.

Steel, M. A, Székely, L. A., and Hendy, M. D. (1994). Reconstructing trees from sequences when sequence sites evolve at variable rates. *Journal of Computational Biology*, **1**, 153–163.

Steel, M., Goldstein, L., and Waterman, M. S. (1996). A central limit theorem for the parsimony length of trees. *Advances in Applied Probability*, **28**, 1051–1071.

Steel, M., Dress, A. W. M., and Böcker, S. (2000). Simple but fundamental limitations on supertree and consensus tree methods. *Systematic Biology*, **49**, 363–368.

Swofford, D. L., Olsen, G. J., Waddell, P. J., and Hillis, D. M. (1996). Phylogenetic inference. In: *Molecular Systematics*, 2nd edn. (ed. D. M. Hillis, C. Moritz, and B. K. Marble). Sinauer, Sunderland, USA, pp. 407–514.

Székely, L., Steel, M. A., and Erdös, P. L. (1993). Fourier calculus on evolutionary trees. *Advances in Applied Mathematics*, **14**, 200–216.

Thomas, A. (1993). A note on the four-colourability of pedigrees and its consequences for probability calculations. *Statistics and Computing*, **3**, 51–54.

Thorley, J. L. (2000). Cladistic information, leaf stability and supertree construction. Unpublished PhD thesis, University of Bristol, UK.

Tuffley, C. and Steel, M. (1997). Links between maximum likelihood and maximum parsimony under a simple model of site substitution. *Bulletin of Mathematical Biology*, **59**, 581–607.

Tuffley, C. and Steel, M. (1998). Modelling the covarion hypothesis of nucleotide substitution. *Mathematical Biosciences*, **147**, 63–91.

Vach, W. (1980). Least squares approximation of additive trees. In: *Conceptual and Numerical Analysis of Data* (ed. O. Opitz). Springer-Verlag, Berlin, pp. 230–238.

Waddell, P. J. and Steel, M. A. (1997). General time-reversible distances with unequal rates across sites: mixing Γ and inverse Gaussian distributions with invariant sites. *Molecular Phylogenetics and Evolution*, **8**, 398–414.

Walter, J. R. (1978). Representations of chordal graphs as subtrees of a tree. *Journal of Graph Theory*, **2**, 265–267.

Wang, L. and Jiang, T. (1994). On the complexity of multiple sequence alignment. *Journal of Computational Biology*, **1**, 337–348.

Warnow, T. (1997). Mathematical approaches to comparative linguistics. *Proceedings of the National Academy of Sciences USA*, **94**, 6585–6590.

Waterman, M. S. (1995). *Introduction to Computational Biology*. Chapman and Hall, London.

Yang, Z. (1994). Maximum likelihood phylogenetic estimation from DNA sequences with variable rates over sites: approximate methods. *Journal of Molecular Evolution*, **39**, 306–314.

Yang, Z. (2000). Complexity of the simplest phylogenetic estimation problem. *Proceedings of the Royal Society of London* (*B: Biological Sciences*), **267**, 109–116.

Yang, Z. and Rannala, B. (1997). Bayesian phylogenetic inference using DNA sequences: a Markov chain Monte Carlo method. *Molecular Biology and Evolution*, **14**, 717–724.

Zaretskii, K. A. (1965). Constructing trees from the set of distances between pendant vertices. *Uspehi Matematiceskih Nauk*, **20**, 90–92.

Commonly used symbols

Chapter 1

$G[V']$	subgraph of G induced by V', 6
$G\backslash e$	graph obtained from G by deleting e, 6
G/e	graph obtained from G by contracting e, 6
$G\backslash v$	graph obtained from G by deleting v, 6
$\mathring{V}(T)$	set of interior vertices of T, 7
$\mathring{E}(T)$	set of interior edges of T, 7
ρ	root vertex of a rooted tree, 8
$\leq_T$	partial order on the vertices of the rooted tree T, 8
$tw(G)$	tree-width of G, 11

Chapter 2

$\mathcal{T} = (T;\phi)$	X-tree, 16
$T(\mathcal{T})$	underlying tree of $\mathcal{T}$, 16
$\phi(\mathcal{T})$	labelling map of $\mathcal{T}$, 17
$\mathcal{L}(\mathcal{T})$	label set of $\mathcal{T}$, 17
$V(\mathcal{T}), E(\mathcal{T})$	vertex and edge sets of $T(\mathcal{T})$, 17
$\mathcal{P}$	set of semi-labelled trees, 17
$\mathcal{L}(\mathcal{P})$	label set of $\mathcal{P}$, 17
$\mathcal{P}(X)$	set of all X-trees, 17
$B(n)$	set of all binary phylogenetic trees on $\{1,\ldots,n\}$, 17
$b(n)$	size of $B(n)$, 17
$\mathrm{lca}_{\mathcal{T}}(x,y)$	greatest lower bound of $\phi(x)$ and $\phi(y)$ in the rooted tree T, 19
$\mathcal{T}^{-\rho}$	X-tree obtained from $\mathcal{T}$ by suppressing ρ, 20
$P(n)$	set of phylogenetic trees on $\{1,\ldots,n\}$, 20
$RP(n)$	set of rooted phylogenetic trees on $\{1,\ldots,n\}$, 20
$RB(n)$	set of rooted binary phylogenetic trees on $\{1,\ldots,n\}$, 20
τ	tree shape, 25
$\mathcal{F}$	phylogenetic forest, 35
$\mathrm{Ext}(\mathcal{T}_1,\mathcal{F})$	set of extensions of $\mathcal{F}$ with kernel $\mathcal{T}_1$, 38

Chapter 3

$A\|B$	split of X, 43
$\Sigma(\mathcal{T})$	set of X-splits induced by $\mathcal{T}$, 43
$\Sigma_{\text{triv}}(X)$	set of trivial splits of X, 46
$\mathcal{T} \leq \mathcal{T}'$	$\mathcal{T}'$ refines $\mathcal{T}$, 47
$\mathcal{H}(\mathcal{T})$	set of clusters of the rooted X-tree $\mathcal{T}$, 51
$G(\Sigma)$	Buneman graph on a set Σ of X-splits, 56
$V(\Sigma)$	vertex set of the Buneman graph on Σ, 57
$E(\Sigma)$	edge set of the Buneman graph on Σ, 57

Chapter 4

χ	character on X, 65
C	set of character states, 65
$\bar{\chi}$	extending χ to the set of vertices of an X-tree, 65
$\pi(\chi)$	partition induced by χ, 69
$\mathcal{C}$	set or sequence of characters on X, 69
g_χ	generalized character associated with χ, 77

Chapter 5

$\text{Ch}(\bar{\chi})$	changing set of $\bar{\chi}$, 84
$\text{ch}(\bar{\chi})$	size of $\text{Ch}(\bar{\chi})$, 84
$l(\chi, \mathcal{T})$	parsimony score of χ on $\mathcal{T}$, 84
$l(\mathcal{C}, \mathcal{T})$	parsimony score of $\mathcal{C}$ on $\mathcal{T}$, 85
$l(\mathcal{C})$	parsimony score of $\mathcal{C}$ on the maximum parsimony tree, 85
$h(\chi, \mathcal{T})$	homoplasy of χ on $\mathcal{T}$, 86
$h(\mathcal{C})$	total homoplasy of $\mathcal{C}$, 86
$\text{Cut}(\chi, \mathcal{T})$	set of all minimum-sized cut-sets for χ on $\mathcal{T}$, 86
$l_d(\chi, \mathcal{T})$	metric parsimony score of χ on $\mathcal{T}$, 93

Chapter 6

$\mathcal{T}\|X'$	restriction of $\mathcal{T}$ to X', 110
$\text{co}(\mathcal{P})$	compatible span of $\mathcal{P}$, 113
$ij\|kl$	quartet tree that induces the split $\{i,j\}\|\{k,l\}$, 116
$a_1 \cdots a_r\|b_1 \cdots b_s$	phylogenetic tree with one non-trivial split $\{a_1, \ldots, a_r\}\|\{b_1, \ldots, b_s\}$, 116
$\mathcal{Q}(\mathcal{T})$	set of restricted quartet trees of $\mathcal{T}$, 116
$\mathcal{Q}$	set of quartet trees, 116

$\Sigma(\mathcal{Q})$	set of splits supported by $\mathcal{Q}$, 116
$\mathcal{R}(\mathcal{T})$	set of restricted rooted triples of $\mathcal{T}$, 119
$[\mathcal{R}, S]$	clustering graph on S induced by a set $\mathcal{R}$ of rooted phylogenetic trees, 119
$\mathrm{scl}_2(\mathcal{Q})$	semi-dyadic closure of $\mathcal{Q}$, 129
$\mathrm{exc}(\mathcal{P})$	excess of $\mathcal{P}$, 134
$\mathrm{Exc}(\mathcal{P})$,	set of excess-free subsets of $\mathcal{P}$, 135
$\mathcal{Q}(\mathcal{P})$	union of $\mathcal{Q}(\mathcal{T})$ over all $\mathcal{T}$ in $\mathcal{P}$, 140

Chapter 7

$P(T; u, v)$	path in T joining vertices u and v, 145
$d_{(T;w)}$	dissimilarity map on $V(T)$ induced by $(T; w)$, 145
$d_{(\mathcal{T};w)}$	dissimilarity map on X induced by $(\mathcal{T}; w)$, 145
δ	dissimilarity map, 145

Chapter 8

$D(\pi)$	diagonal matrix associated with the vector π, 184
$P(e)$	transition matrix associated with e, 185
$p(e)$	substitution probability for e, 185
π^v	distribution of character states at vertex v, 185
$\overline{P}(e)$	reversed transition matrix associated with e, 187
T_ρ	rooted tree obtained by rooting $\mathcal{T} = (T; \phi)$, 188
$p(\chi) = p(\chi \mid \mathcal{T}, \theta)$	probability of generating χ on $(\mathcal{T}, \theta)$, 188
$p(x, y)$	dissimilarity probability for x and y, 188
$(\mathcal{T}, \theta)$	parameters for a Markov model on a phylogenetic tree, 189
$p(\chi \mid \mathcal{T}, \theta, \alpha)$	conditional probability of generating χ on $(\mathcal{T}, \theta)$ given root has state α, 190
P^{xy}	matrix of conditional probabilities of states at $\phi(y)$ given states at $\phi(x)$, 192
J^{xy}	matrix of joint probabilities of states at $\phi(x)$ and $\phi(y)$, 192
$\kappa(e)$	expected number of substitutions on e under stationary process, 195
$\kappa(x, y)$	evolutionary distance between x and y, 195
$\overline{p}_A$	probablity of generating a binary character corresponding to $A \mid (X - A)$, 198

Index